AUTOMATION IN THE STEEL INDUSTRY: CURRENT PRACTICE AND FUTURE DEVELOPMENTS
(ASI'97)

A Proceedings volume from the IFAC Workshop,
Kyongju, Korea, 16 - 18 July 1997

Edited by

W.H. KWON

School of Electrical Engineering, Seoul National University, Seoul, Korea

and

S.WON

SPARC, Pohang University of Science and Technology, Pohang, Korea

Published for the

INTERNATIONAL FEDERATION OF AUTOMATIC CONTROL

by

PERGAMON

An Imprint of Elsevier Science

UK Elsevier Science Ltd, The Boulevard, Langford Lane, Kidlington, Oxford, OX5 1GB, UK
USA Elsevier Science Inc., 660 White Plains Road, Tarrytown, New York 10591-5153, USA
JAPAN Elsevier Science Japan, Tsunashima Building Annex, 3-20-12 Yushima, Bunkyo-ku, Tokyo 113, Japan

First edition 1998

Library of Congress Cataloging in Publication Data

A catalogue record for this book is available from the Library of Congress

British Library Cataloguing in Publication Data

A catalogue record for this book is available from the British Library

ISBN 0-08-043029 5

Transferred to Digital Printing 2009

IFAC WORKSHOP ON AUTOMATION IN THE STEEL INDUSTRY: CURRENT PRACTICE AND FUTURE DEVELOPMENTS

Sponsored by
International Federation of Automatic Control (IFAC)
- Technical Committee on Mining, Mineral and Metal Processing
Steel Processing Automation Research Center (SPARC) at POSTECH in Korea
The Institute of Control, Automation and Systems Engineers (ICASE) in Korea

Supported by
Pohang Iron & Steel Co., Ltd. (POSCO)
Pohang University of Science and Technology (POSTECH)
Graduate School of Iron and Steel Technology (GSIST)
Korea Science and Engineering Foundation (KOSEF)
POSCON Co., Ltd.
Kangwon Industry Ltd.

International Programme Committee (IPC)

General Chair : Mr. Eric Rose (U. K.)

Member :

Prof. Z. Bien (Korea)
Prof. G. C. Goodwin (Australia)
Prof. M. Hadjiyski (Bulgaria)
Prof. A.S. Hauksdottir (Iceland)
Prof. J. Heidepriem (Germany)
Mr. S. J. Lee (Korea)
Prof. J. Paiuk (Argentina)
Mr. W. Don Stepto (South Africa)
Prof. S.-L. Jamsa-Jounela (Finland)
Mr. H. C. Choi (Korea)
Prof. M. J. Grimble (U. K.)
Mr. M. Hague (U. K.)
Mr. T. Iwamura (Japan)
Prof. W. H. Kwon (Korea)
Dr. N. Mort (U. K.)
Dr. R. Takahashi (Japan)
Prof. S. Won (Korea)
Prof. T. J. Williams (U.S.A.)

National Organizing Committee (NOC)

General Chair : Prof. Sangchul Won (POSTECH)

Member :

Mr. D. H. Chung (POSCO)
Prof. B. H. Kwon (POSTECH)
Prof. P. Park (POSTECH)
Dr. J. K. Lee (POSCO)
Prof. K. Nam (POSTECH)
Mr. Y. G. Kim (Kangwon Industry Ltd.)
Prof. W.K. Chung (POSTECH)
Prof. S. W. Kim (POSTECH)
Prof. I. B. Lee (POSTECH)
Prof. J. S. Lee (POSTECH)
Mr. K. N. Paek (RIST)

Advisory Committee

General Chair : Prof. Sooyoung Chang (Korea)

Member :

Prof. T. Y. Chai (China)
Prof. D. H. Chyung (U. S. A.)
Mr. C. S. Shin (Korea)
Prof. K. S. Chang (Korea)
Mr. S. B. Hong (Korea)
Prof. Y. Youm (Korea)

Foreword

The IFAC International Workshop on Automation in the Steel Industry: Current Practice and Future Development(ASI'97) was held in Kyongjoo, Korea from 16 to 18 July 1997, with the Steel Processing Automation Research Center(SPARC) serving as the official host. The objective of the workshop was to bring together engineers and scientists with expertise in applying modern control theories and techniques to industrial problems, particularly those involved in the steel industry.

Progress made in new technologies such as fuzzy controls, AI techniques, neural nets, new advanced controls including predictive controls and discrete event system theories have great potentials for application to the steel industry. Three plenary speakers, Professor P. Albertos, Professor G. Goodwin, and Dr. M. Katebi were invited to speak in these area. Titles of their talks were "Some Issues on AI Techniques in RT Process Control," "Advanced Control Applications in the Steel Industry" and "Predictive Optimal Control of Hot Strip Finishing Mills". In addition, there were three tutorial lectures by Mr. T. Iwamura, Dr. N. Mort and Professor Z. Bien on "Instrument and Control Engineers in the Steel Industry", "Discrete-Event Simulation for Manufacturing Systems" and "Satisfactory Control of Multi-Objective Systems using Fuzzy-Logic Approach".

The extended abstracts from 12 different countries were reviewed by the International Program Committee members. 50 papers were selected and included in the workshop preprint. There were more than 10 papers which dealt with cold rolling mills and hot strip mills indicating that these processes are still hot issues in the steel industry. Furnaces and continuous casting process received considerable attention in the workshop. Sintering process, overhead crane, material handling robot, tube making process, welding process and induction motor were discussed as necessary processes for iron and steel making.

About 10 papers covered modeling of various processes in the steel industry, which indicates that modeling is one of the most important issues even in the steel industry. The proceeding indicate that fuzzy control, intelligent control including neural nets, predictive control, adaptive control, robust and optimal control, and instrumentation and measurement have good applications as seen in these proceedings.

A total 85 engineers and scientists from 12 countries, the majority being from the host country, exchanged information on recent technological advances and practices in the field. The workshop site was close to Pohang Iron and Steel Company(POSCO), one of the leading steel companies in the world, enabling workshop participants to observe real applications in steel making. It is editors' hope that the readers of these proceedings will be able to benefit from the inspiring ideas contained in these papers.

We wish to express our thanks and appreciation to the members of both the International Program Committee and the National Organizing Committee, all of whom gave generously of their time to insure that the workshop would provide an excellent forum for distributing new concepts and methodologies for automation in the steel industry.

Wook Hyun Kwon
Professor of Seoul National University
and
Sangchul Won
Professor of Pohang University of Science and Technology

Editors of ASI'97

CONTENTS

PLENARY PAPERS

TUTORIAL PAPERS

TECHNICAL PAPERS

Some issues on AI techniques in RT process control

P. Albertos, J. Picó, J.L. Navarro and A. Crespo

Dept. of Systems Eng., Computers and Control
Universidad Politécnica de Valencia, Spain
e.mail: pedro@aii.upv.es

Abstract. The use of AI techniques in control raises the problem of implementing operations with not well delimited computing time, in a real time framework. In this paper, the combined used of conventional and AI control strategies at the lower levels of process control is analised. The main time constraints as well as some structures, namely the integrated and the hierarchical control structures, are reviewed. Two applications in the control of industrial kilns are also discussed. *Copyright © 1998 IFAC*

Keywords. Real time control, delays, Intelligent control, Artificial intelligence, kiln control.

INTRODUCTION

To deal with the integral control problem in the process industry, where a number of different tasks are involved, the use of a bunch of control techniques is required. Other than the well established solutions based on the classical control theory, artificial intelligence (**AI**) techniques are also becoming a common practice to develop control systems for local and global control, adaptation, learning, supervision and coordination, optimization, planning and even system management.

Classically, the integral control problem may be decomposed into different levels, as shown in fig. 1, each one dealing with different goals and kind of information. From the real time point of view, the time constraints are also quite different.

At the first level, the local control is implemented based on information directly got from the process. Typically, PID controllers or simple automata cope with most of the goals. The control viewpoint is very close to the process and there is almost one control loop per variable.

Optimal and interactive control have a multivariable structure but the approach to the control problem is quite similar: an algorithm provides the control action based on numeric data. Both, the control structure and the parameters are fixed.

At the second level, the control parameters are automatically tuned by means of adaptation laws, also operating in an algorithmic way. This strategy allows the control of time varying or non-linear processes.

The supervision, also involving local optimization algorithms, may change the controller structure, the set points or the global control strategy. At this level, some heuristic is required. It may be obtained either directly or by reasoning procedures. In the case of multimode operating processes, the supervision system allows the use of the most suitable controller among a set of predefined ones.

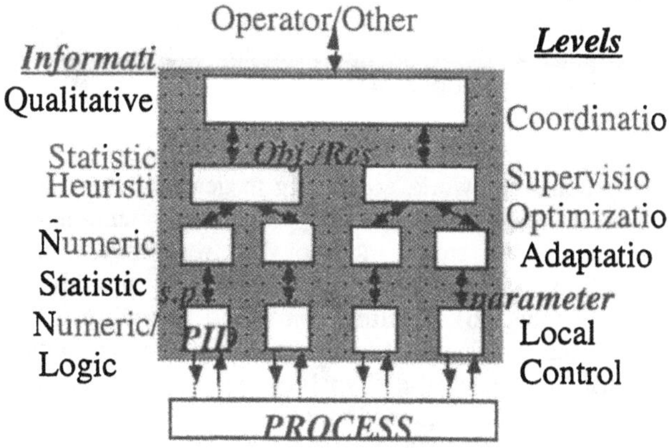

Fig. 1. Integral Control Structure

At the top level, where decisions about the goals, the processes, sequences and so on are taken, a lot of qualitative information is involved.

For the last twenty years, several proposals have come out concerning new techniques trying to improve the performances provided by the controllers by endowing them with facilities so far considered as belonging to the human domain. Learning, symbolic reasoning and pattern recognition are the most representative. Most of these techniques come from the *Artificial Intelligence* field, adapted to the intrinsic characteristics of control systems. This has opened a new broad research line, the so-called *Intelligent Control.*

Typically AI techniques have been applied to activities at the top levels: expert systems, off-line optimization, diagnosis, decision-making, and so on.

In fact, these techniques are suitable to deal with: 1) Fast full information from non-linear and complex models (Neural Nets), 2) Static and dynamic approximated knowledge (Fuzzy Systems), 3) Heuristics in both model and goals, involving reasoning (Expert Systems), 4) Optimization problems with multigoal complex specifications (Genetic Algorithms), as well as a mixture of all of them.

The use of these new control methodologies also reaches the lower levels due to the need of controlling systems with increasing complexity. Basically, in process control one can find three main causes for difficulties, [1]: inherent complexity of the system, presence of nonlinearities, and presence of uncertainties. Facing them requires to modify the traditional concept of control, so as to include features such as decision taking, plan generation, learning, etc., required to design systems with a high degree of autonomy [2].

In this context, control system reaction can be computed as result of either, evaluating rules (reasoning), or ``unconscious" reflex reactions, based on the process behavior. ***Intelligence***, thus, stands for the capability to reason on facts and rules and/or the capability to learn about the system behavior [3]. There are a number of industrial applications already running satisfactorily, [4].

All these control activities should be implemented in a common framework, competing to get control of the system resources and, as previously mentioned, accomplishing some, more or less restrictive, time constraints, including: a) Delays in providing conclusions, b) Getting on time response, c) Performing refinements and learning, d) Dealing with distributed systems, e) Synchronism problems and f) Emergency and degraded functioning.

The structure of this paper is as follows. First RT control issues, in both algorithms and software implementation, are summarized. Then, algorithms and schemes involving AI control are reviewed, focusing the attention on the time requirement as well as their suitability for some basic steel industry processes. Different techniques to implement classical/AI cooperative control systems are discussed. Finally, two already developed applications in the cement and the ceramic industries are reported and some conclusions are drawn.

COMPUTER DIRECT DIGITAL CONTROL

In a complex control scheme the existence of various processes (multiprocess), with many control variables and sensors (multivariable), involving many control loops (multiloop), trying to reach many objectives (multigoal), with different kind of data and decisions (multilevel) and operating at different sampling rates (multirate) must be considered. The general problem can be approached in different ways: 1) assume everything is independent, develop control algorithms and apply them at the scheduled time, trying to minimise delays and interactions, 2) assume the actual situation and develop control algorithms based on the full and detailed process model, 3) generate software mechanisms to make available complementary data, 4) provide different levels of service: the fastest for safety operation, the slowest for routine activities, and an optimized intermediate level for normal system operation, 5) combine all the available data and control strategies in the *best* possible way.

In these processes, the assumption of regular and uniform sampling as well as a simple algorithmic running of the control operations are not more possible. In a recent paper [5], a review of the non-conventional sampling schemes is presented. In particular, algorithms for specific operating conditions are discussed, including variable sampling period, non-synchronous sampling, multirate sampling, missing data, and event driven control.

In summary, the leading issues are: i) to compute the control action based on the latest information obtained from the process and updated to the moment of delivering this control action, by means of the process model, ii) to shorten the time interval or delay between control, and measurement samples (the control is *blind*) by re-structuring the computer implementation of the control algorithms, and iii) to reduce the control updating interval (the control system is open-loop) according to the process dynamics by also minimizing the global control period as a result of a reduction in mandatory tasks per period, as discussed in the next section.

To fulfill these constraints, modified classical control algorithms, like PID or state feedback combined with advanced observers, have been proposed [5].

In most of these cases, the control law is predetermined and the computation time is easy to compute. This is not the case for AI techniques, where optional activities with real time constraints should be considered.

RT Constraints and operation

To develop an integral control application a system design able to guarantee the timing constraints of the different processes is needed. The AI based control techniques, such as NN, FLC, rule-based systems, etc., allow to better capture a system model and apply the appropriated techniques to each problem or subproblem but, in most cases, their execution time is unbounded or unpredictable. So, it is required to modify the classical concepts of tasks or activities to work with more flexible systems.
There are control system activities with different kind of time constraints.

Hard time constraints refer to those for which a solution must be provided at or before a required time, otherwise the system integrity is lost. Soft time constraints refer to activities which are relevant but if they are not executed, an acceptable or degraded system behavior is obtained. Some other activities should be done within a longer period of time but their timing is not so critic.

In a classical real time application several parallel activities have to be defined. Periodic tasks model the activities that must be executed at periodic instants of time. Sporadic tasks model the activities that handle external significant events. All these tasks are concurrently executed under the supervision of a scheduler. In a basic control loop implementation, Audsley [6] proposed the control computation to be composed by three blocks: a mandatory, an optional and a final part. If AI techniques should be considered in the model, a complete control loop implementation would have the following scheme:

1. Computation of the control action. It is mandatory and should be executed as soon as possible. At the end of this part, the basic control action is ready to be sent to the actuators.
2. Solution improvement. During this time, an algorithm or method is able to improve the answer obtained in the first part. This part can be considered as optional.
3. Output the control action. Must be done either as soon as possible or within a fixed time interval.
4. Evaluation of the global state and preparation of the data for the next period. This part should be executed before the next period, but at any time within the period.

These activities can be represented in a graphical way, fig. 2, where δ represents the data-acquisition system delay and Δ the control/measurement one.

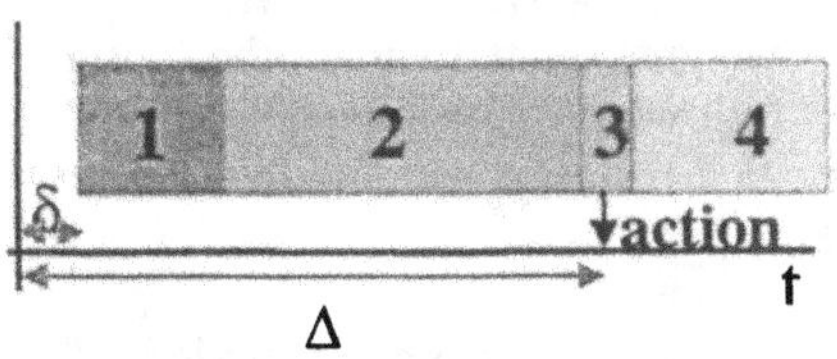

Fig. 2. Actions delay

One of the important aspects in the above model is related to the computation characteristics of part 1, 2 and 3. The first, third and fourth parts can be considered as the classical control loop. The second part introduces the possibility to incorporate new computation depending the time availability to improve the system result.

These aspects are shown in fig. 3.

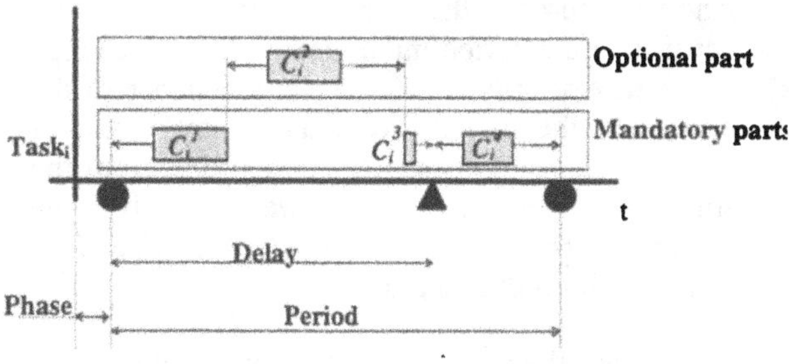

Fig 3. Task scheme

Activities 1 and 3 must be executed before the deadline of the task. Activity 2 is optional and can be executed as far as there is available time.

The main concerns being to fix and keep as short as possible the time used in activities 1 to 3, and the timing between tasks, if they must share data and/or results.

For a number of parallel tasks, competing to get CPU access, some issues should be considered. Each task will have a different priority but many of them will be at the same level. One of the software engineering practices is to determine if the application is schedulable or not based on a few parameters, that is, if the time requirements are guaranteed. However, even if the system is schedulable, some action delays can modify the system behavior. The interval of possible variation of the actuation, the Control Action Interval, (CAI) is defined as the time interval where the action is sent to actuators at different periods.

If the number of tasks increases, the CAI of the low level priority tasks is also increased. Although from an application viewpoint the priorities should be related to the relevance of the activity, from a software viewpoint the priority assignation is basically determined taking into account the period of the task. A higher priority is assigned to more frequent task than to less frequent ones. Because of the CAI depends on the interference level with other tasks, if the priority level of a task is increased or decreased, its CAI will be consequently reduced or increased.

However, from the scheduling theory point of view the priority of a task should not increased or decreased. In order to evaluate the schedulability, a priority is attached to each task depending on its rate or deadline. A first solution is to decrease the deadline of a task maintaining the period. When deadlines are shorter than periods, the deadline monotonic priority assignation can assign a higher priority to the modified task. Using this approach, the system is stressed and the modification can produce a non schedulable system.

Another approach that does not stress the system consists in the period modification. So, the change of the task priority should be a consequence of the change of the period. From the control point of view, we can not change the period, but from the software design point of view, we can apply some techniques to change the period of a task. This change will produce a priority .

To increase the priority of a task, we could split a task into two parts, dividing by two the period and executing alternatively a part each period.

An alternative method consists on increasing the period of a less important task from the user point of view, but more frequent than other which are more important. In any case, the reduction of the CAI of a task produces an increment in the CAI of another one.

BASIC AI TECNIQUES IN CONTROL

Let us review the basic AI techniques as well as some applications in the steel industry.

a) Neural Networks

A broad descriptive definition of *Artificial Neural Network* (NN), inspired in their biological counterparts, may simply formulate a NN as the structure which outcomes from the interconnection of basic processing units, or neurons, each of them performing a nonlinear operation on its inputs.

From the neurocontrol point of view, a NN may be considered as a convenient parameterization of a family of nonlinear mappings, capable of representing general nonlinear systems [7] under certain assumptions. Therefore, they may be considered like another function approximation tool, but endowed with some interesting advantages [8].

Out of the many possible neural network topologies devised over the last thirty years, only three of them are widely used for neurocontrol purposes, for they represent suitable nonlinear function parameterizations:

- Multilayer perceptron, or multilayer feedforward NN (MFNN).
- Radial basis function NN (RBFNN).
- Recurrent, or dynamic network (RNN).

From the point of view of the time requirements, two aspects can be considered regarding the implementation of NN:

- Parallel distributed processing. As most of the operations involved in the NN processing are repeated over many nodes, parallel (vectorial) processors or transputer arrays may lead to very efficient software implementations.
- Dedicated hardware implementations. The referred parallel processing of simple elements (neurons) can be implemented in specific VLSI hardware leading to dramatic speed improvements over its software counterparts (moreover if compared with non-parallel CPU's) when applied to challenging real-time problems. Once a problem is solved with a network, incrementing the number of neurons is the way to solve a similar one with more variables, so the processing time keeps constant even with increased dimensionality of the problem. Processing times in parallel hardware implementations depend on the number of *layers* and *not* on the number of *nodes* in each layer as these are processed all at a time.

Most of the times NNs are implemented as a piece of software code to be run on a sequential processor. In this case the processing time depends both on the number of layers and the number of neurons in each layer.

In any case the processing time is always bounded and can be known for all the implementation described. The only case in which the processing time may be variable is in the case of variable structure NNs implemented on sequential processors.

On the other hand, two different sequences of operations can be distinguished during run-time:

- Forward propagation of information. The NN processes its inputs, forwarding them through the layers, and outcoming its output. The time consumed depends on whether the NN is implemented in dedicated hardware, a parallel processor or a sequential processor. In all cases the processing time is fixed, and forward propagation of information is a compulsory task.
- Weights tuning. The learning of the correct weights requires backpropagating the information through the network, in order to obtain the required derivatives of the network error with respect to the network weights. Afterwards the weights are modified following some adaptation rule. On-line learning can be carried out following several time schedules:

1. Learning at each sampling period. This will require that both forward and backward propagation, and the adaptation rule are executed at every sampling period. The last two are executed jointly, and their execution can be deferred to the final part of the sampling period.
2. Periodic Learning, accumulated learning over a fixed number of periods. Forward and backward propagations are executed at each period, and the results are stored. At the end of the fixed interval, the stored results are used to update de weights. The backward propagation can also be deferred to

the end of the interval. In this case a time-efficient vectorial implementation of the learning algorithm can be used.

3. Sporadic Learning. Whenever a degraded performance of the control system requires re-tuning of the network weights, the backward propagation and adaptation rule are executed. Sporadic learning can also be carried out whenever time resources are available, even if not strictly required. This kind of learning should be carried out over the data corresponding to an interval spanning several sampling periods, to avoid problems caused by using a poor estimation of the gradient of the error with respect to the weights.

b) Fuzzy and neurofuzzy controllers

Fuzzy controllers have become one of the most successful applications of fuzzy logic for the last years, [9]. They can be easily applied to control processes of which we lack a useful mathematical model, or in which vagueness plays a fundamental role and we have qualitative models. The knowledge supplied by an expert human operator is coded in the form of IF-THEN production rules, where the premise and consequent variables are categorized in terms of fuzzy sets. At the controller output, the control action inferred by a set of fired rules is obtained by means of interpolative reasoning methods.

Fuzzy controllers with propositional rules, that is, those in which the consequent of a rule is never the antecedent of another rule are functionally equivalent, under the usual assumptions on the class of inputs and membership functions employed, to NNs with one layer [10,11]. Therefore, if the possibility of adapting the membership functions is assumed, all the considerations mentioned for NNs are valid.

Neurofuzzy controllers are but an implementation of a fuzzy system under the structure of a NN. Therefore, their time requirements are equivalent to those described for NNs.

The simplest control schemes are represented by single intelligent controllers or by estimators implementing a direct mathematical mapping. Fuzzy systems with simple propositional inferences, single NNs and neurofuzzy systems are typical representatives. To get most of the advantages of these new approaches, optionality about knowledge sources and procedures should be considered, introducing additional difficulties in RT control. Some of these new structures are discussed in the next sections.

The natural conception of expert systems as well as evolutionary programming techniques (like genetic algorithms) makes these approaches not suitable for RT applications, and will not further considered.

c) Steel Industry Processes

The steel industry involves a lot of different processes. Some of them are the basic steel production processes, such as the blast Furnace, the Soaking Pits, or the Rolling Mill, but there are also many other complementary processes, such as blending processes (in raw materials), recycling of waste, coating and treatment lines and so on.

Most of these processes fulfill the characteristics for AI techniques suitability: the state knowledge is approximate due to difficult measurement of relevant variables, the behavior is highly nonlinear, and there are joint goals to be reached. Some applications have been already developed, [12].

Blast Furnace

In steel making processes, high temperature and chemical compositions are difficult to measure. They are estimated and sometimes based on operator heuristics. Also the mathematical models are based on parameters which are not well known and changing with time.

There are long time constants, like thermal conditions and hot metal balance, blast volume, furnace heat index, as well as very short time constants like blast furnace pressure. That is, it is a multivariable, multiprocess, multirate system.

Although the common applications refer to diagnosis and monitoring, some RT applications have been already reported, [12]. In [13], an interesting combination of classical and AI techniques is presented: physical models are used to forecast the operating conditions and to optimize the furnace behavior in parallel with fuzzy controllers copying with deviations.

Also, in [14], NN to model the BOF behavior and to predict the finishing time and other characteristics are applied.

Soaking Pits.

In order to prepare the ingots with a uniform temperature before rolling them, they are introduced in batch reheating furnaces, where the saving of energy and the on-time operation are basic in the whole steel making process.

Again, difficulties in measurement and uncertainties in the parameters are the key reasons to use AI techniques, [15].

Hot Strip Finishing Mill

This is a paradigmatic example of complex system where hierarchical control has been applied, [16]. A number of variables, strip gauge, looper angle, and strip tension, are simultaneously controlled and jointly optimized to reach a global objective. Instead of simple decentralized control, interactive control has been widely used, (see, for instance, [17]).

The high nonlinearities of this process leads to the use of NN as a complementary tool to model and implement classical control strategies. In [18], NN are used for both, off-line modeling of the stands, and on-line adaptive tuning of the parameters of the algorithmic model or adding corrective actions.

INTEGRATED CONTROLLERS

As previously described, in a large and complex industrial process there are a variety of tasks that must be performed to keep the system under control. Although based on different data sources, a number of manipulated variables (inputs to the actuators) should be computed. All these tasks impose different requirements to the execution environment, but from the point of view of the control system designer, it is very useful to have a unified approach to reduce the workload in the development process, [19].

The main requirements are:

- The control system can work in several modes to solve different control situations with unlike goals, such as normal operation, start-up, shut-down, failure, etc. For each mode, different actions must be performed that can be solved with any of the techniques outlined above.
- The tasks can be executed with a periodic scheme or can be launched when an event is detected in the control system (event driven).
- The development system must hold different techniques (AI or conventional ones) at the same time, so it is necessary to dispose the appropriated algorithms and data structures.
- For large systems it is desirable to distribute the computational load between various CPUs to increment the execution speed.

Based on these considerations, it is useful to have an architecture able to support different types of techniques for implementing control algorithms matching the RT constraints. This architecture is based on the idea that any of its components, can be seen as an information process that produces a conclusion or output from an environment data set or another source. This base component will be named **Intelligent Regulator Agent (IRA)**, [20] because this module is part of a system that is used to control or regulate a process and it can support algorithms derived from artificial intelligence techniques or conventional ones. Each IRA, as shown in fig. 4, defines a set of problem solvers (rule-based, algorithmic, genetic, etc.) in order to provide solution to a specific subsystem. Also, they are structured for achieving the real time constraints imposed to the control system.

Various components can be considered in an IRA:

Regulator knowledge: It defines the functionality of the IRA by means of the declaration of the structures, parameters or information needed to characterize its behavior. This information is dependent on the technique used in the regulator (real valued parameters for linear discrete regulators, production rules in expert systems, topology and learning methods for neural networks, etc). Also, various levels can be defined to cope with the real time constraints. Each level produces a better response of the system, but it takes more time to compute the final solution, so the scheduler must select the level that produces the best response in the CPU time assigned to evaluate the IRA.

Local data base: A local area to store some temporary and local results produced along the evaluation of each agent is necessary. When the evaluation ends, all or some data is transferred to the global data base and can be shared by another modules.

Evaluation agent: This element implements the chosen algorithm and obtains a result. It depends on the used technique and it is necessary to have a library of different methods that are selected on the design stage.

The scheme of the agent activity depends on the algorithm implemented in the evaluation agent. If a procedural regulator is used (discrete linear controller, neural network or fuzzy controller) then the required execution time is bounded and can be known, so an a priori scheduling can be used and there is no optional part. Moreover, if an anytime algorithm is used then the task is executed until the maximum executed time is elapsed and the current solution is exported to the global data base. In both cases only one level of the regulator knowledge is used.

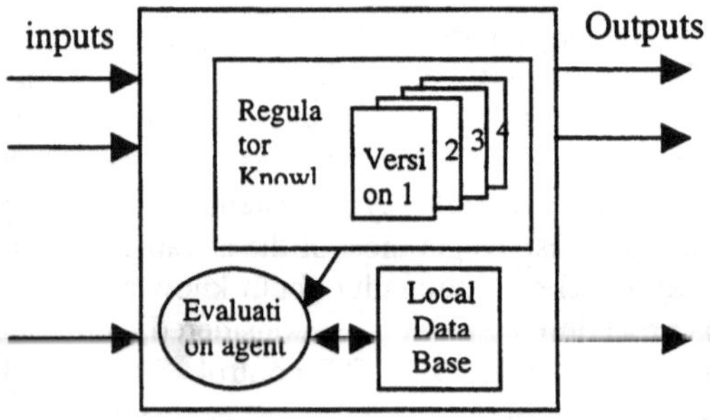

Figure 4. Intelligent regulator agent structure

But if the execution time of the algorithm is not bounded or it may have very large variations depending on the input data presented to the system, different levels or versions must be used in order to allow that a response can be available on-time, even with less accuracy. It is like a discrete anytime algorithm: it improves the response when a level is executed, then the next one is taken and the response is again improved until the execution time is expired and the last response is concluded.

Examples of this approach can be found in expert system for the direct control or supervision of a process. The first level or version of the intelligent regulator should be a system with bounded execution time like a fuzzy controller. It can be modeled like one layer production rules (relates directly the input variables to the output ones), or with general production rules without using computing intensive functions in the antecedents of the rules (like temporal logic functions). The next level should improve the answer taking into account more variables and information. Then, upper levels can include first order predicate calculus and temporal information. The optional tasks of this agent will be the tasks formed at level 2 or higher.

For the design of the global controller it is necessary to start with a decomposition on specialized agents [21] that focuses in different problems or situations that can be met in the process. These agents can be independent or some of them can use information derived from another one. For every agent a technique must be selected.

HIERARCHICAL CONTROL

The main characteristic of the hierarchical schemes is the fact that the final control action on the process is obtained through a chained sequence of intermediate control actions calcula. This is different from the progressive action refinement where successive controllers are run, each one refining the control action provided by the previous one.

In a hierarchical control structure there are different control actions belonging to different levels of decision in the system. Therefore, there is a hierarchy of controllers corresponding to a number of fixed levels. Usually the controllers at the highest levels are AI-based, while the ones at the lowest are mainly based on classical control theory.

The AI-based highest levels in the hierarchy carry out tasks requiring an important degree of reasoning, such as diagnosis and supervision and, in them, progressive refinement techniques may be also possible.

Quality control is one of the cases in which this kind of hierarchical schemes is often encountered. Typically, quality defaults are detected and classified (for instance, by means of a NN-based pattern classification). Depending on the detected defaults and other relevant variables of the process, an inference mechanism on a rule-based system yields new values for the set-points of some process variables. In turn, some lower level controllers take these set-points and manipulate the appropriate process variables so as to keep them within given specifications.

Regarding the control implementation, each of the aforementioned levels can be implemented as a program plan, that is a set of tasks to be executed. An example of the structure of a typical plan is depicted in figure 5, where a plan for production scheduling in an industrial application is shown.

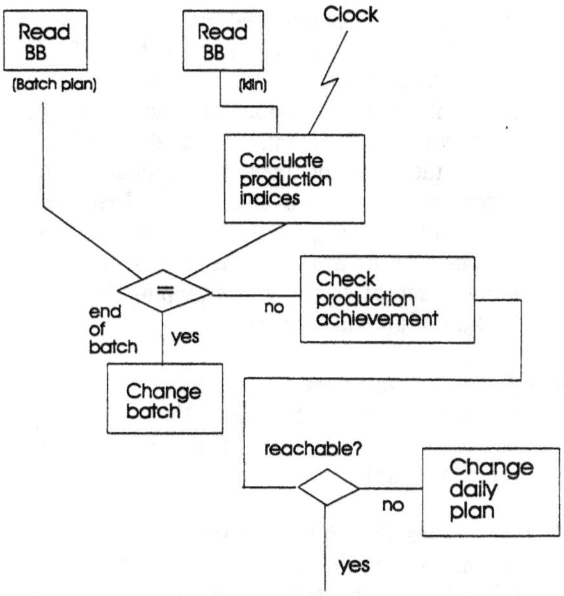

Fig. 5. Production plan

Each of the boxes represents a task to be executed if some conditions are satisfied. The plan is periodically activated. The whole plan can be programmed on a single program thread, or several ones can be used if some of the tasks in the plan can be concurrently executed.

On the other hand, the plans defined in the system architecture interact among them, as depicted in fig. 6, where a blackboard is used to keep and share information.

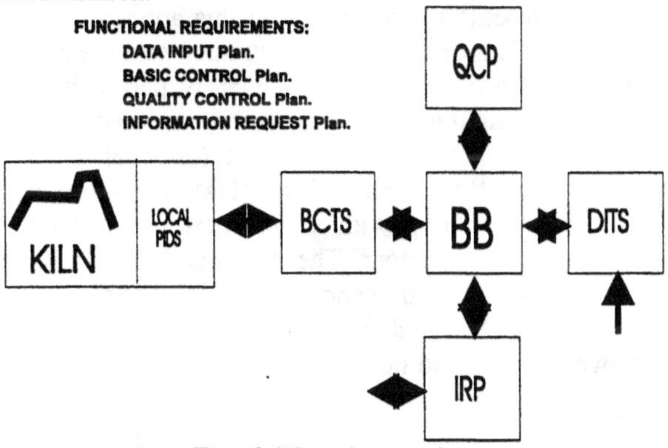

Fig. 6. Plans interaction

From the point of view of the time requirements, two levels can be considered:

- Plans architecture. As said before, in an industrial complex control system, an ordered hierarchy of control functionalities allows for managing different decision levels. Each of these levels is implemented as a set of tasks, so called a plan. Different functioning modes, either set by external events to the plan or by internal ones, may switch on and off some of the facilities of the control system. This introduces optionality among tasks, therefore modifying the available processor time for other tasks. For the number of tasks is finite, there is a finite set of options. For each one of them all the tasks involved must be executed, being the time required known and fixed if it is so for each of the tasks in the plans. This is different from the optionality appearing in progressive refinement, where the optional part duration is not fixed, but depends on the processor time availability. Therefore, optionality among tasks consists of the selection of a `program line' out of a finite, known, set of them. The swapping between `program lines' is sporadic.

- Tasks implementation. As previously mentioned, each of the tasks within a plan may be composed of three parts compulsory, optional and final.

In many cases the high level reasoning takes place with a long time period or even sporadically, while the lower levels are activated with short periods. That is, several time scales may be needed.

RT IMPLEMENTATION OF OPTIONAL TASKS

If an AI technique is used, the way the answer is obtained is relevant. In this sense, two different approaches can be considered:

The same technique or algorithm is applied and depending on the available time the result has a higher quality. Anytime algorithms are one of the main representative technique in this approach. In this kind of algorithm, the computation scheme can be described as follows: a mandatory part with a required computation time known is executed first, followed by a improvement part (optional part) which has the form of a classical iterative algorithm, finishing with a final part which send the output to actuators. The optional part can be interrupted at any time and the solution correspond to the last finished iteration. This behavior is shown in fig. 7 and table I.

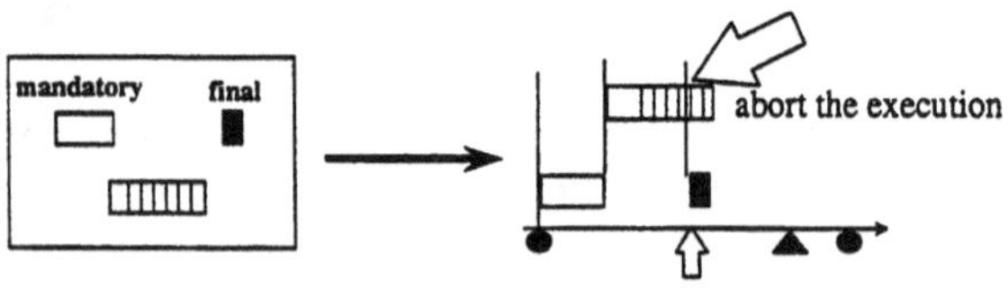

Fig. 7 Anytime option

Table 1. Optional tasks

```
task Control_with_Optional (PERIOD, DELAY,
                            PHASE : NATURAL) is
current, next : TIME := CLOCK ;
output : FLOAT ;
begin
 loop
  -- execution of minimal mandatory part                  1
  -- obtains the output value
  declare
   local_output : FLOAT ;
  begin
   select
    delay until current + DELAY ;                          2
   then abort
    loop
     -- iterative method to improve the output
     local_output := .... ;
     output := local_output ;
    end loop ;
   end select ;
  end ;
  send_to_actuator(output) ;                               3
  --determines the global state of the system             4
  --updates the global variables
  next := next + PERIOD ;
  delay until next ;
 end loop ;
end CONTROL_WITH_OPTIONAL ;
```

Optional part (block 2) is implemented as a control structure provided by the language Ada95. This Asynchronous Transfer Control structure allows the execution of a piece of code limited by the time specified in the first part of the select.

Another approach consists in the definition of different pieces of to solve the same problem. Each piece uses a different model of the problem (more or less detailed) and, as consequence, it requires more or less time. Depending on the available time, the richest approach will be selected. As before, there is a mandatory part providing a basic solution and a number of possible options is ready to be executed, as shown in fig. 8.

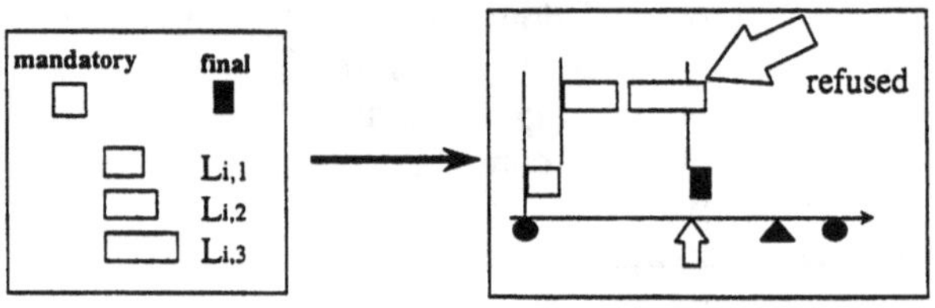

Fig. 8. Selected option

The inclusion of optional parts in control system algorithms allows the use of the remaining CPU time to improve the response to some control demands. There is always some remaining CPU time because some sporadic activities related to alarms, threshold detection and emergencies are seldom executed, and for any task, the a priori scheduler assumes more time than usually required to prevent the worst case, that is, the occurrence of blocking of resources, full data, and so on. The saved time that can be invested in the improvement of some results.

The execution of optional parts will be determined by the availability of remaining time and the demand for soft sporadic load. Optional parts are not executed at the same priority level then that of the mandatory ones. Several approaches to include optional computation in real time systems can be found in the literature. In one of the first works in this line [22] a task model to incorporate imprecise computation was defined.

From the scheduling point of view, optional parts should be handled in a specific way:

- they should be executed after the completion of their related mandatory part
- they should be selected based on their priority, required execution time, and time availability before the deadline of their related task.
- they should be enabled if the data they require from other tasks is available in the common memory (blackboard).

Optional server

All optional components can be included at run-time through a server process (the optional server). Initially, the optional server component does not affect this off-line feasibility, as it is always executed in slack time. However, at run-time some action must be taken if the guaranteed deadlines are affected [23].

In [24] several strategies have been defined to execute optional parts. The optional server executing at the higher priority of the system with a policy based on task importance is shown in fig. 9, where two concurrent tasks are considered. In this case, task 1 has the higher importance on the system and, as a consequence, when there is available time and need of optional computation, the optional part is executed even delaying the execution of task T2 (without breaking the deadline).

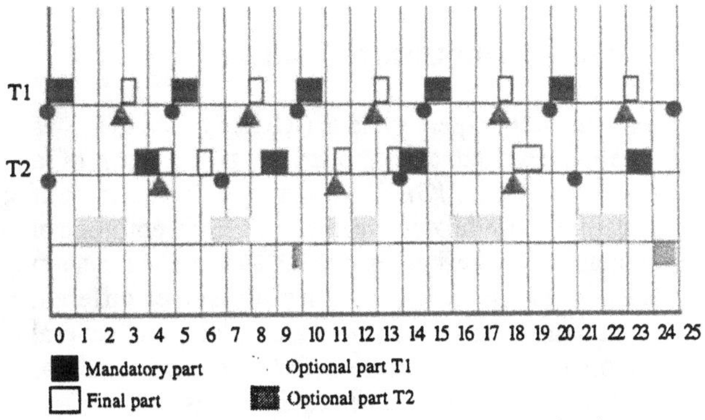

Fig. 9. Optional server using an importance criteria

These strategies have been evaluated and compared. The criteria used for each are:

- based on the user importance. The user importance does not coincide with the system priority.
- sharing the optional part of all tasks to equilibrate the result improvement
- applying the second strategy but selecting the best fitted optional part.

The main conclusion of this evaluation is the difficulty to choose a strategy because in most cases the best solution depends on the task process and the solution is problem dependent. Thus, more than magic general solutions, a trial and error approach with off-line previous simulation will determine the best structure for a given application or operating scenario.

APPLICATIONS

As previously mentioned, RT applications of AI in the process industry are becoming more and more common. Our department has been involved in the development of two typical applications in the control of kilns, one of the basic subsystems in the process industry. They illustrate the real-life application of the control structures proposed above.

Ceramic Tile Production

Like in the steel industry, ceramic tile production critical variables are related to temperatures and compositions which are, many times, difficult to on-line measure. Thus, the production is critically dependent on the firing process at the kiln. Current control systems only apply local control, and product quality is not directly considered. As consequence, there are important material losses and process inefficiency. MARCK (Multi-process Adaptive Real-time Control for a Kiln), is a tool developed under the ESPRIT program aiming at the application of high technology in the production control of quality tiles.

The general architecture of MARCK is depicted in fig. 10. It consists of a set of program threads sharing information through blackboards, and driven by events, [25].

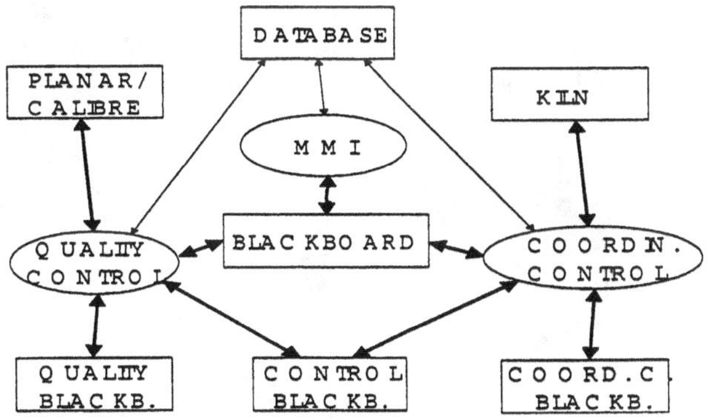

Figure 10: Global threads architecture.

The kiln control devices are the PIDs driving the burners, the PLCs managing alarms and auxiliary devices, and the set of sensors, in particular, the planar and caliber measuring system. They communicate with the application through the coordinating and the quality control sets of tasks (described below), while the kiln operator does through the man-machine interface.

All the current data is stored in a set of blackboards, and the communication among the application tasks is carried out through shared blackboards. Long-term data regarding the production is stored in a database, which also holds all the information about the different tile models.

From the control point of view, the main functionalities sought are:

1. Control of the kiln temperature profile.
2. Adjustment of the kiln to lack of entering material (voids).
3. Transfer between temperature profiles (tile model).
4. Quality control, with main care taken on dimensional problems.

For this purpose, the devised hierarchical control architecture consists of three levels:
1. Low level control: at this level, the local PID controllers at each zone along the kiln drive the associated burners.
2. Coordinating control level: this level uses a dynamic model of the kiln to take into account the interactions among the different kiln zones. Accordingly, the set-points sent to the PID local controllers are modified using a decoupling adaptive feedforward control strategy. Due to the uncertainty of the interaction parameters, an adaptive fuzzy decoupler is well suited.
3. High, rule-based control level : defines high level set-points for the temperature and pressure profiles along the kiln, with the aim of counteracting quality defects appearing in the tiles. Different strategies can be defined by the process engineer, attending to the defect to be controlled and the kind of material.

A basic scheme of the proposed control architecture is depicted in fig. 11, where the local PID controllers are assumed to be constitutive part of the kiln.

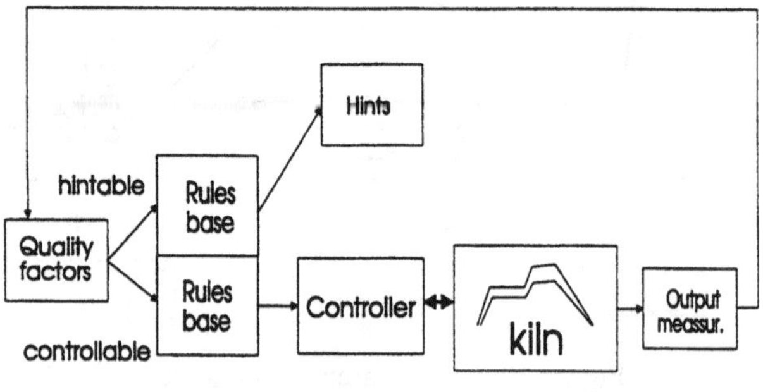

Figure11: Global control scheme.

Different operation modes in the application allow for switching on and off the control levels:

- Monitor: read-only mode at which no control action is taken.
- Manual: local PID set-points are directly provided by the kiln operator through the man-machine interface.
- Automatic low: only the coordinating control is carried out.
- Automatic high: both coordinating and quality controls are switched on.

The goal of the decoupling controllers is to achieve a temperature behavior only defined by the local PID settings at each section, independently from the neighbor.

The coordinating control is implemented as a periodic set of tasks, being the responsible for acquiring data from the kiln through the PIDs, and writing on them.

At the quality control level the main goal is to generate new set-points for the controlled kiln variables so that the quality defects at the output tiles can be corrected. MARCK treats a quality default trying to emulate the steps followed by a human operator when faced with the same problem:

1. Select the kiln sections where an action should be applied. This is referred as the strategy selection.
2. Calculate, by means of a fuzzy rule base inference, the temperature modification to be applied in the selected sections. This modification is considered on the temperature difference between the upper and bottom profiles.
3. Distribute the calculated gradient modification between the upper and bottom temperature profiles.
4. Correct the temperature set-points of left and right neighbor sections if the temperature gradient is too large (sympathy action).

Cement Kiln Control: Rigas

A real time expert system (RIGAS), [26-28], has been developed and applied to the control of a cement kiln. RIGAS allows the definition of Intelligent Regulator Agents, and different control schemes can be implemented. The control system can work in several modes to solve different situations (normal, alarm, heating, etc.) and several levels of solution can be defined in each regulator to cope with the real time requirements.

The main characteristics of the kiln process are : distributed process; different operation situations; large number of variables; measuring delays; heuristic used in the control. These properties justify the use of AI techniques in the control of this system and also real time requirements must be imposed because the direct interaction between the system and the process.

The kiln operation situations can be classified in regular (start up, shutting down, normal production, etc.) and exceptional (coolers obstruction, raw cut off, cyclone obstruction, etc.). All them require a different control policy and this implies that diverse controllers must be used. The control system is organized in three blocks modeling the basic situations *: secure operation*, when the control goal is the production ; *exception operation,* which require a set of specific actions outside normal functioning and *transient operation*, that includes the actions needed to reach the secure operation. The developed system uses an IRA to identify the current situation and the transitions between them, activating the specific blocks designed for that situation. This structure can be seen in fig. 12.

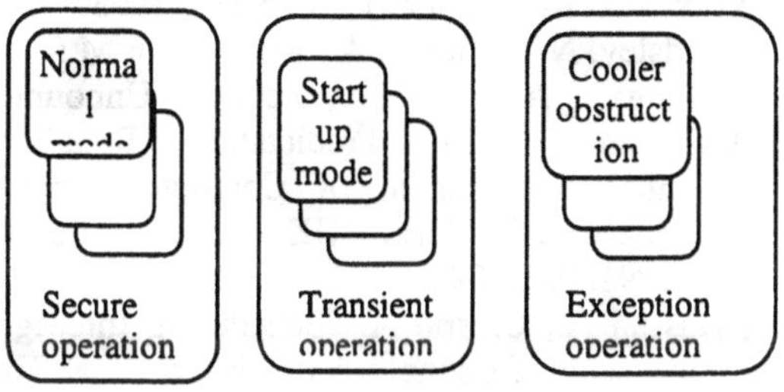

Fig. 12 : Kiln situation modes

In order to show the methodology to implement the control system, the normal mode structure is presented in fig. 13. In this mode, the control goal is to reach maximum production with the minimum coal consumption. The kiln state is divided into three static states (hot, normal, cold) and three dynamic ones (heating, stable or cooling). An IRA is used to conclude the kiln state and a confidence coefficient from the measured values. Next, this information is used in 4 IRA, one for each output variable (coal flow, exhaust aperture, kiln revolutions and raw flow), to obtain the numeric values of the respective control action that is applied to the process.

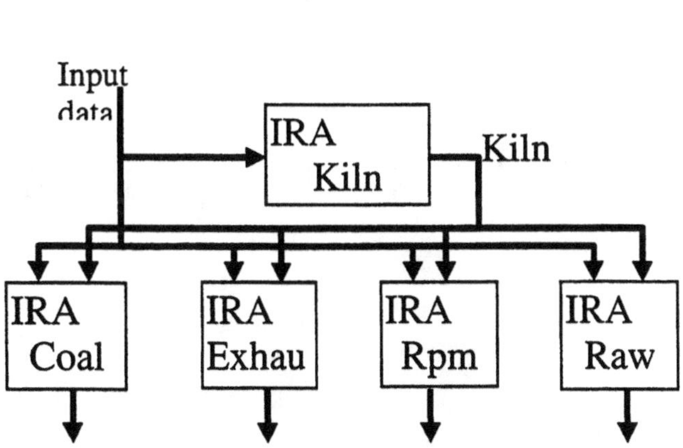

Fig. 13. Kiln normal control mode.

At the first level of each IRA, a fuzzy controller with only one level of rules (the rule antecedent are input variables and the consequent are output variables) has been implemented. These rules model the heuristic behavior of the kiln operators. At the second level, more information is processed, like temporal response of the variables or reasoning with qualitative models. Finally, at the third level an optimization algorithm based on an experimental model is implemented.

CONCLUSIONS

The appealing properties and friendly use of some AI techniques promote their use even at the control levels close to the process, where the time constraints are usually hard. Among these constraints are the sampling period, the delay between sampling and updating the control action, the variation of this interval, and the synchronism for data exchange between processes. Some of them are strongly influenced by the computation of the control action by softcomputing techniques, because there is always some optionality and this should be taken into account in the design of the control system.

In order to get some insight in these problems, a review of timing issues from both viewpoints, control and software implementation, has been made.

To use conventional and AI control techniques in a complementary way, two main strategies have been proposed. The intelligent regulator agents are organized in such a way that to compute their output, the control action, some options are possible. If there is a minimum time to execute each one of them, the one to be used will be determined each time by the CPU time availability. These integrated controllers keep all the information they need to run the different control options. This solution has been successfully applied to control the cement kiln in a cement factory.

A classical control approach, the hierarchical control, has been also proved to be suitable to combine both techniques. In order to allow an easy data exchange between tasks, local and global blackboards are implemented in connection with the different control activities. Although AI techniques are more appropriated for higher levels in the control hierarchy, they can be easily implemented for intermediate and lower levels of control. This hierarchical structure allows a better control of the execution time by means of a modular computation. This strategy has been applied in the control of a ceramic kiln where the operating scenario often changes.

REFERENCES

[1] Narendra, K. Adaptive Control using Neural Networks. In *Neural Networks for Control,* MIT Press, 1990.

[2] Antsaklis P.J. Neural Networks in Control Systems. *IEEE Control Systems*, 10(3), Apr. 1990.

[3] Verbruggen H.B. and Krijgsman, A.J. and Bruijin, P.M. Towards Intelligent Control: Integration of AI in Control. In *Application of Artificial Intelligence in Process Control,* Pergamon Press, 1992.

[4] Lu, Y-Z. *Industrial Intelligent Control.* John Wiley and Sons. 1996.

[5] Albertos, P. and Crespo, A. Real-Time Control of Unconventionally Sampled Data Systems. *IFAC Workshop on AARTC'97*. Vilamoura, pp 1-13. 1997.

[6] Audsley N.C. "The Deadline monotonic scheduling". Real-Time Systems Research Group. Department of Computer Science. University of York, UK. Report number YCS146. 1991

[7] Narendra, K. And Mukhopahyay, S. Intelligent Control using Neural Networks. *Procs ACC,* pp.1069-1074, Boston, MA, 1991.

[8] Ljung, L. and Sjoberg ,J. A System Identification Perspective on Neural Nets. Tech. Report, Dept. Of Electrical Eng., Linkoping U., Sweden, 1992.

[9] Albertos, P. Fuzzy Logic Modeling and Control. In *Identification Adaptation, Learning.* S. Bittanti and G. Picci (Ed.) Springer pp 479-513. 1996.

[10] Picó, J. and Moya, P. Localized BF-type Networks for Identification and Adaptive Control of Discrete.time Nonlinear Systems, *Proc. IFAC Symp SICICA'97*, Annecy (France), 1997.

[11] Wang, L.-X. *Adaptive Fuzzy Systems and Control,* Prentice-Hall, 1994.

[12] Lu, Y.Z. 1996a. "Meeting the Challenge of intelligent system technologies in the iron and steel industry". *Iron and Steel Engineer.* September 1996. pp 139-148.

[13] Iida, O., Ushijima, Y and Sawada, T. Application of AI techniques to blast furnace operations. *Iron and Steel engineer.* pp 24-28. October 1995.

[14] Yun, S. Y. Chang, K.S. and Buyn, S. M. Dynamic Prediction using NN for Automation of BOF Process in Steel Industry. *ISSM* Aug. 1996. pp 37-42.

[15] Lu, Y-Z. And Williams, T.J.Computer control strategies ... for the control of steel Mill Soaking Pits. *ISS Trans.* Vol. 2 pp 35-43. 1983

[16] Singh, M. G. *Dynamical Hierarchical Control.* North Holland. 1980.

[17] Okada, M. *et al* . Optimal Control System for Hot Strip Finishing MillSession 7b-04, pp 493-498 *IFAC World Congress* 1996.

[18] Portman, N.F. Lindhoff, D. Sorgel, G. and Granckow O. . Applications of NN in rolling mill automation. *Iron and Steel Eng.* Pp 33-36, Feb 1995.

[19] Morant, F., P. Albertos, J.L. Navarro, and A. Crespo, Intelligent regulators: Design and Evaluation. *Proc. 12th IFAC World Congress,* Vol. 7, pp. 479-484, Sydney 1993.

[20] Albertos, P. A. Crespo, F. Morant, and J.L. Navarro. Intelligent Controllers Issues. *Proc. IFAC Symp. SICICA'92*, pp. 321-334, 1992.

[21] A. Crespo, J.L. Navarro, R. Vivó, A. García, A. Espinosa, A real time environment for process control. *Proc. Int. Workshop on Artificial Intelligence in Real Time Control,*1991.

[22] Burns, A. "Preemptive Priority Based Scheduling: an Appropriate Engineering Approach" Real-Time Systems Research Group. Department of Computer Science. University of York, UK. Report number YCS214, 1993.

[23] Audsley, N.C., Burns, A., Richardson M.F. and Wellings, A.J. "Incorporating Unbounded Algorithms Into Predictable Real-Time Systems". Department of Computer Science. University of York, UK. Report number RTRG/91/102. 1991.

[24] Hassan H., Crespo A. "Scheduling Intelligent Tasks with Optional Parts in a Real-Time Environment". *3th IFAC/IFIP Workshop on Algorithms and Architectures for Real-Time Control.* May-95.

[25] Bondia, J, Moya, P. Picó, J. and Albertos, P. MARCK: An Intelligent Adaptive Real-Time System for Ceramic Kiln Control. *Interfaces 97.* Montpellier. May 1997.

[26] A. Crespo, J.L. Navarro, R. Vivó, A. García, RIGAS: an expert server task in real-time environments. *Proc. Int. Symp. on Art. Intel. In Real-Time Control*, pp. 631-636, Delft, 1992.

[27] Navarro, J.L., P. Albertos, F. Morant, M. Martinez, and R. Vivó, The Cement kiln: AI Approach to Model and Control, *IFAC Int. Symp. ADCHEM'91*, pp.331-334, Toulouse, 1991.

ADVANCED CONTROL APPLICATIONS IN THE STEEL INDUSTRY [1]

Graham C. Goodwin and **Richard H. Middleton**

Centre for Integrated Dynamics and Control
University of Newcastle
NSW 2308, Australia

All industries are under increasing pressure to remain competitive in the face of global market forces. To remain competitive in this environment it is important that industries keep up with the field in the area of product quality.

Advanced control has long been proposed as a mechanism for achieving significant quality improvements. However, the success in using advanced control has been varied with some spectacular successes matched by apparent failures in other areas.

There is now, however, a marked maturity in this area leading to renewed confidence in the application of advanced control. In part this is due to the widespread availability of computer aided design tools which allow the user to focus on the requirements of the particular application rather than being distracted by the details of the design methodology. Also, rapid prototyping tools have become available which greatly simplify the step from the design phase to implementation on the process for evaluation and testing.

These tools also allow "forensic" investigation of control difficulties. The issue here is that it is sometimes more important and difficult to pinpoint the intrinsic nature of a problem than it is to solve the problem once it has been identified.

This talk will illustrate the use of these new tools by applications drawn from the steel industry. Topics to be covered include combined tension and gauge control in rolling mills, eccentricity compensation, mould level control in continuous steel casting, zinc coating mass estimation in a continuous galvanizing line, and shape control in rolling mills. It will be shown how each of these applications can benefit from the use of advanced control ideas.

It will be argued that in many of these applications the solution to a control problem at one level opens the way for a more detailed understanding of the problems at deeper levels. This allows a staged approach to solving advanced control problems in which advances in sensor technology, actuator response times, modelling accuracy and control understanding can be combined to successively improve the performance of systems.

---

[1] Full paper not received

Predictive Optimal Control of Hot Strip Finishing Mills

M.J. Grimble & M.R. Katebi,
Industrial Control Centre, Strathclyde University, Graham Hills Building,
50 George St., Glasgow G1 1QE, UK.
fax : +44 141 548 4203 and e-mail : m.grimble@eee.strath.ac.uk

Abstract

The use of supervisory control in hot strip tandem rolling mills is discussed. A new model, which extends to distributed variables, is described and a controller using a predictive control methodology is designed. The controller is used in a supervisory capacity and takes into account constraints. A technique for controlling gauge and crown over the last two stands is described and simulation results are presented. A new result for the constrained control of gauge is also shown. *Copyright © 1998 IFAC*

1. Introduction

A hot strip tandem finishing mill takes in a bar of metal (at roughly 1100 °C in temperature for mild steel) and puts it through a series of rolling stands, typically decreasing its average thickness by an order of ten. As the width does not change much there is also a corresponding increase in length of the bar by an order of ten. Figure 1 shows a side-on view of the last two stands of a tandem finishing mill; there are typically six or seven in a steel rolling mill.

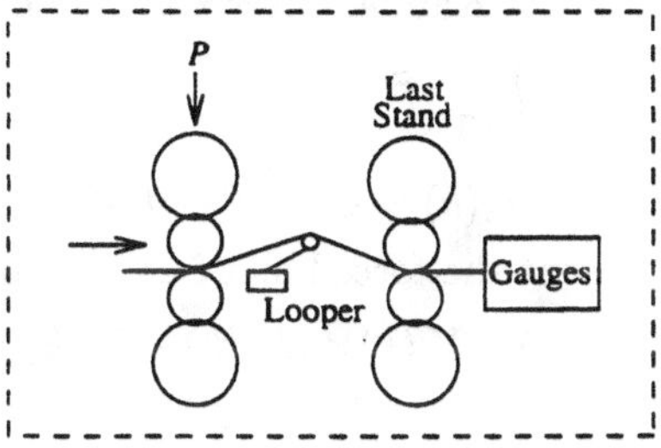

Figure 1: The Last Two Stands of A Tandem Mill

At each stand are actuators to change the roll gap setting between the work rolls, which act on the strip by generating a rolling force, P. The majority of the power to reduce the thickness of the strip comes from one main drive per stand which turns the work rolls, and whose speed is used as a manipulated variable. The purpose of the looper is to store some strip between stands in order that disturbances do not cause the strip to become too loose or tight. A perfect looper keeps the strip tension constant and hence decouples upstream stands from downstream ones. Also present at each stand and interstand are sensors to measure rolling force and looper height, from which the strip gauge and tension are estimated.

A 3-D view of the mill, such as in figure 2 introduces an additional process variable: profile. Profile is shorthand for the thickness profile across the strip width. Figure 2 shows the third main actuation at a stand, the bending force J. This, in its simplest usage, ameliorates the bowing effect that the two rolling forces P exert on the rolls and, consequently, the strip.

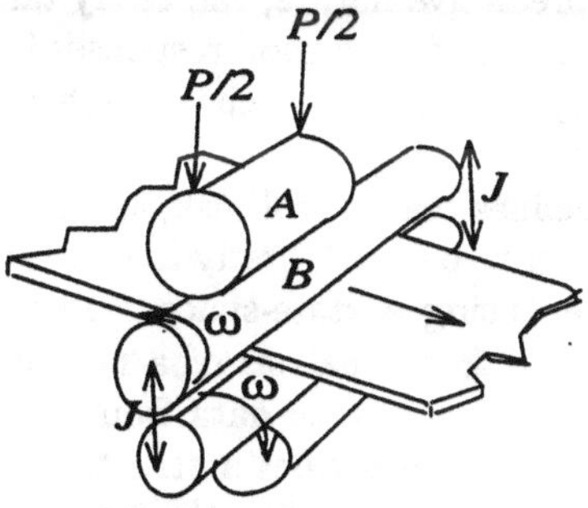

A - back-up roll
B - work roll

Figure 2: Three-Dimensional View of a Rolling Stand

At the exit of the mill are various gauges to measure the process variables themselves. These are gauge (on the strip centre-line), profile, temperature (which indicates strip surface quality). A metric for profile is *crown*, which is the difference between the centre-line thickness and the mean edge thickness.

During the rolling of a bar the control tasks carried out in a finishing mill can be split up into two broad groups:

- Local regulation of such variables as gauge, strip tension and looper height.
- Manipulation of regulator set-points across the mill in response to error readings at the exit of the mill.

The set-point manipulation requires the coordination across several stands in response to a signal at one stand. It is an example of a *supervisory* control scheme, as it can be represented in a hierarchical layer above the regulation layer. In Europe, reports suggest that set-point manipulation schemes are mainly multi-loop rather than multi-variable. However, in Japan the dynamic control of a whole mill by use of optimisation techniques has been described ((Okada *et al.*, 1996) and (Mizuno *et al.*, 1989)). In both of these cases a decentralised approach to controller design was adopted, albeit after matrix transformations that rendered the states of the models non-physical. An optimisation approach to a multi-stand design that assembles it from controllers designed on a single stand basis is also reported ((Hearns *et al.*, 1996) and (Anbe *et al.*, 1996)). In the latter case the designs were commissioned on a real plant. All these projects have concentrated on the process variables that do not vary widthways i.e. centre-line gauge, tension and the looper height.

This paper describes a modelling extension that includes profile as a controlled variable. This model is used to formulate a coordinated controller using the Generalised Predictive Control (GPC) methodology (in state-space form) which, because of its *receding horizon* and predictive nature, can easily take into account constraints and give good responses in the presence of transport delays, which are inherent in the rolling process.

Section 2 defines the control task, and section 3 sets down the model to be used, firstly in transfer function form and then using a state-space realisation. The model is derived from analysis of a non-linear simulation, which is verified with data from Llanwern hot strip mill. Section 4 describes a structure for coordinating the control of gauge over the last two stands of the mill, and contains some simulation results. There are results shown for the control of gauge and crown at the mill exit, using actuations at the last two stands, and the control of only gauge, taking into account a constraint at the last stand.

2. Control Design Methodology

The first part of the control task is to find a model for the variables of interest such that it can be put in the following standard form, indexed by discrete time k: As usual, Y, X and U represent the vectors of output, state and manipulated variables, and V represents measured or modelled disturbances.

$$X(k+1) = AX(k) + BU(k) + EV(k) \qquad (1)$$

$$Y(k) = CX(k) + DU(k) \qquad (2)$$

This leads to the second part of the control task which is to solve for a trajectory $[U^T(k), U^T(k+1), ..., U^T(k+NU-1)]^T$ in the GPC cost functional written down in (3). Note that only $U(k)$ is actually *applied* since the procedure is repeated at the next time step.

$$J(k) = \sum_{j=N1}^{N2} (R(k+j) - Y(k+j))^T \Gamma(R(k+j) - Y(k+j)) + \sum_{j=1}^{NU} (U(k+j-1))^T \Lambda (U(k+j-1)) \qquad (3)$$

$N1$ and $N2$ are the system deadtime and prediction horizon respectively, NU is the control horizon, Γ is the error weighting and Λ is the matrix of control penalties. $N2$ and NU together are used for gross tuning of the system, and trade of robustness versus control activity (a long prediction horizon gives a gentler response, whereas a long control horizon increases control activity). Λ is then used for fine-tuning, and so does not have quite the same role as in a linear quadratic cost function.

The procedure for solving a GPC problem in state-space is to first estimate the system state using a Kalman filter incorporating the system model. An optimiser then finds the future outputs and controls that minimise the objective function.

The above cost function, and the system modelled by 1 and 2, are now represented as follows, including an expression for constraints.

$$Y_s(k) = [Y^T(k+N1), ..., Y^T(k+N2)]^T$$
$$R_s(k) = [R^T(k+N1), ..., R^T(k+N2)]^T$$
$$U_s(k) = [U^T(k), ..., U^T(k+NU-1)]^T$$
$$f(k) = [(CA^{N1}\hat{X}(k))^T, ..., (CA^{N2}\hat{X}(k))^T]^T$$

where $\hat{X}$ is the Kalman filter estimate of the state

$$G = \begin{cases} g_{ij} = CA^{j-i-1}B & j > 1 \\ 0 & j \le i \end{cases}$$

$$Y_s(k) = GU_s(k) + f(k)$$

G and f represent the forced and free systems respectively

$$J(k) = (R_s(k) - Y_s(k))^T (R_s(k) - Y_s(k)) + U_s^T \Lambda U_s$$
$$U_s \le b \qquad (4)$$

If the expression for Y_s in terms of G, U_s and f is substituted into J(k) it is a simple matter, in the absence of constraints, to analytically minimise the expression with respect to U_s to give:

$$U_s = (G^T G + \Lambda I)^{-1} G^T (R_s - f) \qquad (5)$$

With constraints, in general, an optimisation problem must be solved at each time step (typically using a quadratic programming method). Consequently, this results in a time-varying controller with no fixed structure.

3. Derivation of the Model

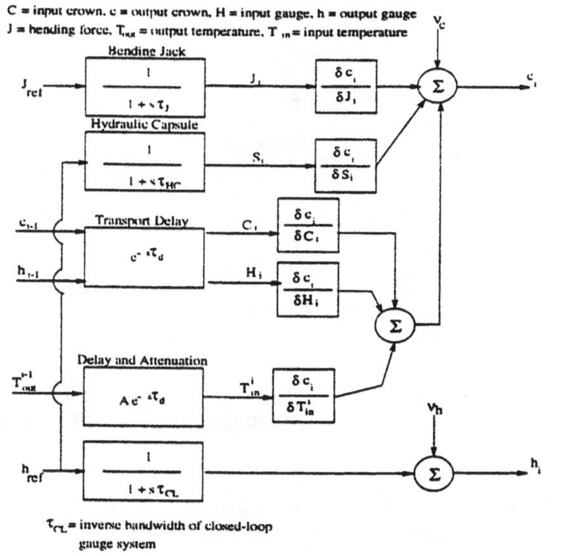

Figure 3: Gauge and Crown Model of One Stand Unit

The model in figure 3 is formulated on the basis of tests on a non-linear simulation and heuristic knowledge. The operating points around which it is based are defined by the *schedule* of the mill. This is the stand by stand requirement for strip quality parameters and other important variables like rolling force and strip speed. It can be realised in state space form by approximating the time delays as lags $\frac{1}{1+s\tau_l}$. For a given stand unit i, this is written thus:

$$\dot{x}_i = A_{ii}x_i + B_{ii}u_i + E_{ii}v_i + s_i \qquad (6)$$

$$y_i = C_{ii}x_i + w_i \qquad (7)$$

The assignments $y_i = [c_i, h_i]^T$, $u_i = [J^i_{ref}, h^i_{ref}]^T$, $x_i = [J_i, S_i, C_i, H_i, T^i_{in}, h_i]^T$, $s_i = [\frac{c_{i-1}}{\tau_l}, \frac{h_{i-1}}{\tau_l}]^T$, $v_i = T^{i-1}_{out}$ and $w_i = [\nu_c \nu_h]^T$ are made, where $k_i1 = \frac{\delta c_i}{\delta J_i}$, $k_i2 = \frac{\delta c_i}{\delta S_i}$, $k_i3 = \frac{\delta c_i}{\delta C_i}$, $k_i4 = \frac{\delta c_i}{\delta H_i}$ and $k_i5 = \frac{\delta c_i}{\delta T^i_{out}}$. The system matrices are defined as

$$A_{ii} = \begin{bmatrix} \frac{-1}{\tau^i_J} & 0 & 0 & 0 & 0 & 0 \\ 0 & \frac{-1}{\tau^i_{HC}} & 0 & 0 & 0 & 0 \\ 0 & 0 & \frac{-1}{\tau^i_l} & 0 & 0 & 0 \\ 0 & 0 & 0 & \frac{-1}{\tau^i_l} & 0 & 0 \\ 0 & 0 & 0 & 0 & \frac{-1}{\tau^i_l} & 0 \\ 0 & 0 & 0 & 0 & 0 & \frac{-1}{\tau^i_{CL}} \end{bmatrix}$$

$$B_{ii} = \begin{bmatrix} \frac{1}{\tau^i_J} & 0 \\ 0 & \frac{1}{\tau^i_{HC}} \\ 0 & 0 \\ 0 & 0 \\ 0 & 0 \\ 0 & \frac{1}{\tau^i_{CL}} \end{bmatrix}$$

$$C_{ii} = \begin{bmatrix} k_i1 & k_i2 & k_i3 & k_i4 & k_i5 & 0 \\ 0 & 0 & 0 & 0 & 0 & 1 \end{bmatrix}$$

$$E_{ii} = [0, 0, 0, 0, \frac{1}{\tau^i_l}, 0]^T$$

The interactions s_i are a function of y_{i-1}, and consequently x_{i-1}. Hence, this system can be represented in the standard form, seen in equations 1 and 2, where the vectors and matrices in the latter are stacked and blocked forms of the entities of a single stand model.

4. Experimental Results

Figure 4 shows the structure of the control scheme to be applied in the simulation experiment. The gauge and crown are regulated at the exit of the finishing mill using a single GPC controller that supplies increments to the gauge and crown references at stands 6 and 7 as the manipulated variables.

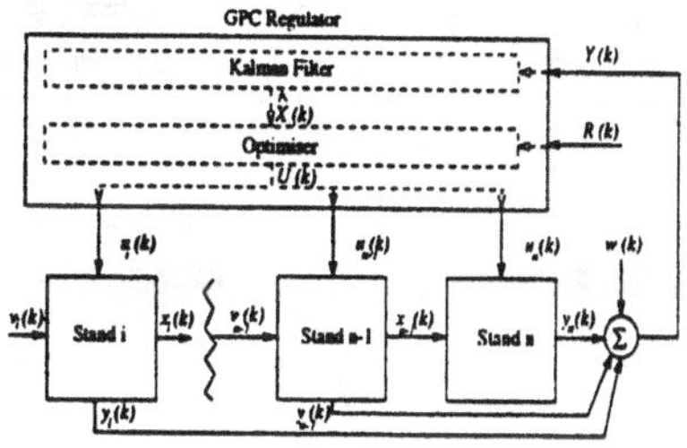

Figure 4: Predictive Control Structure

A step disturbance of 1mm was injected at the stroke input of stand 6 after 1 seconds, and the controller was switched on after 4 seconds. The intention was to test against steady-state bias, so this condition was allowed to be reached before allowing control action. Representative results are shown in figure 5 which depicts the outputs, crown and gauge, along with the actuations, roll-gap setting stroke and bending at stands 6 and 7. In test 1 all the weightings (Γ and Λ) are set to unity. It can be seen that after four seconds some rejection of the crown and gauge disturbances occur. In test 2 the control penalties are reduced by a factor of 10, making the control more active. This results in the gauge and crown nearly being returned to zero (as desired), however the stroke at stand 6 is momentarily 5mm, which is excessive. In order to curb this the inputs at stand 7 are penalised with respect to stand 6 in test 3. It can be seen that the stroke at stand 7 is reduced to an acceptable level whilst the output rejection does not return to the un-tuned level of test 1. This illustrates the ability of the predictive control methodology to tune and apply soft constraints.

Figure 6 shows the results of a simpler experiment where the output is gauge and the inputs are the stroke at stands 6 + 7. The units are normalised. A hard constraint is placed on the control activity at stand 7.

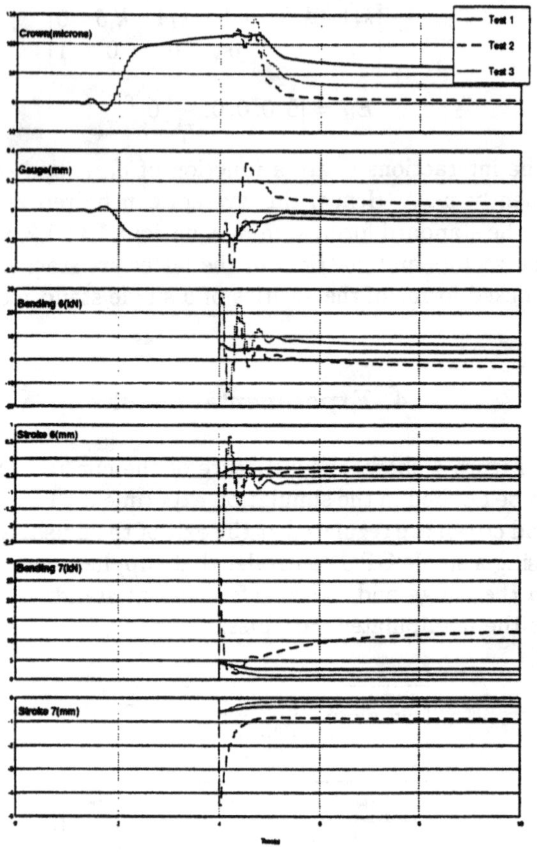

Figure 5: Multivariable Results for Various Settings of Λ

Roll gap changes may be constrained because of acuator limitations or implied deterioration of the output quality. In this case the output at the gauge sensor is made up mostly from the response at stand 6, which is delayed by approximately a second with respect to stand 7. This illustrates the constraint handling of predictive control, for the controller dynamically assigned a higher control gain to stand 6: without taking into account constraints various undesirable responses can occur from a loss of performance (offsets and overshoots) to instability.

5. Conclusions

This paper has described the background to a control problem in hot strip tandem rolling. It has then outlined a design and modelling methodology and presented results which showed that predictive control is suitable for upper-level control, and can use information about constraints to give improved designs. To improve the quality of of rolled strip, metallurgical and mechanical engineering properties must be controlled and the rolling load applied on stands must therefore satisfy strict requirements. In conventional control of multi-variable systems it is difficult to impose hard constraints on variables. The predictive control approach offers a mechanism whereby limits can be pre-specified and guaranteed behaviour obtained. The use of predictive optimal control for the supervisory level should therefore offfer greater control over product quality and thereby provide competitive advantage.

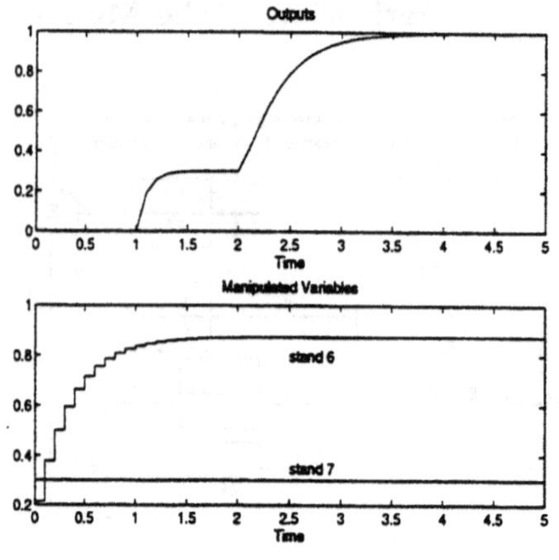

Figure 6: Two Stand Gauge Control: With Constraints

6. Acknowledgements

The authors are grateful to the ACME directorate for their support on the project EPSRC GR/J/54864, and for the support of the project's industrial partners Alcan, British Steel, Cegelec and Davy. The substancial help provided by Gordon McNeilly in producing the paper and developing the simulation results is also gratefully acknowledged.

References

Anbe, Y., K. Sekiguchi and H. Imanari (1996). Tension control of a hot strip mill finisher. In: *13th Triennial World Congress.* San Francisco.

Hearns, G., M.R. Katebi and M.J. Grimble (1996). Robust control of a hot strip mill looper. In: *13th Triennial World Congress.* San Francisco.

Mizuno, K., Y. Morooka and Y. Katayama (1989). Multivariable nonlinear control system design for multistand rolling mill. In: *Production Control in the Process Industry.* IFAC. Osaka.

Okada, M., Y. Iwasaki, K. Murayama, A. Urano, A. Kawano and H. Shiomi (1996). Optimal control system for hot strip finishing mill. In: *13th Triennial World Congress.* San Francisco.

Discrete-Event Simulation for Manufacturing Systems [1]

Neil Mort

Department of Automatic Control and Systems Engineering
University of Sheffield
Mappin St., Sheffield S1 3JD, U.K.

The use of software for the simulation of systems with continuous dynamics is widespread among engineers in universities and industry. A pre-requisite for using this software is that the systems dynamics be described in mathematical form (for example, differential equations or state-space models). In the case of manufacturing systems, the use of such mathematical models is often not appropriate and so the question of how to simulate these systems must be asked. This tutorial provides an introductory look at some of the issues involved in looking at approaches to simulation of discrete-event manufacturing systems. No previous knowledge of the topic will be assumed.

After a brief introduction to the material to be covered, the distinction between probabilistic and deterministic process is made. The basic ideas of statistical simulation are then illustrated using an inventory control example. Following this, the principles of discrete-event simulation languages are described with a brief coverage of the special attention required for output data analysis, e.g. dealing with initial transients and variance reduction methods. Finally, an example from a steel processing plant, where the objective is to investigate the operational effectiveness of a soaking pit/rolling mill process, is presented which illustrates some of the topics referred to earlier in the tutorial.

[1] Full paper not received

Satisfactory Control of Multi-objective Systems using Fuzzy-Logic Approach[1]

Zeungnam Bien

Department of Electrical Engr., KAIST
Taejon, Korea
E_mail : zbien@ee.kaist.ac.kr

Controlling a complex uncertain dynamic system often entails a heuristic approach of directly forming the control laws from experience. Recently, it is well accepted that the control law of fuzzy if-then rule type is a practical alternative for such large-scale uncertain systems for which mathematical models are either unavailable or futile. When multiple performance objectives are involved, however , the fuzzy-logic based controller design becomes quite problematic because the multiple objectives may render logically inconsistent control rules due to their competing nature.

In this talk, we first address the issues of inconsistent rule base. We revise the definition of inconsistency between two fuzzy if-then rules and then refine the measures of inconsistency between the rules. Also, we briefly review the FLC design methods in consideration of the levels of inconsistency of the rule set.

Next, we investigate the Multiple-Objective System Control Problem(MOSCP) for the case when the plant is complex and uncertain. It is known that, if the mathematical model is available, the multiplicity of the control objectives often results in mathematical intractability for many types of optimal problems. For a given complex uncertain dynamic system with multiple objectives, there arise two types of difficulty in the design of a control system: one is that no convincing mathematical control law can be derived due to lack of certainty model and the other is that the concept of optimality would not make sense because there would be no way of knowing the optimality due to the uncertainty of the plant as well as because the total ordering of the vector values of the performance indices is not possible.

[1]Full paper not received

As a means of circumventing the first difficulty, we may obtain the fuzzy if-then control rules by interviewing expert operators, studying relevant literature and/or incorporating information from system design engineers. In this case, with multiple objectives and with multiple view points regarding the system environment, multiple operators may produce rules that are logically inconsistent. In this regard, the many MOSCP problems should be handled in consideration of logical inconsistency issue for realizing an effective controller. In analyzing multiple objective control systems, it is also important to understand the relationships between the objectives. One objective may be (or may not be) independent of the other objectives. An objective may be conflicting with some other objective. We analyze the nature of the objective functions in relation to decision making and discuss several ways of handling them for problem formulation.

The second difficulty involved in the MOSCP problem is that, when the plant is complex with uncertainty , the concept of "Optimal Control", or the superlative control, is not meaningful in most cases. As an alternative, we propose a new control solution called "Satisfactory Control". The concept of satisfactory control is a subjective one which is hardly defined in a mathematical way. The term roughly means that the worst case of the control performances is acceptable with respect to the specifications and its average performance is competent in comparison with the performances of any existing control systems. To be more specific, we propose that each of the negotiable performance objectives is transformed to a fuzzy set whose membership function is called a satisfaction degree function. Then, instead of solving the original optimal control problem for multi-objective system, we try to find a satisfactory solution for the given plant under the hard and soft constraints. As an practical approach, a systematic method is presented for finding a satisfactory control for a complex system for which a set of possibly inconsistent rules is assumed to exist. With the newly defined level of inconsistency, the fuzzy c-means technique is used for grouping the rules and the max-min concept is employed to improve the solution.

Several simulation examples as well as experimental results are illustrated to show the validity of the methods.

FUZZY ADAPTIVE CONTROL OF THE HYDRAULIC BENDING ROLLS

Qiao Jun-fei Guo Ge Chai Tian-you

Research Center of Automation, Northeastern University
Shenyang, 110006, China

Abstract: Fuzzy adaptive strategy is suggested for the control of hydraulic bending roll system. A simple but useful fuzzy adaptive controller is developed which is composed of parameter self-optimization mechanism and structure self-tuning mechanism. The effectiveness of this control strategy was proved by experiments. *Copyright © 1998 IFAC*

Keywords: Shape control, hydraulic system, fuzzy adaptive control

1. INTRODUCTION[1]

Force control of hydraulic bending rolls is a basic and most important link in automatic flatness control system (AFC). Its dynamic characteristics play a primary role for the performance of the whole system. So it's the first problem to be solved in shape control system.

PID method is commonly used in hydraulic bending roll systems (Anders, G.C., et al., 1991; Hoshno, I., et al., 1993). But it is unsatisfactory in meeting the need of control performance, owning to the nonlinear time-varying charateristic of the hydraulic systems and some random disturbances which makes precisely modeling very difficult. A self-tuning method was used by Y. Xu (1995) to control the bending roll system but ended up with a performance not very ideal. A more superior but relatively simple fuzzy adaptive control approach is presented in this paper whose effectiveness is proved by experiments.

[1] This research was supported by Chinese National Natural Science Fund

2. HYDRAULIC BENDING ROLL SYSTEM

The research work was conducted with the shape control system of a three-stand continuous rolling mill in Chinese National Key Lab. of Rolling Technology and Continuous Rolling Automation, Northeastern University. The hydraulic bending roll system of this rolling mill comprises the following main parts: a servo amplifier, a elctro-hydraulic servo valve and a hydraulic cylinder.

2.1 Nonlinearity of the system

The sliding valve is a positive-opening three-passageway valve. Its flow characteristic equation can be written as

$$Q = C_1(\Delta + X)\sqrt{P_s - P_l} - c_1(\Delta - X\sqrt{P_l}) \quad (1)$$

where

C_1 is a throttling coefficient, Δ is the positive opening value of the valve, X is position shift of the valve, P_s and P_l are oil source pressure force and load pressure respectively.

The nonlinear throttling characteristic of the sliding valve is linearized as the following common used form to simplify our analysis

$$Q=\frac{\partial Q}{\partial X}\Big|_P\cdot X+\frac{\partial Q}{\partial P}\Big|_X\cdot P_l=K_qX+K_lP_l \qquad (2)$$

where

K_q and K_l are the coefficients of flow gain and flow pressure of the valve orifice respectively, which will change with X and is time-independent. This will be described later.

2.2 Time varying characteristic of the system

When the compressive characteristic aroused by the pressure of oil and the leakage flow are considered, we obtain the flow equation according to the continuity principle of flow

$$Q-Q_p-Q_l=A\frac{\partial y}{\partial t} \qquad (3)$$

where

A and y are the cross sectional area of the piston and its position shift respectively. namely the inflow of the system equals to a sum of flow lost when the oil is compressed, the leakage flow and the volume increasing rate of the system. Here the lost flow is

$$Q_p=\frac{V}{\beta}\frac{\partial P}{\partial t} \qquad (4)$$

The leakage flow is

$$Q_l=L_pP_l \qquad (5)$$

where

V is the volume of oil, β is the modulus of elasticity of oil volume and L_p is the coefficient of oil leakage. Hence equation (5) reads as

$$Q=A\frac{\partial y}{\partial t}+\frac{V}{\beta}\frac{\partial P}{\partial t}+L_pP_l \qquad (6)$$

From this equation we can see that the volume of oil changes with pressure.

2.3 Uncertainty of the system

There are inevidently changes of temperature and feeding in the production of strip. This results in rough random disturbances in hydraulic bending roll system. In addition, some parameters such as the modulus of elasticity of oil capacity and the elastic force are slow time-varying with the temperature of oil. Thus the estimation of the dynamic parameters of this system is inaccurate. Moreover, precise system model is inaccessible in that the relevant factors such as the inertia force, the viscous drag, the elastic force and the constant disturbance force etc., the accurate obtaining of all of which are very difficult, should all be dealt with in the force analyzing process of the hydraulic piston.

3. FUZZY ADAPTIVE CONTROL

The structure of the fuzzy adaptive control system is shown in Fig.1 where the two inputs are the error between the set point P_t of bending force and its output P and the variance of this error.

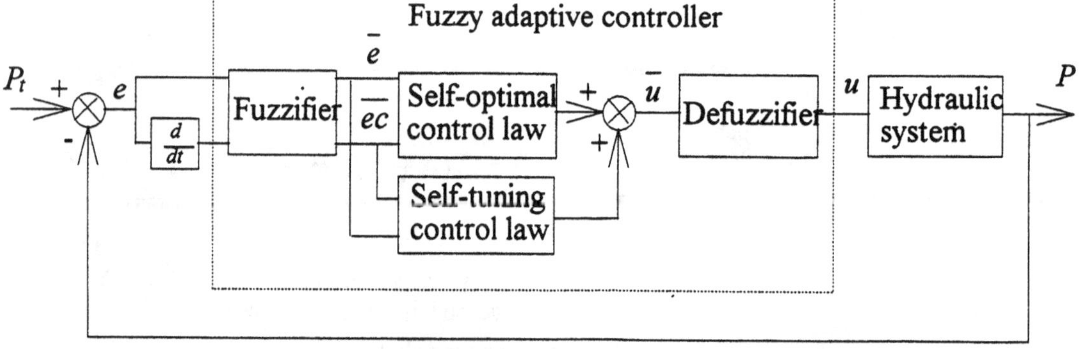

Fig. 1 The fuzzy adaptive control system

The control value, system error and the change of error are respectively in the ranges $[-x_e, x_e]$, $[-x_{ec}, x_{ec}]$ and $[-y_u, y_u]$. They are fuzzified in the same universe of discourses

$$[-6,-5,-4,-3,-2,-1,0,1,2,3,4,5,6]$$

3.1 Fuzzy self-optimal control law

A fuzzy self-optimal control law is introduced into the force control system to ensure stability when error is relatively small and, otherwise, delete system error. The control law reads

$$\overline{u_c} = -[\alpha\overline{e} + (1-\alpha)\overline{ec}] \tag{7}$$

where

$\alpha = \frac{1}{n}|\overline{e}|$ and $n = \max|\overline{e}|$.

3.2 Fuzzy self-tuning control law

In order to obtain perfect dynamics and system robustness, the above optimal control law is compensated by self-tuning control method.

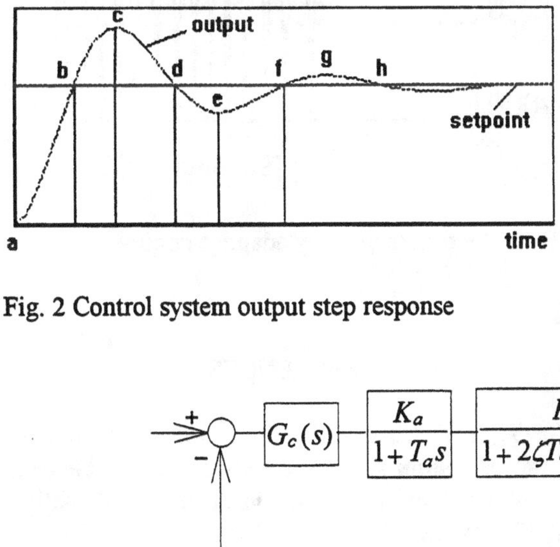

Fig. 2 Control system output step response

An output response of the control system is depicted in Fig.2. It's obvious that within the ranges a-b, c-d and e-f, there is "good" extent of output error decreasing; and in the segments b-c, d-e and f-g, "bad" extent of error increasing takes place. Define a cost function

$$J = e \cdot ec \tag{8}$$

Subsequently we obtain the self-tuning fuzzy control law

$$\overline{u_{tc}} = -\alpha K_{tc} e \qquad J < 0 \tag{9}$$

$$\overline{u_{tc}} = -K_{tc} e \qquad J > 0 \tag{10}$$

where

K_{tc} is the gain coefficient of the self-tuning control value.

Finally the fuzzy adaptive control law is obtained

$$\overline{u} = \overline{u_c} + \overline{u_{tc}} \tag{11}$$

4. SIMULATING EXPERIMENTS OF THE CONTROL SYSTEM

The brief structure of the whole bending roll control system as depicted in Fig.3. (Yang, J., et al., 1996) in which $G_c(s)$ is the transfer function of the system controller.

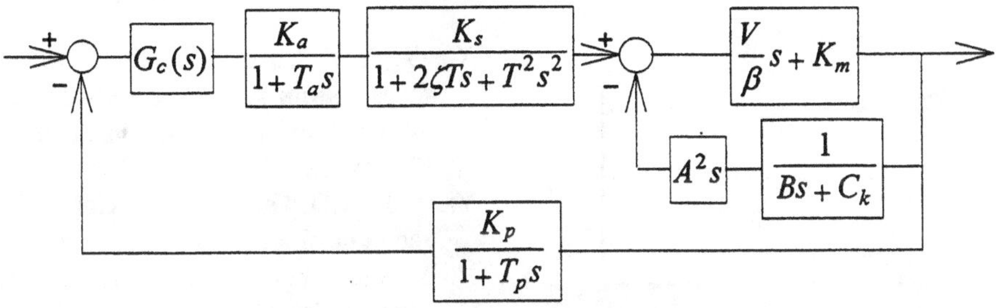

Fig. 3 The structure of bending roll control system

When practical plant data are sent to all the parameters, we get the close loop transfer function of the controlled bending roll system

$$G(s) = \frac{5.728\times10^{-3}s+68.2}{(3.46\times10^{-4}s^2+0.329s+160)(2.5\times10^{-3}s+1)(1.96s+1)(2\times10^{-3}s+1)} \tag{12}$$

In discrete domain, sampling time is supposed to be 10 ms, the corresponding expression of equation (12) is

$$G(z^{-1}) = \frac{z^{-1}(1.2\times10^{-3}+1.3\times10^{-3}z^{-1}+10^{-3}z^{-2}}{1-0.8063z^{-1}-0.1703z^{-2}} \tag{13}$$

From equation (13) we can write out the CARMA model of the hydraulic bending roll system

$$A(z^{-1})y(k) = z^{-d}B(z^{-1})u(k) + C(z^{-1})\xi(k)$$

where

$d=1$, $y(k)$、$u(k)$ and $\xi(k)$ are system output, input and noise term respectively, d is transport delay and here

$$A(z^{-1}) = 1 - 0.8063z^{-1} - 0.1703z^{-2}$$

$$B(z^{-1}) = 0.0012 + 0.0013z^{-1} + 0.001z^{-2}$$

$\xi(k)$ is a independent random noise series satisfying

$$E\{\xi(k)\} = 0 \quad \text{(14-a)}$$

$$E\{\xi(i)\xi(j)\} = \begin{cases} \sigma^2 & i = j \\ 0 & i \neq j \end{cases} \quad \text{(14-b)}$$

$$\lim_{N\to\infty} \frac{1}{N}\sum_{k=1}^{N} \xi(k)^2 < \infty \quad \text{(14-c)}$$

Equation (14-a) through (14-c) show us that the mean of random noise is zero, its variance is a finite positive value and its sampled mean square is bounded. $C(z^{-1})$ is a stable polynomial which is assumed to be

$$C(z^{-1}) = 1 - 0.25z^{-1} - 0.5z^{-2}$$

Some simulation experiments are conducted with the following parameters

x_e=0.8, x_{ec}=0.1, y_u=1200, K_{tc}=5

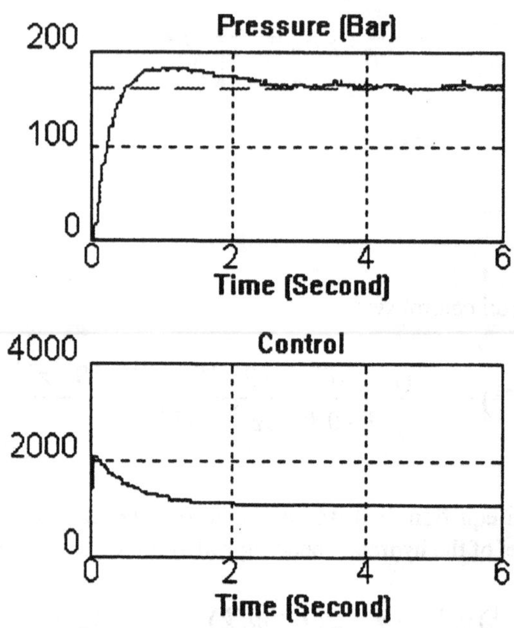

Fig. 4 The results of self-tuning control

As the conventional PID controller can not perform well in regulating bending roll force, a self-tuning algorithm based on generalized least variance (Xu, Y., et al., 1995) was used in place of PID method. The control performance of this method is still not very satisfactory, which is shown in Fig. 4. Thus the fuzzy adaptive control method presented in current paper is used and its dynamic behavior (see Fig. 5) is remarkably improved when compared with that of the self-tuning controller.

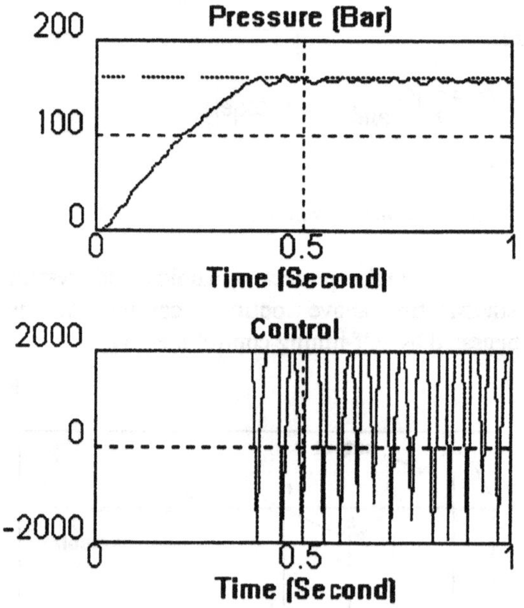

Fig. 5 The results of fuzzy adaptive control

REFERENCES

Xu, Y., T.Y. Chai, Z.Z. Mao (1995). Research on shape control system of four-high reversible cold rolling mill. Metallurgical Automation, 19(1):8~12.

Yang, J., X.D. Guan (1996). Characteristic analysis and control algorithm design of shape control in hydraulic bending roll system. Proceedings of Chinese Control and Decision Conference, 1105-1108.

Chai, T.Y. (1985). Research of Multivariable Self-tuning Controller. Dr. Essay of Northeastern University, China.

Anders, G.C., O. Keijser (1991). Modern approach to flatness measurement and control in cold rolling. Iron and Steel Engineer, 68(4): 34~37

Mason, J.D. (1990). The measurement and control of flatness. Proc. of 5th International Rolling Conference, 205~209.

Hoshno, I., M. Kawai, M. Kokubo, et al. (1993). Obersver-based multivariable flatness control of the cold rolling mill. IFAC, 149~156

APPLICATIONS OF MIN-MAX GENERALIZED PREDICTIVE CONTROL TO SINTERING PROCESSES

Yong Ho Kim* and **Wook Hyun Kwon***

* *Control Information Systems Lab.,*
School of Electrical Engr., Seoul National Univ.,
San 56-1 Silim-dong Kwanak-gu, Seoul 151-742, Korea
TEL.: +82-2-873-2279, FAX.: +82-2-878-8933,
E-mail: yongho@isltg.snu.ac.kr

Abstract: This paper presents applications of min-max generalized predictive control (MMGPC) to burnthrough point control in industrial sintering processes. A simple model is used as a plant model for the burnthrough point control. It is derived for a complicated sintering process system by introducing events into a continuous variable system. As a control algorithm, the MMGPC is used. It is robust to disturbances and has guaranteed stability. Simulations of the burnthrough point control using the MMGPC are performed using a derived POSCO (Pohang Steel Company, Korea) sintering plant model and real burnthrough time data of sinter packages. The simulation results of the MMGPC are compared with those of PID control and are satisfactory. *Copyright © 1998 IFAC*

Keywords: Generalized Predictive Control, Processes, Stability, Simulation

1. INTRODUCTION

The sintering process is a pre-process for blast-furnace materials. The sintering of iron ore is an essential part of modern iron-making, since a sinter has a great effect on the operating conditions of the blast furnace. In the sintering process, there are several control loops. The major control loops consist of the feed control, return fine/coke control, water control, burnthrough point (BTP) control, and bed height control. There are close relations among these control loops. The BTP control, which is treated in this paper, is the most important, since this control affects other control loops directly and results in major sources of disturbance in them. The purpose of BTP control is to keep the burnthrough point near a desirable set-point at approximately the end of the strand and to reduce speed changes of the strand. The existing mathematical models (Dash and Rose, 1977; Hansen *et al.*, 1992; Lewis *et al.*, 1983; Rose *et al.*, 1993; Young, 1979) are, however, too complicated for control purposes. Moreover, there exist few works in the open literature on BTP control methods and related models (Dash and Rose, 1977; Hansen *et al.*, 1992; Kwon *et al.*, 1997; Lewis *et al.*, 1983; Rose *et al.*, 1993; Young, 1979).

The current BTP of a sinter package needs to be calculated from the patterns of the already-passed-over wind box temperatures after the sinter package has passed over the highest temperature. The information about BTPs is old for the newly arriving sinter packages which are to be controlled. Hence some kinds of delay exist in calculating BTPs. When average strand speed is applied to a motor drive system as a command input signal, there may be some errors between the actual motor speeds and the average strand speeds due to the characteristics of the motor drive system. In this case, the errors exist in the input and act as input disturbances. The temperature is measured in multiple positions of each wind box and the interpolation method is used to get the temperature profile at all positions between temperature sensors. In this case, the errors between calculated temperatures and real temperatures exist and act as output disturbances. Due to these delays and disturbances, a burnthrough time control algorithm which is robust to disturbances and has prediction concepts is needed.

In this paper, a min-max generalized predictive control (MMGPC) algorithm is used, which is robust to disturbances and stable under some conditions. The MMGPC minimizes a cost function while disturbances maximize the cost function. The original form of the MMGPC was first proposed without stability results in Noh *et al.* (1996). The MMGPC is obtained from the well known results of LQ discrete-time min-max game theory (Basar and Oldser, 1982) with terminal output weightings (Kim *et al.* 1997). In Kim *et al.* (1997), the stability conditions of the MMGPC were presented and converted to a LMI (Linear Matrix Inequality) form to obtain appropriate parameters for the stability easily.

In Section 2, an event-based model of the burnthrough point control system in the sintering processes is presented. In Section 3, the MMGPC is presented and its stability properties are presented based on the paper (Kim *et al.*, 1997). In Section 4, the BTT prediction algorithm is given. In Section 5, the simulation results are shown. Finally, some concluding remarks are given in Section 6.

2. AN EVENT-BASED MODEL FOR SINTERING PROCESSES

There are some complex mathematical models (Dash and Rose, 1977; Hansen *et al.*, 1992; Lewis *et al.*, 1983; Rose *et al.*, 1993; Young, 1979). Since those models are too complex to be used in the control of burnthrough point, the simple model is used. A simple dynamic model for a burnthrough point control system is very important for real applications. The simple model used in this paper is derived using events in Kwon *et al.* (1997). Before some definitions are introduced, it is assumed that the typical temperature history of a sinter package follows a pattern like Figure 1. *The burnthrough time (BTT) for a sinter package is defined as the time duration for the sinter package to reach the highest temperature from the instant when it was loaded onto the strand.* The temperature history and the burnthrough time of a sinter package are both dependent on the characteristics of the sinter package. *The burnthrough point (BTP) of*

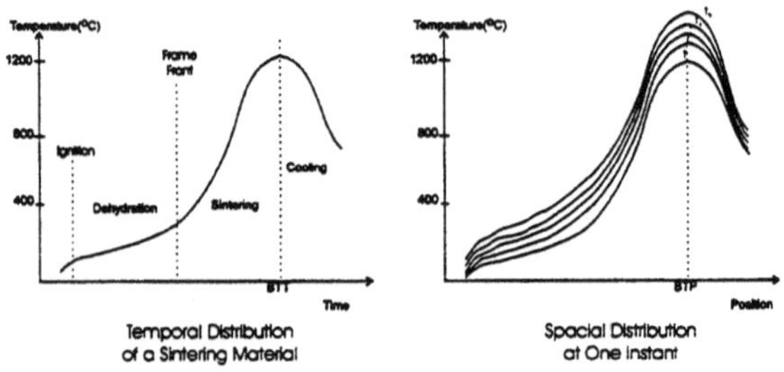

Fig. 1. Typical Temperature Patterns of Sintering Materials

a given sinter package is defined as the point on the strand where the temperature of that sinter package reaches its highest value. The BTP of each sinter package is indicated as a distance from the charging end of the strand. The burnthrough point control problem is to keep the burnthrough point of each sinter package at a desired fixed point near the end of the strand. The BTP of each sinter package can be changed by the strand speed. Hence, for burnthrough point control problems, the control variable is the strand speed and the controlled variable is the burnthrough point of each moving sinter package. The BTT is also involved in the dynamics for the BTP control problem. The relationship between these variables is very complicated.

Let l be a certain length, the length of the strand be $N \times l$, and the desired position of the burnthrough point be $n \times l$ distances from the charging end of the strand. In this plant, the distance of the strand is measured from the charging end. *Here an event is defined to occur whenever the strand moves a distance l from the starting time.* Let the 1-th event occur at the starting time when the strand has just begun to run. The second event occurs when the strand moves the distance l from the starting time. Then, the k-th event occurs when the strand moves a distance l after the occurrence of $(k-1)$-th event. Let the k-th material be the materials loaded on the strand between the k-th event and the $(k+1)$-th event. Since the strand is a long straight belt of finite length, the $(k+1)$-th event occurs when the $(k-1)$-th material moves the next distance l after the occurrence of the k-th event. *Let $u(k)$ be the time duration between the k-th and $(k+1)$-th event.* That is, $u(k)$ is the time duration during which the strand moves a distance l after the occurrence of the k-th event. It is noted that the time duration $u(k)$ is not constant but is varying. The state $x_i(k)$ is defined as the time duration which the sinter package at the position of $i \times l$ distance from the charging end of the strand at the occurrence of the k-th event has spent from the instant when it was loaded onto the strand. Since the sinter package at the position of $i \times l$ distance from the charging end of the strand at the occurrence of the $(k+1)$-th event lay at the position of $(i-1) \times l$ distance from the charging end of the strand at the occurrence of the k-th event and will spend $u(k)$ between the k-th event and the $(k+1)$-th event, the following relationships are given:

$$
\begin{aligned}
x_1(k+1) &= u(k) \\
&\vdots \\
x_i(k+1) &= x_{i-1}(k) + u(k) \\
&\vdots \\
x_N(k+1) &= x_{N-1}(k) + u(k).
\end{aligned}
\tag{1}
$$

$x_i(k)$ for $i \geq k$ is undefined from the definition of $x_i(k)$. The desired burnthrough point is assumed to be at the position of $n \times l$ distance from the charging end of the strand. Let the value $y(k)$ be the time duration which the sinter package at the desired BTP at the occurrence of the k-th event has spent from the instant when it was loaded onto the strand. The controlled output $y(k)$ is taken as

$$y(k) = x_n(k). \tag{2}$$

Using the input $u(k)$ and the output $y(k)$, the input-output relation is

$$y(k) = u(k-1) + \cdots + u(k-n). \tag{3}$$

If disturbances $w(k)$ are added to the above model (3) using the change of inputs, the I/O model can be converted to

$$
\begin{aligned}
A_p(q^{-1})y(k) &= B_p(q^{-1})\Delta u(k-1) \\
&\quad + C_p(q^{-1})w(k), \qquad (4)\\
A_p(q^{-1}) &= 1 - q^{-1}, \\
B_p(q^{-1}) &= 1 + q^{-1} + \cdots + q^{-n+1}, \\
C_p(q^{-1}) &= c_0 + c_1 q^{-1} + \cdots + c_{p-1} q^{-p+1},
\end{aligned}
$$

where q^{-1} is the delay operator, $\Delta = 1 - q^{-1}$, and $w(k)$ is the disturbance.

3. A CONTROL ALGORITHM

The control algorithm used in this paper is the min-max generalized predictive control which has been introduced in Noh *et al.*, (1996) and Kim *et al.* (1997). It has characteristics such as robustness to disturbances and guaranteed stability under some conditions. First, output predictor equations have to be investigated to present controller equations. The general relations of the vector forms between outputs and inputs are given by the following matrix equation:

$$\overline{y}(k) = G\overline{u}(k) + D\overline{w}(k) + \overline{f}(k). \tag{5}$$

Using the constraint horizon N_F and the output horizon N_y, the above matrix equation can be partitioned to

$$
\begin{bmatrix} \overline{y}_1(k) \\ \overline{y}_2(k) \end{bmatrix} = \begin{bmatrix} G_1 \\ G_2 \end{bmatrix} \overline{u}(k) + \begin{bmatrix} D_1 & 0 \\ D_F & D_2 \end{bmatrix} \begin{bmatrix} \overline{w}_1(k) \\ \overline{w}_2(k) \end{bmatrix} + \begin{bmatrix} \overline{f}_1(k) \\ \overline{f}_2(k) \end{bmatrix}, \tag{6}
$$

where input, output and disturbance vectors are defined as

$$
\begin{aligned}
\overline{y}_1(k) &= [\, y(k+1) \cdots y(k+N_y) \,]^T, \\
\overline{y}_2(k) &= [\, y(k+N_y+1) \cdots y(k+N_y+N_F) \,]^T, \\
\overline{u}(k) &= [\, \Delta u(k) \cdots \Delta u(k+N_u-1) \,]^T, \qquad (7)\\
\overline{w}_1(k) &= [\, w(k+1) \cdots w(k+N_y) \,]^T, \\
\overline{w}_2(k) &= [\, w(k+n+1) \cdots w(k+N_y+N_F) \,]^T,
\end{aligned}
$$

N_u is the control horizon, and the gain matrices G_1, G_2, D_1, D_F, D_2 and the vectors $\overline{f}_1$, $\overline{f}_2$ have appropriate dimensions. It is assumed that $C(q^{-1}) = 1$. This assumption means that disturbance sequences are uncorrelated and the current disturbance $w(k)$ gives influence only on the current output. $\overline{f}_1(k)$ and $\overline{f}_2(k)$ represents zero-input response and can be rewritten as

$$
\begin{aligned}
\overline{f}_1(k) &= F_{1y}\underline{y}(k) + F_{1u}\underline{u}(k), \\
\overline{f}_2(k) &= F_{2y}\underline{y}(k) + F_{1u}\underline{u}(k),
\end{aligned}
\tag{8}
$$

where $\underline{y}(k) = [y(k)]^T$, $\underline{u}(k) = [\Delta u(k-1)\ \Delta u(k-2) \cdots \Delta u(k-n)]^T$, and the matrices ($F_{1y}$, F_{2y}, F_{1u}, F_{2u}) have appropriate dimensions.

The burnthrough point control has two direct objectives:

(1) If possible, make the output $y(k)$ equal to the burnthrough time $BTT_n(k)$ for all k.
(2) Keep the variation in the strand speed as small as possible.

The second objective is necessary since variation in the strand speed is the main source of disturbance in other control loops. This control problem can be seen as a tracking problem, where the reference signal is $BTT_n(k)$. It is well-known to engineers that BTP control is difficult. Based on the above control objectives, consider the following cost function:

$$
\begin{aligned}
&\min_u [\max_{w_1} (J_k(u(k), w(k)))] \qquad (9)\\
&= \min_u [\max_{w_1} \{ J_1(k) - \gamma^2 \overline{w}_1^T(k)\overline{w}_1(k) + J_2(k) \}],
\end{aligned}
$$

where

$$J_1(k) = \sum_{i=1}^{N_y} [\lambda \Delta u(k+i-1)^T \Delta u(k+i-1)]$$

$$= \lambda \overline{u}(k)^T \overline{u}(k),$$

$$J_2(k) = \sum_{i=N_y+1}^{N_y+N_F} [\{\hat{y}(k+i|_{k+N_y}) - y_r(k+i)\}^T q_2 \cdot \{\hat{y}(k+i|_{k+N_y}) - y_r(k+i)\}] = (\hat{\overline{y}}_2(k) - \overline{y}_{r2})^T Q_2 (\hat{\overline{y}}_2(k) - \overline{y}_{r2}), \quad (10)$$

$$Q_2 = q_2 I, \; q_2 \geq 0,$$
$$\hat{\overline{y}}_2(k) = [\hat{y}(k+N_y+1|_{k+N_y}) \cdots \hat{y}(k+N_y+N_F|_{k+N_y})]^T,$$

and $\hat{y}(k+i|_{k+N_y})$ is the value of some predictions at time $k+i$ using data before time $k+N_y$ *without disturbances after time* $k+N_y$.

The following proposition modified from the result in Kim *et al.* (1997) can be used as a control law in this paper.

Proposition 1. For the given system(4), the MM-GPC law which satisfies the equations (9), (10), and (10) can be described by the following control law:

$$\begin{aligned}
&\Delta u_{MM}(k) \\
&= \overline{e}_1[\lambda I + G_2^T(Q_2 + Q_2 D_F \Sigma_2^{-1} D_F^T Q_2) G_2]^{-1} \\
&\quad \cdot G_2^T(Q_2 + Q_2 D_F \Sigma_2^{-1} D_F^T Q_2)(\overline{y}_{r_2} - \overline{f}_2(k)) \\
&= \overline{e}_1[\lambda I + G_2^T(Q_2^{-1} - \frac{1}{\gamma^2} D_F D_F^T)^{-1} G_2]^{-1} G_2^T \\
&\quad \cdot (Q_2^{-1} - \frac{1}{\gamma^2} D_F D_F^T)^{-1}(\overline{y}_{r_2} - \overline{f}_2(k)) \qquad (11) \\
&= \overline{e}_1 \frac{1}{\lambda} G_2^T [\frac{1}{\lambda} G_2 G_2^T + Q_2^{-1} - \frac{1}{\gamma^2} D_F D_F^T]^{-1} \\
&\quad \cdot (\overline{y}_{r_2} - \overline{f}_2),
\end{aligned}$$

where $\overline{e}_1 = [1 \;\; 0 \;\; \cdots \;\; 0]$ and $\overline{y}_{r_2}$ is the reference vector, if the inequality:

$$\Sigma_2 \triangleq \gamma^2 I - D_F^T Q_2 D_F \; > 0 \qquad (12)$$

holds. While the maximizing disturbance is

$$\overline{w}^* = (\gamma^2 I - {D_F}^T Q_2 D_F)^{-1} {D_F}^T Q_2 (G_2 \overline{u} + \overline{f}_2 - \overline{y}_{r_2}). \qquad (13)$$

The above control law can stabilize the given system if the inequalities presented in the theorem of Kim *et al.* (1997) are satisfied.

4. PREDICTION OF REFERENCE SIGNALS

In the equation (11), the reference $\overline{y}_{r2}$ is needed in calculating control inputs. The reference is burnthrough time of each sinter package. Hence, burnthrough times must be decided before the control law is activated. That is to say, $BTT_{n-N_y}(k)$ in the equation (11) has to be calculated or estimated. $BTT_{n-N_y}(k)$ is the burnthrough time of the material at the position of $(n - N_y) \times l$ distances from the charging end of the strand at the occurrence of the k-th event,

The burnthrough time depends mainly on the ingredients of materials, permeability, amount of water, bed height, and suction speed, where bed height and suction speed are usually constant during the operation. In practice, permeability is known to be the most important factor, and thus the burnthrough time is calculated by measuring the permeability of the raw mix at the first wind box. Permeability, however, is very difficult to measure and costly. Since the ingredients of the raw mix or the permeability change slowly, the burnthrough time is predicted from the history of the measured burnthrough times. The least-square algorithm is adopted with covariance resetting (Goodwin and Sin, 1994) for predicting $BTT_{n-N_y}(k)$. $BTT_{n-N_y}(k)$ is predicted from measurable $BTT_n(k)$ and $BTT_{n+N_y}(k) = BTT_n(k - N_y)$. In order to compute the burnthrough time $BTT_l(k)$ for $l > n$, the temperature history of the sinter package is measured, from which the burnthrough time is computed from the definition. Let $z(k) = BTT_{n-N_y}(k), z(k-1) = BTT_n(k)$, and $z(k-2) = BTT_{n+N_y}(k)$. It is assumed that the burnthrough time follows the stochastic auto-regressive model:

$$\begin{aligned}
z(k) - \overline{BT} &= \alpha(k)(z(k-1) - \overline{BT}) \\
&\quad + \beta(k)(z(k-2) - \overline{BT}) + w_{BTT}(k) \\
&= \phi^T(k-1)\theta(k) + w_{BTT}(k) \qquad (14)
\end{aligned}$$

where

$$\begin{aligned}
\phi^T(k-1) &= \left[z(k-1) - \overline{BT}, \; z(k-2) - \overline{BT}\right], \\
\theta^T(k) &= [\alpha(k), \; \beta(k)], \qquad (15)
\end{aligned}$$

and $w_{BTT}(k)$ is a white noise sequence, $z(k-1)$ and $z(k-2)$ are known burnthrough times of processed sinter packages, and $\overline{BT}$ is the mean value of the known burnthrough times. Here, $\hat{\theta}(k)$ is calculated by the least-square algorithm with covariance resetting (Goodwin and Sin, 1994):

$$\hat{\theta}(k) = \hat{\theta}(k-1) + \frac{P(k-2)\phi(k-1)}{1 + \phi^T(k-1)P(k-2)\phi(k-1)} \cdot \left[z(k) - \overline{BT} - \phi^T(k-1)\hat{\theta}(k-1)\right] \qquad (16)$$

with $\hat{\theta}(0), P(-1) > 0$ given. Let $\{K_s\} = \{k_1, k_2, \cdots\}$ be the times at which resetting occurs, then for $k \notin \{K_s\}$ an ordinary sequential least-square update is used; that is

$$P(k-1) = P(k-2) - \frac{P(k-2)\phi(k-1)\phi^T(k-1)P(k-2)}{1+\phi^T(k-1)P(k-2)\phi(k-1)}. \quad (17)$$

Otherwise, for $k = k_i \in \{K_s\}$, $P(k_i - 1)$ is reset. Then, the predicted burnthrough time is given by

$$\widehat{BTT}_{n-N_y}(k) = \hat{z}(k) = \phi^T(k-1)\hat{\theta}(k) + \overline{BT}. \quad (18)$$

The predicted signal $\widehat{BTT}_{n-N_y}(k)$ in (18) will be used for the reference $\overline{y}_{r2}$.

5. SIMULATIONS

To compare the results of the MMGPC, a PID algorithm is adopted and applied. Control inputs of the PID algorithm are calculated using the following equation:

$$\begin{aligned} u(k) = u(k-1) &+ G_i(BTT_n(k) - x_n(k)) \\ &+ G_d(BTT_n(k-k_d) \\ &- BTT_n(k - 2\times k_d)), \quad (19) \end{aligned}$$

where G_i and G_d are the gain, and k_d is a time constant.

The control algorithms are applied to the POSCO sintering plant model (1) and (2). A diagrammatic representation of the POSCO sintering plant is given in Figure 2. The length of the sinter strand

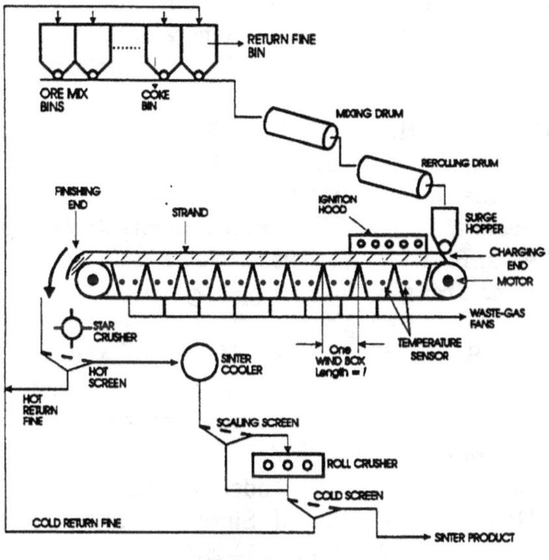

Fig. 2. A Systematic Diagram of POSCO Sintering Plant IV

is 100m. There exist 25 wind boxes, each of which has a 4m width. The temperature is measured in multiple positions of each wind box at the occurrence of each event. The desired position of the burnthrough point is 92.2 m from the charging end of the strand. Then $N = 50$, $n = 47$. N_y is chosen

Table 1. Parameter Values Used in the Experiment

Parameter	value	Reference
$\hat{\alpha}(0)$	0.6	Eq. (16)
$\hat{\beta}(0)$	0.3	Eq. (16)
Init. Cov.	$\begin{bmatrix} 0.003 & 0.001 \\ 0.001 & 0.002 \end{bmatrix}$	Eq. (16)
Cov. Reset Period	20 Steps	Eq. (17)
# of Windows	50	2 × # of W.B.
BTP	92.2	
Horizon N_y	11	Eq. (11)
Horizon N_F	1	Eq. (11)
Horizon G_i	0.0005	Eq. (19)
Horizon G_d	0.0001	Eq. (19)
Horizon k_d	11	Eq. (19)

W. B.: Windboxes.

as 11. The parameters used in the experiments are listed in Table 1.

The system block diagram is shown in Figure 3. And the flow of the control algorithm is as follows:

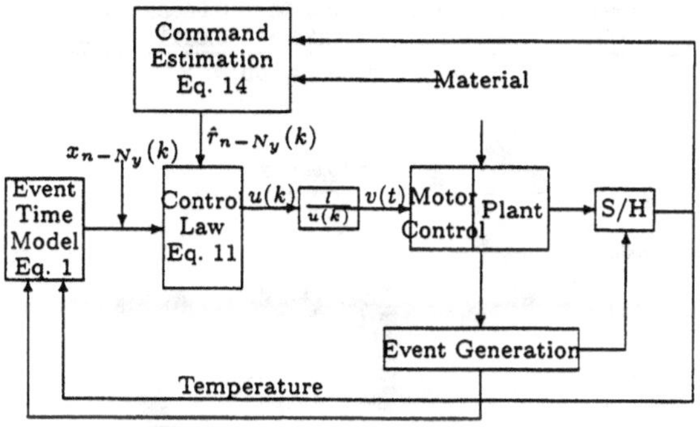

Fig. 3. The System Block Diagram

(a) Calculate $x_i(k)$, $i = 1, 2, ..., N$, in Eq. (1) and (2), $z(k-1)$, $z(k-2)$, and $\overline{BT}$ in Eq. (14) using the measured data: the strand speed and the temperature history.
(b) Predict the burnthrough time $\widehat{BTT}_{n-N_y}(k)$ using the least-square prediction algorithm with covariançe resetting: Equations (16)-(18).
(c) Calculate the control $u(k)$ of each algorithm.
(d) Calculate the strand speed $v(k) = \frac{l}{u(k)}$.
(e) Calculate $x_i(k+1)$, $i = 1, 2, ..., N$, using the model (1) and (2).
(f) Check that the next event occurs, measure and calculate the temperature with the interpolation method, and calculate the burnthrough time of the sinter material at the desired burnthrough point: $k = k + 1$ and go to (b).

It is assumed that the temperature of air sucked into the wind box i, at the occurrence of the k-th event, represents the average temperature of the sinter package at the wind box i. And the average temperature of a sinter package follows the pattern shown in Figure 1 while moving on the strand. Since the suction speed of the fan is much faster than the strand speed, it is assumed that no delay exists in the measurement.

Once $u(k)$ is determined from the control algorithms, the average strand speed between the k-

th event and $(k+1)$-th event is obtained by the equation $v(k) = \frac{l}{u(k)}$. This average strand speed is applied to the motor drive system as a command signal. The strand speed is proportional to the actual motor speed which is controlled by the voltage input. Generally, the actual motor speed depends on the characteristics of the motor drive system. There can be some errors in following the command signal. These can act as small input disturbances.

The comparison results are shown in the table 2 and figure 4. In the table 2, results using MMGPC

Table 2. Statistics of Simulation Results

		PID		MMGPC	
Var. of $w(k)$		0	50	0	50
Error in	†M	0.0732	0.1233	0.0254	0.0299
BTP (m)	‡S	1.1774	2.9201	0.9460	2.7463
Speed	†M	0.0505	0.0505	0.0505	0.0506
(m/s)	‡S	3.525♭	2.995♭	0.0011	0.0017

†: Mean, ‡: Standard Deviation, ♭: $\times 10^{-4}$.

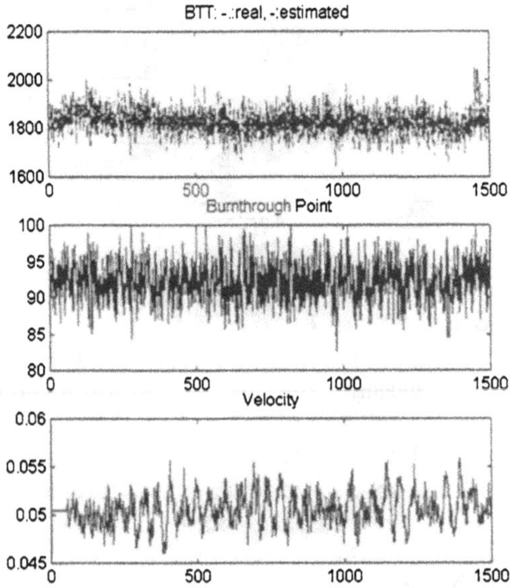

Fig. 4. Simulation Results Using MMGPC with var$(w(k))$=50.

are better than those using PID control in all aspects. But if disturbances are applied to the system in Equation (4), mean of BTP errors in the case using MMGPC is much less than that using PID control. Since the variance of the strand speed is too small considering the mean of the strand speed, it could be negligible.

6. CONCLUSIONS

This paper treats the applications of the MMGPC to the burnthrough point in the sintering processes. Specifically, a simple CARIMA (Controlled Auto-Regressive Moving Average) model is presented for the burnthrough point control system by defining events. In the model, the reference signal is the burnthrough time and the controlled output is the time which a sinter package has taken to reach the desired burnthrough point. For a known reference signal, a robust and stable control law (11) is used. It is a robust GPC in one shot form using discrete-time min-max game theoretic approaches. For the unknown reference signal, a prediction method of the BTT is suggested, which is based on the least-square algorithm with covariance resetting. This alleviates the necessity of making the very difficult permeability measurement, which cuts down the expense. The MMGPC with the BTT prediction was applied to the BTP control problem. The simulation results of the MMGPC using real BTT data shows better performance than those of PID control.

REFERENCES

Basar, T. and G. J. Oldser (1982). *Dynamic Noncooperative Game Theory,* New York: Academic.

Dash, I. R. and E. Rose (1977). Simulation of a sinter strand process, *Ironmaking and Steelmaking,* pp.321-328.

Goodwin, G. C. and K. S. Sin (1994). *Adaptive Filtering, Prediction, and Control,* Prentice Hall, New York.

Hansen, J. D., R. P. Rusin, M. H. Teng and D. L. Johnson (1992). Combined Stage Sintering Model, *Journal of the American Ceramic Society,* Vol. 75, Iss. 5, pp. 1129-1135.

Kim, Y. H., Y. I. Lee and W. H. Kwon (1997). A Stabilizing Min-max Generalized Predictive Control, Will be published in *the proceedings of ACC'97.*

Kwon, W. H., Y. H. Kim, S. J. Lee and K. N. Baek (1997). Event-based Modelling and Control for the Burnthrough Point in Sintering Processes, *Submitted to IEEE Control Systems Technology.*

Lewis, R. W. *et al.* (1983). *Numerical Methods in Heat Transfer,* Vol. II, pp. 485-510, John Wiley and Sons, New York.

Noh, S. B., Y. H. Kim, Y. I. Lee and W. H. Kwon (1996). A robust generalized predictive control with terminal output weightings, *Journal of Process Control,* Vol. 6, Iss. 2-3, pp. 137-144.

Rose, E., W. R. M. Anderson and I. M. Orak (1993). Simulation of Sintering, *IFAC World Congress,* Vol. 10, pp. 289-294.

Young, R. W. (1979). Dynamic model of sintering process, *Ironmaking and Steelmaking,* pp.25-31.

APPLICATION OF NEURAL NETWORK TO SUPERVISORY CONTROL OF REHEATING FURNACE IN STEEL INDUSTRY

Young Il Kim*, Ki Cheol Moon**
Byoung Sam Kang, Chonghun Han[1] and Kun Soo Chang

**Environmental Catalysis Team, Research Institute of Industrial Science and Technology*
***Facilities Technology & Engineering Dept., POSCO*
Automation Research Center and Dept. of Chemical Eng., POSTECH

Abstract: The efficient and reliable control of a reheating furnace is a challenging problem due to: a) many different types of billets to process, b) strong intercorrelation among process variables, c) large dimension of the input and output space, d) the strong interaction among process variables, e) a large time delay, and f) highly nonlinear behavior. Thus, the conventional reheating furnace operation has been heavily dependent upon the look-up table which lists the optimal set points. We have developed a modified modular neural network for the supervisory control of the reheating furnace. Based on the divide-and-conquer concept, a modular network is capable of dividing a complex task into subtasks and modeling each subtask with an expert network. To model such activities, a gating network is used for the classification and allocation of the input data to the corresponding expert network. To overcome the correlation effects among process variables and the dimension problem, principal component analysis (PCA) has been employed to remove the correlation and reduce the problem dimension. From PCA analysis, we were able to decide the optimal dimension for the problem to describe the dynamic behavior of the furnace. The proposed neural network has been trained and tested using operation data from the reheating furnace and has been implemented on the wire rod mill process of POSCO™. *Copyright © 1998 IFAC*

Keywords: Reheating furnace, supervisory control, PCA, PLS, HME

INTRODUCTION

Due to increasing global competition among steel companies, the productivity enhancement of the process has become an essential task to survive and succeed in the world trade market. Thus, it has been a natural move to automate the operation of many unit processes to improve the productivity and the safety for the steel making processes.

In steel works, a reheating furnace is a key unit process of a rod mill plant where a billet, a steel bar, is heated to have uniform target temperature. After a billet is charged into a reheating furnace, the billet goes through three heating zones from inlet to outlet. At the outlet, the billet should be heated as close as possible to the target temperature so that the billet is milled without any damage to the rollers to produce a wire.

The supervisory control problem of the reheating furnace can be formulated as follows: given a billet type and the billet target temperature at the furnace outlet, determine the amount of fuel flow rate for each heating zone in an optimal manner. Traditionally, the mathematical models (Bobrie and

[1] Author to whom all correspondence should be addressed
Tel:+82-(0562)279-2279: e-mail : chan@vision.postech.ac.kr

Louis, 1986; Farmer et al., 1994) have been developed to predict the temperature profile of a billet and consequently determine zone temperatures. However, such models are only valid for localized operating regions. Furthermore, in case of a real furnace, we do not have enough amount of data on physical properties and model parameters. As a result, the furnace has been usually operated based on the experience and knowledge of the operators.

Neural networks have been proven by many researchers as universal approximators of any continuous function with arbitrarily desired accuracy. This property has been widely used to make the nonlinear map of the input to output space. However, when input space consists of several different classes of input data, it becomes very difficult to converge the network during the training phase. This phenomenon, called *temporal crosstalk* (Sutton, 1986), cannot be efficiently handled by a single multilayer perceptron neural network. The training may require too many hidden nodes to resolve the temporal crosstalk and capture the mapping relationship among the input and the output space. A modular neural network (MNN) and a hierarchy of mixture of expert networks (HME) have been recently developed to handle such problems. It has been reported (Jacobs et al., 1991) that MNN/HME are very efficient to handle large and unrelated input space. Jordan and Jacobs (1994) have extended the modular system to a hierarchical system, made links to the statistical literature on classification and regression trees (Brieman et al., 1984), and developed an Expectation-Maximization(EM) algorithm for the architecture.

However, even HME cannot work efficiently when there are many measurements which are strongly correlated each other. To remove the correlation effects among many measurement variables and reduce the problem dimension, PCA/PLS has been widely used recently. PCA/PLS are multivariate analysis techniques whose main advantage is to reduce the dimension of the data set identifying the collinear relationships among the measured variables. In this work, we have applied PCA and HME to develop an inverse model that gives the set points for three heating zones given the billet type at the furnace inlet and the billet target temperature at the furnace outlet. Learning the operation of the human behavior, the developed model was able to predict the set point temperatures for the furnace.

The organization of this paper is as follows: first, this paper introduces the theoretical concepts behind modular network, PCA and PLS at section 2 and 3. At section 4, the application of HME and PCA to modeling of the supervisory control behavior for a reheating furnace will be presented. The reheating furnace and the current control system will be described in detail. The application results and discussions will be presented followed by conclusions.

2. MODULAR NEURAL NETWORK

A modular neural network that learns to partition a task into two or more functionally independent tasks and allocates different networks to learn different training patterns will be able to learn and compute different and complex functions. The modular architecture, as shown in Fig. 1, consists of two types of networks: expert networks and a gating network. The expert networks compete to learn the training patterns and the gating network mediates this competition. The gating network is restricted to have as many output units as the number of expert networks, and the activation of these output units must be nonnegative and sum to one. Generally, Jacobs and Jordan (1993) have introduced the "*softmax*" function and the activation of the ith output of the gating network, denoted g_i is

$$g_i = \frac{e^{s_i}}{\sum_{j=1}^{n} e^{s_j}} \tag{1}$$

where, s_i denotes the weighted sum of unit ith inputs $s_i = \sum \mathbf{W}x_k$ and n denotes the number of expert networks. The output vector of the entire architecture, denoted $\mathbf{y}$, is

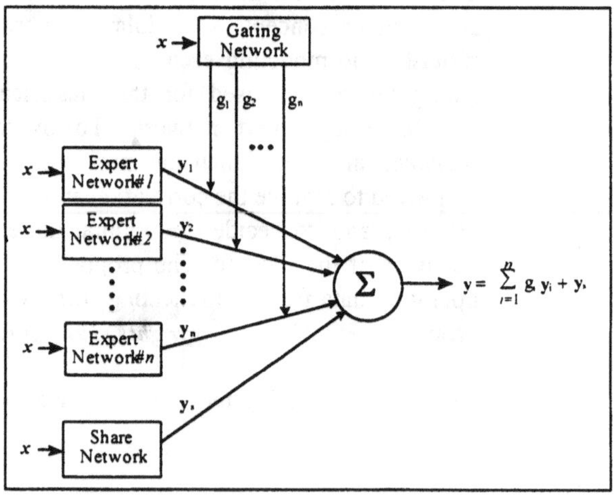

Fig. 1. Schematic diagram of modular neural network

$$\mathbf{y} = \sum_{i=1}^{n} g_i \mathbf{y}_i + \mathbf{y}_s \tag{2}$$

where $\mathbf{y}_i$ denotes the output vector of the ith expert network. During training, the weights of the expert and gating networks are adjusted simultaneously using the backpropagation algorithm so as to maximize the cost function, $\ln L$,

$$\ln L = \ln \sum_{i=1}^{n} \frac{g_i}{\sigma_i} \exp\left(-\frac{1}{2\sigma_i^2} \left\| \mathbf{y}* - \mathbf{y}_i - \mathbf{y}_s \right\| \right) \tag{3}$$

where $\mathbf{y}*$ denotes the target output vector and denotes a scaling parameter associated with the ith expert network. The goal of the architecture is to model the distribution of training patterns. This is achieved by gradient ascent in the log likelihood function as follows:

$$\frac{\partial \ln L}{\partial s_i} = h_i - g_i \tag{4}$$

where h_i is the a posteriori probability that the ith expert network generates the target vector:

$$h_i = \frac{\frac{g_i}{\sigma_i}\exp\left(-\frac{1}{2\sigma_i^2}\|\mathbf{y}^* - \mathbf{y}_i - \mathbf{y}_s\|\right)}{\sum_{j=1}^{n}\frac{g_j}{\sigma_j}\exp\left(-\frac{1}{2\sigma_j^2}\|\mathbf{y}^* - \mathbf{y}_j - \mathbf{y}_s\|\right)} \tag{5}$$

Differentiation of $\ln L$ with respect to $\mathbf{y}_i$ yields

$$\frac{\partial \ln L}{\partial \mathbf{y}_i} = \frac{h_i}{\sigma_i^2}(\mathbf{y}^* - \mathbf{y}_i - \mathbf{y}_s) \tag{6}$$

Finally we compute the derivative of the log likelihood with respect to σ_i^2, the variance associated with the ith expert network. Differentiation of $\ln L$ yields

$$\frac{\partial \ln L}{\partial \sigma_i^2} = \frac{h_i}{2\sigma_i^4}\left(\|\mathbf{y}^* - \mathbf{y}_i - \mathbf{y}_s\|^2 - \sigma_i^2\right) \tag{7}$$

This expression implies that the parameter σ_i^2 is adjusted toward the sample variance $(\mathbf{y}^* - \mathbf{y}_i - \mathbf{y}_s)^2$, with a step size correction that is weighted by the a posteriori probability (Jacobs and Jordan, 1993)

3. PCA and PLS

3.1 *Principal Component Analysis (PCA)*

Principal component analysis is basically multivariate analysis technique. The central idea of principal component analysis is to reduce the dimension of a data set within which there are a large number of interrelated variables, while retaining as much as possible the variations present in the data set. This reduction is achieved by transforming to a new set of variables, *principal components*, which are uncorrelated, and which are ordered so that the first few retain most of the variations present in all of the original variables. Computation of the principal components was reduced to the solution of an eigenvalue-eigenvector problem for a positive-semidefinite symmetric matrix.

PCA provides an approximation of a data matrix $\mathbf{X}$, in terms of the product of two small matrices $\mathbf{T}$ and $\mathbf{P}^{\mathrm{T}}$. These matrices, $\mathbf{T}$ and $\mathbf{P}^{\mathrm{T}}$ capture the essential data patterns of $\mathbf{X}$. The N rows in the data set are termed "*objects*". These often correspond to chemical or geological samples. The K columns are termed "*variables*" and comprise the measurements made on the objects. The projection of $\mathbf{X}$ down on a A-dimensional subspace by means of the projection matrix $\mathbf{P}^{\mathrm{T}}$ gives the object coordinates in this plane, $\mathbf{T}$. The columns in $\mathbf{T}$, $\mathbf{t}_a$ are called *score vectors* and the rows in $\mathbf{P}^{\mathrm{T}}$, $\mathbf{p}_a$ are called *loading vectors*. The latter comprises in the direction coefficients of the PC plane. The vectors $\mathbf{t}_a$ and $\mathbf{p}_a$ are orthogonal, i.e.,

$$\mathbf{p}_i^{\mathrm{T}}\mathbf{p}_j = 0, \quad \mathbf{t}_i^{\mathrm{T}}\mathbf{t}_j = 0, \quad i \neq j \tag{8}$$

The deviations between projections and original coordinates are termed the "*residuals*". These are collection in the matrix $\mathbf{E}$. PCA in matrix form is the least squares model:

$$\mathbf{X} = \mathbf{1}\mathbf{x}_m + \mathbf{T}\mathbf{P}^{\mathrm{T}} + \mathbf{E} \tag{9}$$

Here the mean vector $\mathbf{x}_m$ is explicitly included in the model formulation. But this is not mandatory. The data may be projected onto a hyperplane passing through the origin. PCA can be extended to data matrices divided into two or more blocks of variables and then called partial least squares (PLS) analysis (partial least squares projection to latent structures).

3.2 *Projection to Latent Structure (PLS)*

The PLS model is built on the properties of the NIPALS algorithm. It is possible to let the score matrix represent the data matrix. A simplified model would consist of a regression between the scores for the $\mathbf{X}$ and $\mathbf{Y}$ block. The PLS model can be considered as consisting of outer relations ($\mathbf{X}$ and $\mathbf{Y}$ block individually) and an inner relation (linking both blocks). The outer relation for the $\mathbf{X}$ block is

$$\mathbf{X} = \sum_{h=1}^{A}\mathbf{t}_h\mathbf{p}_h^{\mathrm{T}} + \mathbf{E} = \mathbf{T}\mathbf{P}^{\mathrm{T}} + \mathbf{E} \tag{10}$$

One can build the outer relation for the $\mathbf{Y}$ block in the same way:

$$\mathbf{Y} = \sum_{h=1}^{A}\mathbf{u}_h\mathbf{q}_h^{\mathrm{T}} + \mathbf{F} = \mathbf{U}\mathbf{Q}^{\mathrm{T}} + \mathbf{F} \tag{11}$$

One can describe all the components and thus make $\mathbf{E} = \mathbf{F} = \mathbf{0}$ or not. But we describe $\mathbf{Y}$ as well as possible and hence to make $\|\mathbf{F}\|$ as small as possible and at the same time, get a useful relation between $\mathbf{X}$ and $\mathbf{Y}$. PLS algorithm is one of the most complete and elegant ones when prediction is important.

4. Application

We have developed a HME for the reheating furnace at POSCO™. We will describe application results and following discussions in this section.

4.1. *Process description*

A schematic drawing of the furnace is shown in Fig. 2(a). The furnace is composed of four combustion zones: charge, preheating, heating, and soaking zone. The soaking zone is divided into three subzones. As it is very difficult to control the charge zone temperature, we exclude the charge zone from control region. Each control zone has radiant burners at the roof as shown in Fig. 2(b) which are controlled in a group, not individually. The burners in each zone are controlled independently from the burners for the other zones. The input to the reheating furnace is a billet which is a steel bar whose dimension is about 12x12x180 cm. After a billet has been charged into the reheating furnace, the billet goes through all the heating zones one by one in a sequential manner as shown in Fig. 2. At the end of the reheating furnace, the billet, which has been heated to the target temperature for the rolling operation at next stage, is pushed out of the furnace. The reheating furnace processes many different types of billets. Each type of billet has its own index temperature and processing time. Index temperature is the average temperature at the roof of the corresponding zone. It is the index temperature that governs the radiation heat transfer between the furnace and the billet and mainly depends upon the roof temperature. Thus these index temperatures for all the zones are controlled variables for the furnace control problem.

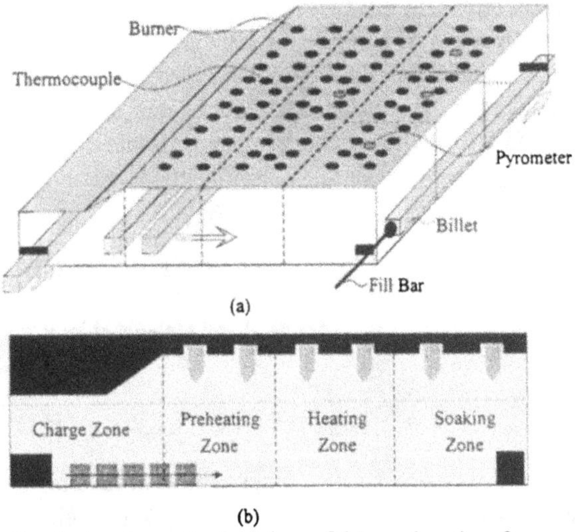

Fig. 2 A schematic drawing of (a) a reheating furnace, (b)the burner types and their locations

For the operation of the reheating furnace, as development of mathematical models for the reheating furnace are difficult due to : a) many different types of billets to process, b) the strong interaction among process variables, c) a large time delay, and d) highly nonlinear behavior, a lookup table reference method has been widely used in industry. However, since this lookup table keeps set point values for only discrete conditions as shown in Table 1, process operators have used interpolation based on this table and decided the set points. However, the interpolation method is not always accurate and reliable enough to calculate the set points. Table 1 shows a lookup-table which lists the typical temperature set points for various steel types during continuous operation. The information flow for the reheating furnace control is shown in Fig. 3. Material flow control system, at the top of the control hierarchy, gives to the supervisory control system the information on billet types, billet target temperatures, the billet locations, element contents, etc. Supervisory control system determines the set points which are downloaded into the distributed control system (DCS). DCS regulates the process (FIELD) to satisfy the downloaded set points.

Table 1 Typical set point values (temperature, °C) for two line continuous operation mode. P, H, S*k*: Preheat, Heat, Soak *k* zone, RM: rimmed steel MR: mild hardness, HR: slow cooling, Killed : killed steel

Type	P	H	S 1	S 2	S 3
RM MR	1120	1120	1120	1120	1110
Killed 1	1120	1120	1120	1120	1110
Killed 2	1140	1130	1140	1150	1120
Killed 3	1140	1140	1150	1160	1130
RM HR	1120	1120	1120	1120	1110
CHQ	1140	1140	1150	1160	1130
PSPC	1150	1150	1160	1170	1140
RH	1160	1160	1170	1180	1140

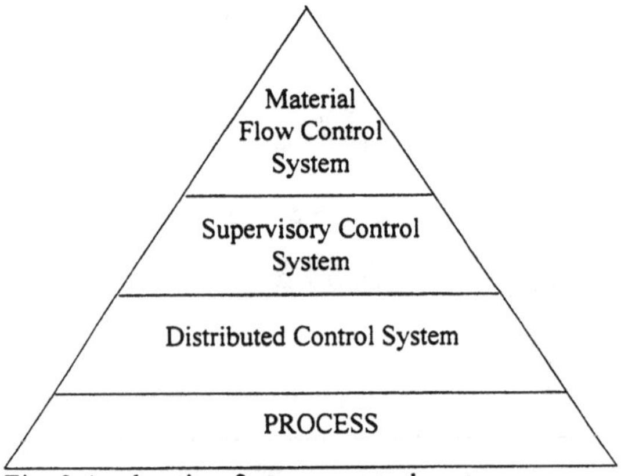

Fig. 3 A reheating furnace control system architecture

4.2 Results and Discussion

Data for the variables shown in Table 2 are scaled and normalized into the range between 0.0 and 1.0. The data collected from DCS and the supervisory control system contain noise, thus they need to be preprocessed before they are used as input data for the training of neural networks. The contents of carbon, manganese, and silicon are read as weight percent and they do not need to be preprocessed. The steel type variable is recorded as cooling code plus JIS system notation. The set point mode has a discrete value which represents the method for setting up the set point values of DCS for each zone: 0 for set points for gas flow rates, 1 for set points for temperatures, 2 for one line operation, so on. When new type of billet (experimental or test billet) is charged or the set point mode does not properly work, mode 0 for gas flow rates begins to work. Zone hold time is the time average of processing time at heating zone.

Table 2 Summary of variables; *:Target -:Not used

Variables	No.	Index name
Carbon	12	$C_1 \cdots C_{12}$
Manganese	12	$Mn_1 \cdots Mn_{12}$
Silicon	12	$Si_1 \cdots Si_{12}$
Steel type	12	$Kind_1 \cdots Kind_{12}$
Set point mode	5	$DCS_1^{con} \cdots DCS_5^{con}$
Temperature SV*	5	$T_1^{SV} \cdots T_5^{SV}$
Temperature PV	5	$T_1^{PV} \cdots T_5^{PV}$
Gas SV-	5	$G_1^{SV} \cdots G_5^{SV}$
Gas PV	5	$G_1^{PV} \cdots G_5^{PV}$
Zone Hold Time	1	WBC
Pressure	1	P_{plant}

We have compared the convergence rate of a HME with one of BP (Back Propagation neural network) as shown in Fig. 4. The HME is a binary tree with 3 layers which has 4 expert nets, and 3 gating nets without hidden layer. BP has two hidden layers and 7 nodes for each layer. To find the optimal number of hidden layers for BP, we have tested several numbers for the number of hidden layers. A BP with 9 layers have been found to be the optimal one with SSE (Sum of Squared Error) value equal to 9.58. Compared with HME, BP has about 1.6 times more weights than HME. The input dimension is dim $\mathbf{X}$ = 1000x66 and the target dimension dim $\mathbf{Y}$ = 1000x5, respectively. Expectation maximization (EM) algorithm with IRLS (Iterative Recursive Least Square) learning algorithms has been used for HME and momentum and learning rate rule for BP. Training continued until there is no more improvement in SSE. HME stopped after 6 epochs and showed faster convergence rate than BP. In case of BP, the training has not been successful due to the temporal crosstalk phenomena as shown in Fig. 4 and Fig. 5.

However, one epoch for HME took much more time than one for BP because the internal IRLS loop takes a significant amount of time when input space is huge like our case($\approx$4min/epoch).

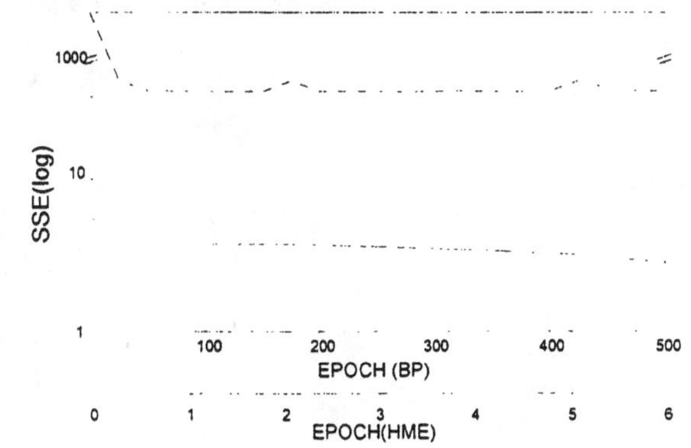

Fig. 4 Comparison of SSE (Sum of Squared Error) from BP with the one from HME; ····:BP, —: HME.

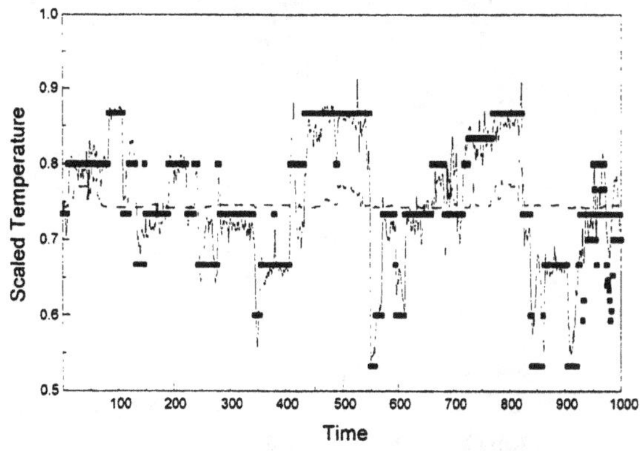

Fig. 5 Comparison of the prediction results from HME and BP with the real data; ····:BP, —: HME (with PCA-based model reduction), • :for real data.

To use a neural network for supervisory control, the prediction from the network should be fast. To make the HME faster, we have used a model reduction technique based on principal component analysis (Wold et al.,1987; Jackson, 1991). Removing the correlation effects among process variables, we have reduced the model dimension form the original 66 variables to 12 principal components and described about 80% of the dynamic behavior of the reheating furnace as shown in Fig. 6 using 12 key principal components.

Fig. 7 shows the prediction results of HME trained with raw data from DCS (dim X=1000x66) and the prediction results of HME trained with preprocessed data using PCA (dim X=1000x23). A HME with 3 layers of binary tree has been used. During the learning phase, sum of squared errors are; 4.19 for HME without PCA, 5.50 for HME with PCA. During the testing phase, SSEs are: 6.33 for HME without PCA, 8.24 for HME with PCA. Therefore, despite large dimension reduction using PCA, the results are almost same. However, the training and the testing of HME with PCA took significantly less time.

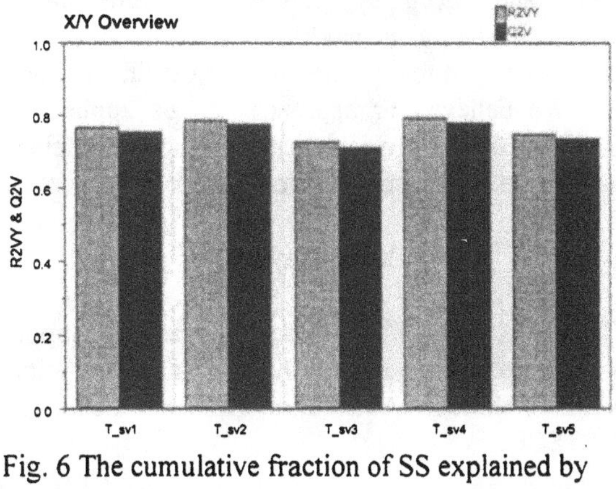

Fig. 6 The cumulative fraction of SS explained by PLS model developed using the selected principal components for set point variables.

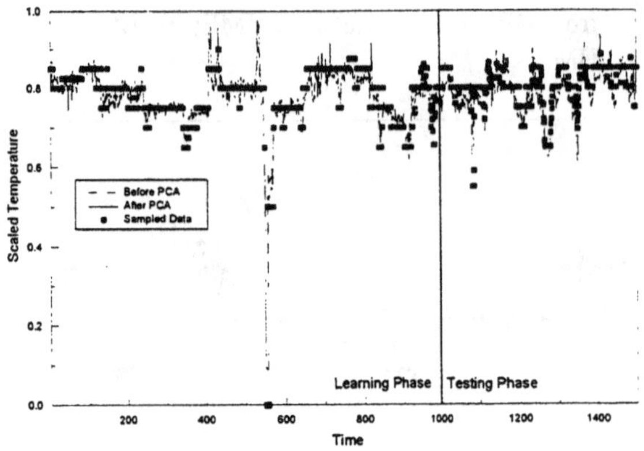

Fig. 7 Comparisons of the prediction results from HME and HME (with PCA-based model reduction) with the real data for heating zone; • for sampled data, ⋯: HME prediction , —: HME prediction (with PCA-based model reduction).

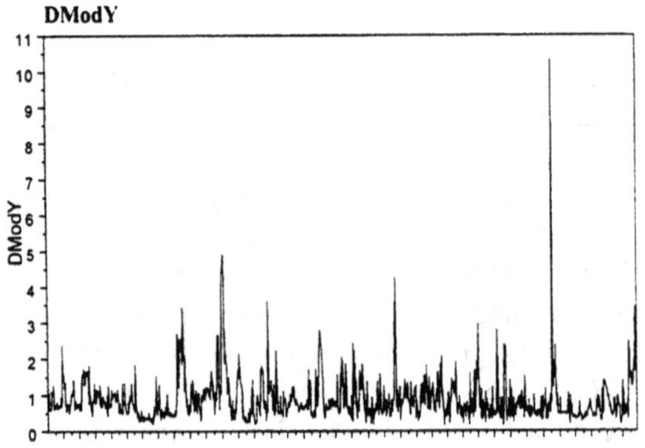

Fig. 8 Plot of residual between PLS prediction values and the measured data

Fig. 8 shows the residual plot where a high peak indicates the occurrence of gross errors such as sensor failure and controller failure.

5. Conclusions

We have developed a HME network with dimension reduction for the supervisory control of the reheating furnace of the wire rod mill plants at POSCO™. The network has shown excellent performance in terms of speed and accuracy, compared with back-propagation neural network. The network has been fully tested with real plant data both off-line and on-line. We believe our approach can be applied to many similar multivariable systems for the efficient and reliable supervisory control.

Acknowledgment

The authors would like to thank the financial support of Korea Science and Engineering Foundation (KOSEF) through Automation Research Center at POSTECH.

REFERENCES

N. Bhat and T. J. McAvoy (1990), Use of neural nets for dynamics modeling and control of chemical process systems *Computers chem. Engng*, **14**, **4/5**, pp. 573-583.

M. Bobrie and D. Louis(1986), Computer control of billet furnaces based on physical models, In: *Process control in the steel industry*, 11-12 Sep., MEFOS.

L. Breiman, J.H. Friedman, R.A. Olshen, and C.J. Stone (1984), Classification and regression Trees, Belmont, CA: Wadsworth International Group.

L. K. Farmer, I. S. Chan, J. G. Nelson (1994), Development of rapid gas heating process for simifinished steel products, *Iron and Steel Engineer*, Sept.

J. E. Jackson (1991), *A User's Guide to Principal Components*, John Wiley & Sons Inc.

R. A. Jacobs, M. I. Jordan, and A.G. Bartom (1991), Task decomposition through competition in a modular connectionist architecture: The what and where vision tasks, *Cognitive Sci.*, **15**, **2**.

R. A. Jacobs, M. I. Jordan, S.J. Nowlan, and G.E. Hinton (1991), Adaptive mixtures of local experts, *Neural Computation*, **3**, pp. 79-87.

R. A. Jacobs and M. I. Jordan (1993), Learning piecewise control strategies in a modular neural network architecture, *IEEE Transactions on Systems, Man, and Cybernetics*, **23**, **2**, March/April.

M. I. Jordan and R. A. Jacobs (1994), Hierarchical mixtures of experts and the EM algorithm, *Neural Computation*, **6**, pp. 181-214.

R. S. Sutton (1986), Two problems with backpropagation and other steepest descent learning procedures for networks, *Proc. Eighth Annual Conf. Cognitive Sci. Soc.*, pp. 823-831.

S. J. Qin and T. J. McAvoy (1992), Nonlinear PLS modeling using neural networks, *Computers chem. Engng.*, **16**, **4**, pp. 379-391.

S. J. Qin and T. J. McAvoy (1996), Nonlinear FIR modeling via a neural net PLS approach, *Computers chem. Engng.*, **20**, **2**, pp. 147-159.

Fred Shenvar (1994), Walking beam furnace supervisory control at Inlands 80-in hot strip mill, *Iron and Steel Engineer*, July.

S. Wold, K. Esbensen and P.Geladi (1987), Principal Component Analysis, *Chemometrics and Intel. Lab Sys.*, **2**, pp. 37-52.

Application of Fuzzy Logic on Flatness Control of Cold Rolling Mill

XueJun Pan and TianYou Chai

*Research Center of Automation, Northeastern University,
Shenyang, 110006,P.R.China*

Abstract: This paper focuses on the flatness control of cold rolling mill using fuzzy logic. Strip flatness is described by an orthogonal polynomial regression based on measurement of output stress distribution. Two fuzzy logic controllers are proposed , the skewing compensation controller to adjust the linear flatness error and the bending controller to eliminate parabolic flatness error. *Copyright © 1998 IFAC*

Keywords: Fuzzy logic, Rules, PI controller, Regression, Simulation.

1. Introduction[1]

Recently, the need for more accurate measurement and control of quality has increased. One important quality characteristic in cold rolling sheet production is the product shape or flatness. Automatic flatness control (AFC) is an important part of strip rolling process control. Various actuators, such as work rolling bending, skewing and local cooling , have been developed for flatness control of cold rolling mills and many control algorithms using these actuators have been proposed. From the first closed-loop flatness control system introduced in the mid-1970's, many flatness closed-loop control approaches have been proposed. The conventional flatness control system for 4-high cold mill employs the traditional PI controllers for adjusting parabolic flatness error by bending the work roll and linear flatness error by skewing the backup roll. Some modern control theory approaches also are used in flatness control system, such as the observer-based multivariable flatness control method by using the rolling process linearized model(Ikuya Hoshino et al. 1993). The sheet steel rolling process is a complex nonlinear

[1] Supported by National Science funds of China

process. The strip flatness is influenced by a large number of phenomena: thermal camber of the roll; roll wear; work or backup bending; roll skewing; work roll shifting ; roll deflection; work roll and strip temperature; material flow; etc., see (Bryant, 1973). Because it is difficult to model the stripe rolling process, many control approaches are limited in flatness control of cold rolling mill.

Fuzzy set theory, established by Zadeh(Zadeh, 1965), have been developed for 30 years. It has been found applications in a lot of industrial process . One important advantage lies in that fuzzy logic is close in inference process to human thinking and natural language. It provides an effective means of converting a linguistic control strategy based on expert knowledge into an automatic control strategy . The linguistic control algorithms which are the generation of human experience show effectiveness especially when precise model of a system is unavailable. On the other hand Fuzzy control system is easy to implement by conventional microprocessor.

In this paper, two fuzzy logic controllers are presented for automatic control system of the 4-high laboratory rolling mill, from which one is to control work roll bending and the other is compensation of hydraulic automatic gauge control system by making backup roll skewing.

2. Strip Shape Pattern

The fuzzy control system , for the 4-high laboratory cold rolling mill is shown in figure 1.

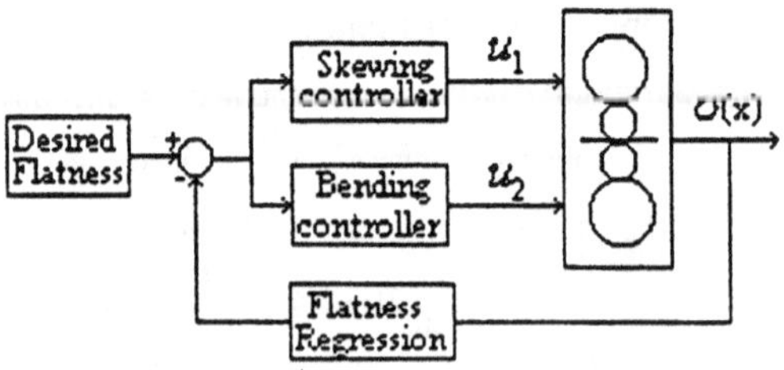

Fig. 1. Flatness control of cold rolling mill

According to the rolling theory, The flatness of output strip can be presented by rolling mill output stress distribution across the strip width. In the rolling mill, the stress distribution of outcoming strip is measured by a contacting shapemeter in the width direction. A orthogonal polynomial regression algorithmic is adopted, and the derivation equation of stress distribution can be get :

$$\Delta\sigma(x) = a_0\theta_0 + a_1\theta_1 + a_2\theta_2 \qquad (1)$$

$$(\theta_i \cdot \theta_j) = \begin{cases} 1 & (i = j) \\ 0 & (i \neq j) \end{cases} \qquad (2)$$

$$\begin{aligned} \theta_0 &= b \\ \theta_1 &= cx + d \\ \theta_2 &= ex^2 + f \end{aligned} \qquad (3)$$

$$a_i = \frac{\sum(\Delta\sigma(x) \cdot \theta_i)}{\sum(\theta_i \cdot \theta_i)} \qquad (4)$$

From the equations (1)-(4), it can be considered that the backup skewing has mainly effect on the first order term coefficient a_1 and the work roll bending has mainly an effect on the second order coefficient a_2. The a_0 is mean stress coefficient. The desired output stress can described as follow:

$$\Delta\sigma^*(x) = a_0^*\theta_0 + a_1^*\theta_1 + a_2^*\theta_2 \qquad (5)$$

The aim of flatness control is to minimize the flatness error, $\Delta\sigma^*(x) - \Delta\sigma(x)$.

3. Fuzzy Control

In this section, the fuzzy controllers, fuzzy skewing compensation controller and fuzzy bending controller, are proposed. The fuzzy skewing compensation controller employs two inputs: the mean stress error signal, $e_0 = a_0^* - a_0$ and the linear flatness error signal, $e_1 = a_1^* - a_1$. and its output is backup skewing compensation value. The input and output membership functions of the fuzzy skewing controller are shown in Fig. 2- Fig 4. The two input signals of the fuzzy bending controller are the mean strain error, e_0 and the parabolic flatness error, $e_2 = a_2^* - a_2$, and its input and output membership functions are shown in Fig. 5 - 7. In the Fig. 2- 7., the NB, NM, NS, Z, PS, PM, and PB are negative big, negative medium, negative small, zero, positive small, positive medium and positive big, respectively. For examples, for skewing control, the term positive means positive skewing and the term negative means negative skewing, for the bending control, the term positive means positive bending and the term negative means negative bending.

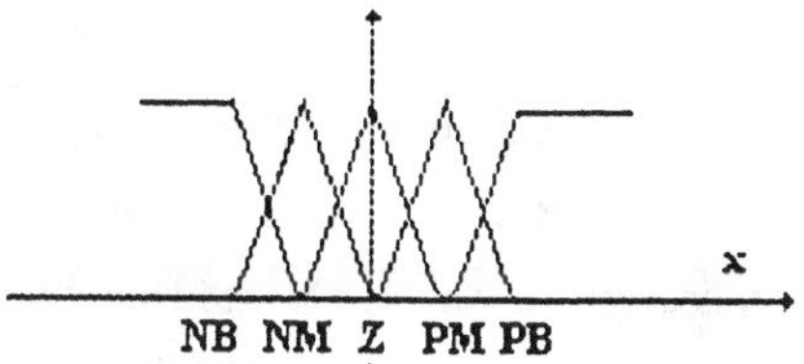

Fig. 2. The input, e_0, membership function for fuzzy skewing controller

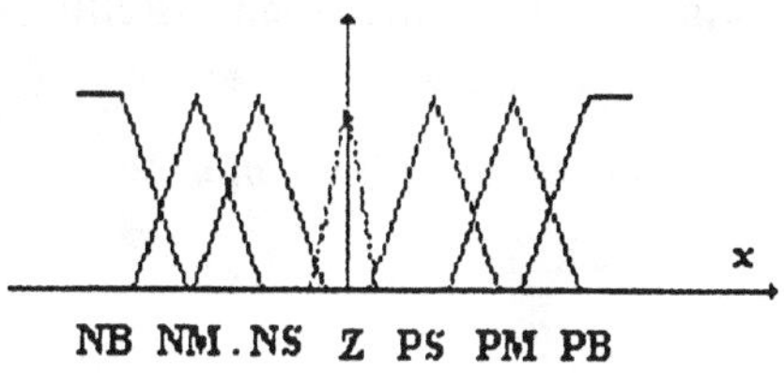

Fig. 3. The input, e_1 , membership function for fuzzy skewing controller

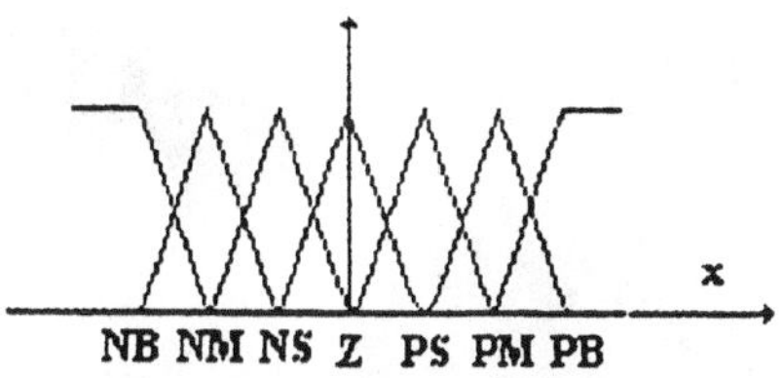

Fig. 4. The output membership function for fuzzy skewing controller

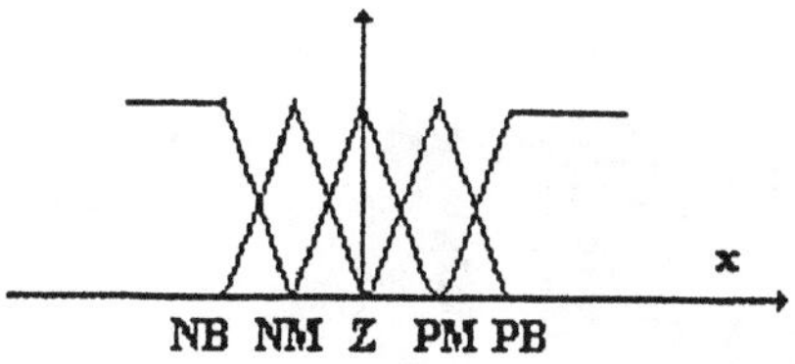

Fig. 5 The input, e_0 , membership function for fuzzy bending controller

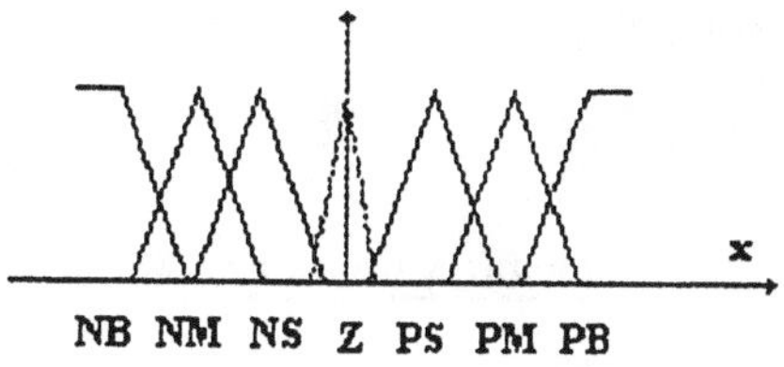

Fig. 6. The input, e_2 , membership function for fuzzy bending controller

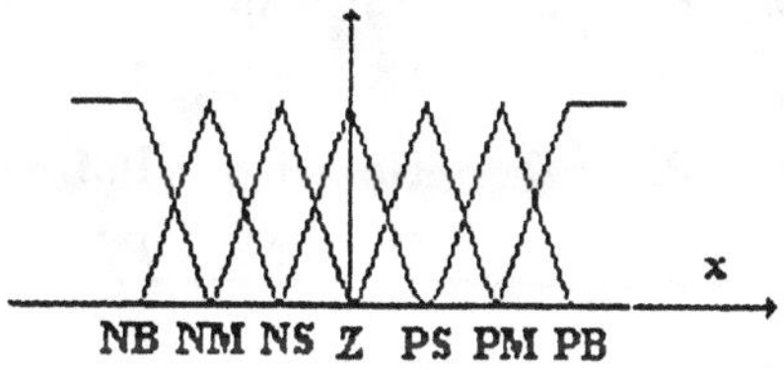

Fig. 7. The output membership function for fuzzy bending controller

All rules of the two fuzzy controllers can be found in table 1 and table 2 and the fuzzy inference results of them are shown in Fig. 8 and Fig. 9. The method of defuzzification uses singletons for each consequent, which is suitable for computational speed.

Table 1 The rules of the fuzzy skewing controller

e_1 / e_0	NB	NM	NS	Z	PS	PM	PB
NB	PM	PS	Z	Z	Z	NS	NM
NM	PB	PM	PS	Z	NS	NM	NB
Z	PB	PM	PS	Z	NS	NM	NB
PM	PB	PM	PS	Z	NS	NM	NB
PB	PM	PS	Z	Z	Z	NS	NM

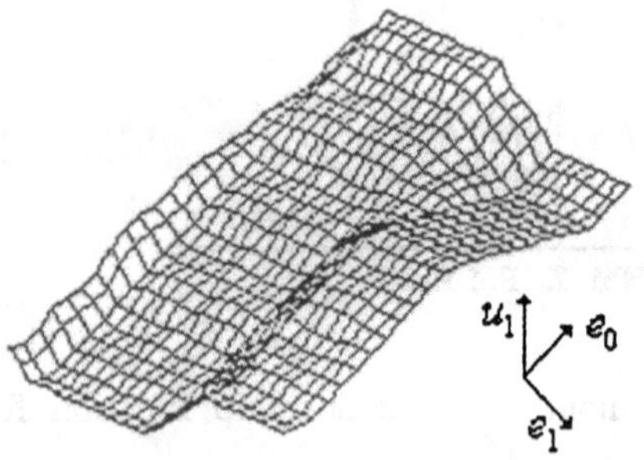

Fig. 8. The fuzzy inference results for the fuzzy skewing controller

Table 2 The rules of the fuzzy bending controller

e_2 / e_0	NB	NM	NS	Z	PS	PM	PM
NB	PB	PM	Z	Z	Z	NM	PB
NM	PB	PM	PS	Z	NS	NM	PB
Z	PM	PS	PS	Z	NS	NS	PM
PM	PS	PS	PS	Z	NS	NS	PM
PB	PS	PS	Z	Z	Z	NS	PS

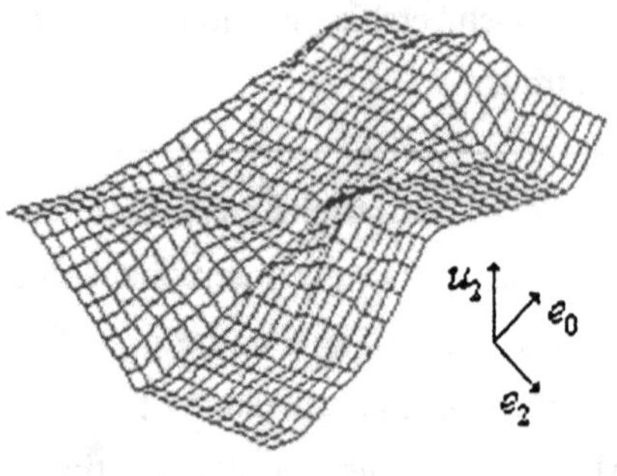

Fig. 9. The fuzzy inference results for the fuzzy bending controller

4. Simulation

For a 4-high laboratory rolling mill, the proposed fuzzy flatness control is implemented by simulation. The simulation shows that proposed Fuzzy flatness control method is effective for flatness control of cold rolling mill, and that excellent flatness control system performance can be achieved using relatively simple and qualitative rules.

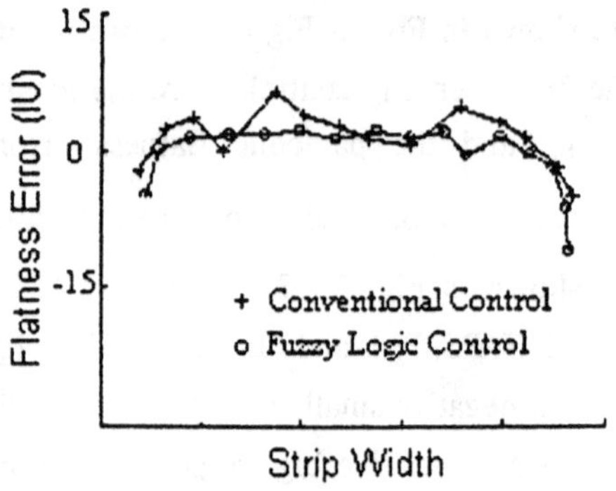

Fig. 10 Simulation result

Reference

Bryant, G.F. (1973). *Automation of Tandem Mill*, pp. 176-212, Iron and Steel Institute, London.

Ikuya Hoshino, Masareru Kawai and Misao Kokubo, (1993). Observer-based multivariable flatness control of the cold rolling mill. *12th world Congress, IFAC*, Vol. 6, pp. 149-156.

Mcdonald, I.R, K.M. Finn and P.D. Spooner, (1990). Improvements in Shape Control for a 4-High Mill. 5th internation Rolling Conference, London, pp. 217-225.

Zadeh, L.A. (1965). Fuzzy sets. *Information control*. Vol. 8, pp. 338-353.

A FUZZY LEAKAGE ALARM METHOD OF LIQUID STEEL

Ge Guo, Junfei Qiao, Wei Wang, Tianyou Chai

Research Center of Automation, Northeastern University
Shenyang, 110006, China

Abstract: A fuzzy pattern recognition method based on an improved thermocouple layout is presented to forecast breakout of liquid steel from the mould in stead of the conventional logic analysis method. A fuzzy decision making network is set up for this purpose. Experiments show that the fuzzy prediction method not only can accurately recognize temperature patterns and forecast steel leakage but also can point out possible reasons and positions of steel leakage. *Copyright © 1998 IFAC*

Keywords: fuzzy network, pattern recognition, leakage prediction.

1. INTRODUCTION[1]

Leakage of liquid steel owning to bulging, corner crack and sticking of the slab with the copper plate of the mould is usually found in continuous casting process. It's not only dangerous to workers and the equipment but also very harmful to the surface of newly-casted slab. So it's very beneficial and important to forecast and prevent possible leakage of liquid steel.

Steel leakage prediction according to temperature measurements on the base of thermocouple system has been used for some years. There are usually two methods to predict possible leakage. One is exciting the alarm by temperature increases of two neighboring upper-lower pairs of thermocouples simultaneously. The other is that leakage alarm is stimulated if there are temperature increases of no less than two in the three neighbors of a thermocouple whose temperature arises abruptly.

Leakage prediction of molten steel in itself is essentially a problem of pattern recognition. The manner of temperature change is different whether it's in the event of normal casting or there is sticking or split. Thus a temperature monitoring method based on fuzzy pattern recognition algorithm is presented in this paper. It involves such tasks as rebuilding of thermocouple system, obtaining of standard pattern base and fuzzy patterns for all the relevant thermocouples, pattern recognition by fuzzy classifying algorithm and finally predicting if there will be steel leakage.

2. IMPROVEMENT OF THERMOCOUPLE SYSTEM

Two rows of thermocouples are usually laid in thermocouple systems as shown in Fig.1 (Haers. F, et al., 1994). There is a high rate of wrong alarm when this method is used in leakage prediction. So the thermocouple system is rearranged in this paper (see Fig.2.) and the number of thermocouples remains the same as in Fig.1. Narrow-face bulging is an indicator of potential leakage of steel. In order to prevent the accruement of narrow-face bulging, an additional center thermocouple is laid in the copperplate which directly contacts the narrow side of the slab.

[1] This research was supported by the Ninth Five-year Plan and Research Fund of the Ministry of Metallurgical Industry of China.

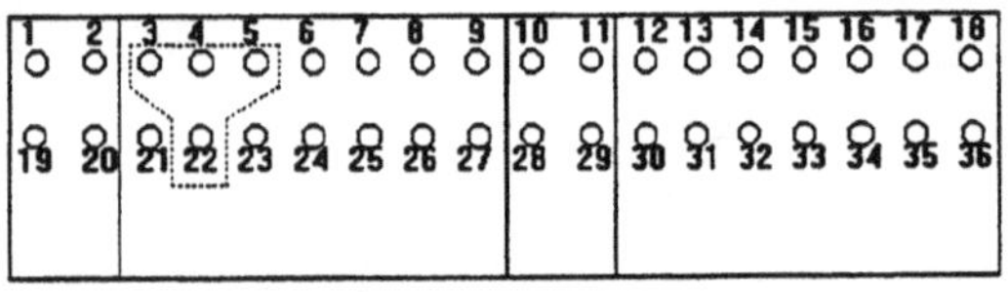

Fig. 1 Conventional layout of thermocouples in the copperplate.

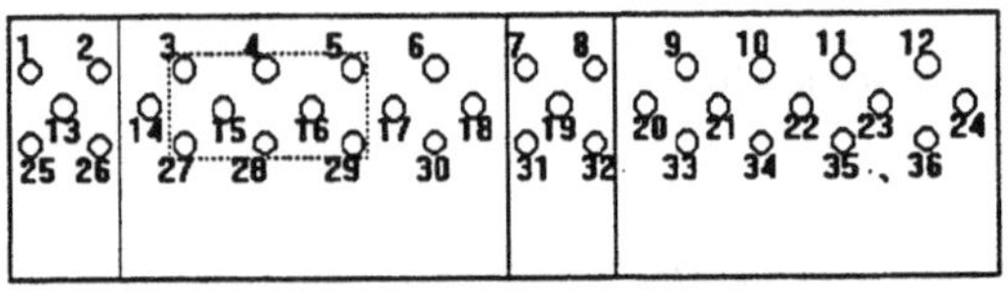

Fig.2 New layout of thermocouples in the copperplate.

The thermocouple system is divided into twelve subsystems with each of the thermocouples in the top row depicts the position of each subsystem. Leakage alarm of molten steel in this subsystem is excited or not depends on the temperature patterns of all the thermocouples within this range.

3. TEMPERATURE PATTERN RECOGNITION WITH A FUZZY ALGORITHM

Since the most important thing in predicting steel breakout is not the precise temperature-time behavior but the over-all tendency of change in temperature. That means the standard temperature patterns are obtained from normalized temperature sampling series. Fig.3 shows the standard reference patterns.

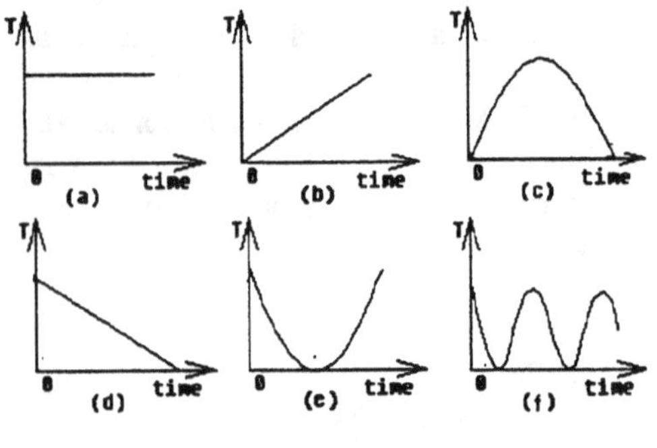

Fig. 3 Six reference temperature patterns

The temperature samples of each thermocouple are written as a vector $T=(t_1,t_2,t_3,\cdots,t_{n_s})$. In order to use fuzzy pattern recognition method in breakout predicting and alarm, the following normalization process is introduced for the temperature samples

$$\sigma t_i = \frac{t_i - \min\limits_{k=1}^{n_s} t_k}{\Delta T} \tag{1}$$

where n_s is the number of samples of each thermocouple, t_i is the i-th temperature sample value, ΔT is the maximum temperature variance in the mould. After this treatment, the temperature measurements become into temperature pattern $P=(\sigma t_1,\sigma t_2,\cdots,\sigma t_{n_s})$.

There are two kinds of fuzzy pattern recognition methods: direct method based on the maximum of membership function selected and indirect method making use of the concept of neighborhood between two fuzzy sets (Zadeh, 1978). These methods are too crude to use in our case. So a more effective fuzzy clustering method is introduced (Hathaway, et al, 1994). Suppose $P_1,P_2,\cdots,P_{n_t}$ are the processed temperature measurements of all the thermocouples in a subsystem, with n_t=8; and $M_1,M_2,\cdots,M_{n_p}$ are the prototype reference patterns, n_p=6. A pattern family $X=(X_1,X_2,\cdots,X_l)$ is set up with $l=n_p+n_t$ and

$$X_k=\begin{cases} M_k & 1\le k\le n_p \\ P_k & k=n_p+1,\cdots,n_p+n_t \end{cases}$$

where X_k is a n_s dimension vector.

There exists a matrix $\mathbf{R}=[r_{ij}]_{l\times l}$ which can describe the relationship between any two fuzzy patterns in this pattern family. Here

$$r_{ij}=\frac{1}{n_s}\sum_{k=1}^{n_s}[X_i(k)\cong X_j(k)] \tag{2}$$

in which "$\cong$" is a fuzzy operator (Pedrycz, et al, 1993) defined as follow

$$x\cong y=\frac{1}{2}[(x\ \varphi\ y)\wedge(y\ \varphi\ x)+(\bar{x}\ \varphi\ \bar{y})\wedge(\bar{y}\ \varphi\ \bar{x})] \tag{3}$$

where

$$x\ \varphi\ y=\min(1,1-x+y)$$

The main task is to find a pattern-class matrix $C=[c_{ij}]_{l\times n_p}$ which can decompose the relation matrix above mentioned

$$R=C\circ C^T \tag{4}$$

c_{ij} describes the extent to which the i-th pattern belongs to the j-th class (or the reference pattern). It is situated in the range [0,1]. Signal "$\circ$" stands for the s-t convolution of C with its transpose. Making

use of the fuzzy relation, the decomposition can be written as

$$r_{ij} = \mathop{S}_{k=1}^{n_p} (c_{ik} \; t \; c_{jk}) \tag{5}$$

where $i, j = 1, 2, \cdots, l$.

A standard gradient-based method, mean square error method, is introduced for the solution of optimal pattern-class matrix. For this purpose, the following performance index is used

$$J = \left\| R - C \circ C^T \right\|^2$$

It can also be expressed in terms of the membership functions of R and C as

$$J = \sum_{i=1}^{l} \sum_{j=1}^{n_p} [r_{ij} - \mathop{S}_{k=1}^{n_p} (c_{ik} \; t \; c_{jk})]^2$$

The goal of the optimization is to obtain c_{ij} so that the performance function is minimized. By using gradient method, the following updating process is obtained

$$c_{ij}^{k+1} = c_{ij}^{k} - \xi \frac{\partial J}{\partial c_{ij}} \tag{6}$$

where $i = 1, 2, \cdots, l$, $j = 1, 2, \cdots, n_p$; ξ is the learning rate which belongs to (0,1]. It's value will affect the pace and stability of the optimization and is selected through prudently organized experiments. And the gradient of the performance index is obtained as follow (Hathaway, et al, 1994).

$$\frac{\partial J}{\partial c_{ij}} = \frac{\partial}{\partial c_{ij}} \left\{ \sum_{u=1}^{l} \sum_{v=1}^{n_p} [r_{uv} - \mathop{S}_{k=1}^{n_p} (c_{uk} \; t \; c_{vk})]^2 \right\}$$

$$= -2 \sum_{u=1}^{l} \sum_{v=1}^{n_p} [r_{uv} - \mathop{S}_{k=1}^{n_p} (c_{uk} \; t \; c_{vk})] \cdot \frac{\partial}{\partial c_{ij}} [\mathop{S}_{k=1}^{n_p} (c_{uk} \; t \; c_{vk})]$$

The detailed computation of the gradient depends on the kind of operators selected for *s-t* convolution. In this paper, they are assumed to be

$$\alpha \; s \; \beta = \alpha + \beta - \alpha\beta$$

and

$$\alpha \; t \; \beta = \alpha\beta$$

After this prolonged updating process, the pattern-class matrix is finally obtained.

4. FUZZY DECISION MAKING AND LEAKAGE ALARMING SYSTEM

When the temperature patterns of all eight thermocouples in a subsystem are recognized, we can subsequently get a overall knowledge of whether there will be any sticking, splits or even leakage of molten steel in the mould by the temperature dynamics (or patterns) obtained above (Reay, et al, 1995).

4.1. Description of the fuzzy alarm system

An alarm function in the following form is defined in this paper

$$y = \frac{\sum_{i=1}^{6} w_i \theta_i}{\sum_{i=1}^{6} \theta_i} \tag{7}$$

where w_i is the weight of each group of thermocouples in a subsystem, and

$$\theta_i = \frac{1}{m_d} \sum \theta_{ij} \tag{8}$$

where m_d denote the number of "dangerous" patterns. In our simulation m_d=4; $\theta_{ij} = \max(c_{uv})$ represents the maximum entries in the pattern-class matrix with respect to all thermocouples in a single group whose temperature characteristics belong to the j-th "dangerous" patterns. For the purpose of simplicity, let

$$z_i = \frac{\theta_i}{\sum_{i=1}^{6} \theta_i} \tag{9}$$

here z_i is called the generalized output of the fuzzy decision making system. Then the alarm function (7) can be written as follow

$$y = \sum_{i=1}^{6} w_i z_i$$

Or in the form

$$y = \mathbf{z}^T \mathbf{w} \tag{10}$$

where

$$\mathbf{z} = [z_1, z_2, \cdots, z_6]^T$$
$$\mathbf{w} = [w_1, w_2, \cdots, w_6]^T.$$

A threshold y_τ is set for the output of this alarm function. When the threshold is exceeded, the alarm is excited for possible leakage of liquid steel. At the same time, the reasons and position of steel leakage are suggested. The position depends on which thermocouple reads the maximum temperature and the reasons depend on the practical temperature patterns of all the six groups in the subsystem.

4.2. Learning of the weights in the alarm function

In order to perform prediction quickly and precisely, a learning algorithm is introduced to adjust the weights of the alarm function. This is reasonable because the fuzzy decision making system works like an artificial neural network, and it can be represented as a feed-forward multilayer network in which all parameters are adjustable (Theocharis, et al, 1994).

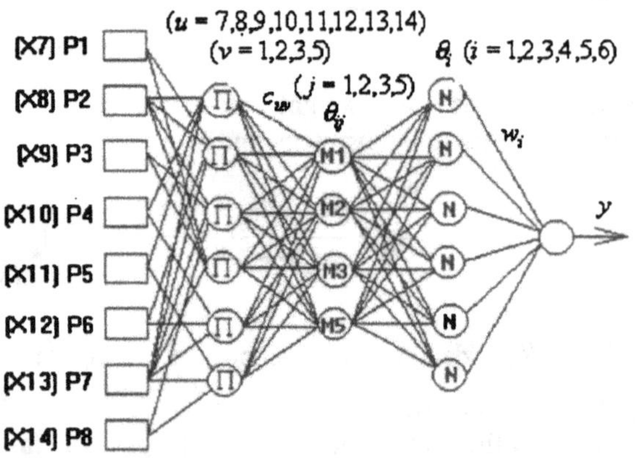

Fig. 4 Structure of the fuzzy-ruler-net

In this paper it is named as the fuzzy-ruler-net as depicted in Fig. 4. The fuzzy network is trained with all possible combinations of temperature patterns before being used for decision making. In this process, a fixed prototype output ψ is set for each specific situation of combinations and an approximation error is defined as $\varepsilon = \psi - y$. The learning approach used in this paper is a standard gradient algorithm (Seo, et al, 1994). At each iteration of the learning process, the gradient estimation is in the form

$$\nabla = \frac{\partial \varepsilon^2}{\partial \mathbf{w}} = 2\varepsilon \frac{\partial \varepsilon}{\partial \mathbf{w}} = 2\varepsilon \mathbf{z}$$

Thus the iteration of the weights can be performed as

$$\mathbf{w}^{k+1} = \mathbf{w}^k - \mu \nabla \tag{11}$$

where μ is the gain constant that regulates the speed and stability of the learning algorithm.

5 EXPERIMENTS AND CONCLUSIONS

In the simulation experiments, it's assumed that the samples of each thermocouple is maintained as n_s=10. After each sampling process, a new measure is added and an old one is deleted. The number of standard reference patterns is n_p=6, and n_t=8 denotes that there are eight thermocouples in a subsystem. Let's take subsystem No. 4 composed of thermocouples (3, 4, 5, 15, 16, 27, 28, 29) as an example. The temperature measurements of this subsystem are listed as follows.

$$\begin{aligned}
\mathbf{T}_1 &= (200,201,203,200,199,198\ 200\ 201\ 199\ 204),\\
\mathbf{T}_2 &= (200\ 205\ 210\ 220\ 225\ 228\ 227\ 235\ 240\ 250),\\
\mathbf{T}_3 &= (203\ 202\ 200\ 199\ 202\ 203\ 201\ 198\ 197\ 201),\\
\mathbf{T}_4 &= (170\ 169\ 174\ 178\ 180\ 190\ 200\ 205\ 215\ 210),\\
\mathbf{T}_5 &= (168\ 172\ 175\ 180\ 185\ 185\ 190\ 200\ 210\ 215),\\
\mathbf{T}_6 &= (145\ 147\ 150\ 148\ 146\ 145\ 147\ 149\ 150\ 151),\\
\mathbf{T}_7 &= (142\ 146\ 148\ 149\ 150\ 150\ 152\ 148\ 151\ 149),\\
\mathbf{T}_8 &= (146\ 147\ 145\ 148\ 150\ 151\ 150\ 149\ 148\ 150).
\end{aligned}$$

And the reference patterns are written in the form of a matrix as

$$\mathbf{M} = \begin{bmatrix}
1 & 1 & 1 & 1 & 1 & 1 & 1 & 1 & 1 & 1\\
0 & 0.1 & 0.3 & 0.4 & 0.5 & 0.6 & 0.7 & 0.8 & 0.9 & 1\\
0 & 0.3 & 0.5 & 0.7 & 0.9 & 1 & 0.9 & 0.6 & 0.4 & 0\\
1 & 0.9 & 0.8 & 0.7 & 0.6 & 0.5 & 0.4 & 0.3 & 0.1 & 0\\
1 & 0.7 & 0.5 & 0.3 & 0.1 & 0 & 0.1 & 0.4 & 0.6 & 1\\
1 & 0.8 & 0.2 & 0.1 & 0.6 & 0.9 & 0.9 & 0.4 & 0.2 & 0
\end{bmatrix}$$

The fuzzy clustering algorithm given in this paper is used and the learning rate is selected as ξ=0.42. The pattern-class matrix C is obtained as follow.

$$\mathbf{C} = \begin{bmatrix}
0.9447 & 0.4395 & 0.1789 & 0.0851 & 0.1016 & 0.3616\\
0.4816 & 0.7280 & 0.5146 & 0.0294 & 0.4719 & 0.4858\\
0.2789 & 0.0003 & 0.6534 & 0.4763 & 0.0174 & 0.7201\\
0.0118 & 0.7880 & 0.0874 & 0.5595 & 0.2978 & 0.7084\\
0.0766 & 0.1078 & 0.4105 & 0.2730 & 0.8287 & 0.5526\\
0.0078 & 0.5616 & 0.1451 & 0.6210 & 0.3143 & 0.7296\\
0.2291 & 0.3542 & 0.2927 & 0.3169 & 0.4169 & 0.3293\\
0.3425 & 1.0000 & 0.6460 & 0.2019 & 0.5924 & 0.4827\\
0.2114 & 0.3318 & 0.2978 & 0.3583 & 0.3839 & 0.3694\\
0.2379 & 0.9190 & 0.5875 & 0.1233 & 0.5499 & 0.4577\\
0.2595 & 0.9396 & 0.5976 & 0.1925 & 0.6221 & 0.4105\\
0.2296 & 0.3753 & 0.3214 & 0.3263 & 0.4189 & 0.3390\\
0.1346 & 0.4327 & 0.4099 & 0.3936 & 0.4025 & 0.4084\\
0.2180 & 0.3724 & 0.3327 & 0.3656 & 0.3689 & 0.3753
\end{bmatrix}$$

In order to decide whether there are possible sticking or splits on the shell of the newly solidified slab in the mould, the fuzzy decision making system presented in section 4 is used. After a prolonged training process, the weights of the alarm function are optimized as w_1=0.95, w_2=0.87, w_3=0.84, w_4=0.84, w_5=0.62, w_6=0.62; the gain constant μ is selected by try and error as μ=0.5. For the six thermocouple groups (4, 15, 16, 28), (3, 4, 27, 28), (4, 5, 28, 29), (3, 4, 5, 28), (15, 27, 28) and (16, 28, 29) in this subsystem, their θ_i's are calculated according to (8) and the results are substituted into the alarm function. The output of the function is y=0.7954. Since it was shown in lots of experiments that the threshold is y_T=0.65, alarm of possible steel breakout is then excited. And the position of leakage is also shown as in the triangular area in which thermocouples 4, 15 and 16 are located. In addition, possible reasons causing the leakage are suggested by the fuzzy system as there are sticking or transverse splits happened with in this triangular area.

In practice, leakage of liquid steel can also be predicted according to the traditional alarm strategy using the temperature samples in the experiment.

To show the effectiveness of the results obtained with the new clustering method, the indirect fuzzy clustering method was used and the degree of similarity between any two patterns are obtained as shown in Table 1.

Table 1 Results of indirect fuzzy clustering method

Thermocouples		M_1	M_2	M_3	M_4	M_5	M_6
3	P_1	0.0600	0.5400	0.5200	0.4900	0.5600	0.4900
4	P_2	0.5000	1.0000	0.8000	0.4800	0.8000	0.6800
5	P_3	0.0600	0.5100	0.5200	0.5200	0.5100	0.5200
15	P_4	0.4600	0.9400	0.8000	0.4600	0.8000	0.7200
16	P_5	0.4700	0.9700	0.8000	0.4800	0.8200	0.6200
27	P_6	0.0600	0.5600	0.5500	0.5000	0.5600	0.5000
28	P_7	0.1000	0.6000	0.6000	0.5300	0.5100	0.5300
29	P_8	0.0600	0.5500	0.5500	0.5300	0.5000	0.5300

It's evident from this table that there are about at lest three thermocouples whose measurements have characteristic of slope-like pattern, which is believed to be one of the most dangerous temperature patterns. This coincides with the results of the new clustering algorithm.

Many experiments have also shown that the new method is a reliable superior leakage alarm method. More over, it can be used for the purpose of quality evaluation and system diagnosis in continuous casting process.

REFERENCES

Haers, F., S.G.Thornton (1994). Application of mould thermal monitoring on the two strand slab caster at Sidmar. Ironmaking and Steelmaking, 21(5), pp.390-398.

Zadeh, L.A. (1978). Fuzzy sets as a basis for a theory of possibility. Fuzzy Sets and System 1, pp.3-28.

Hathaway, R.J., J.C. Rezdek (1994). NERF c-means non-Euclidean relational fuzzy clustering. Pattern Recognition 27, pp.429-437.

Donald, S.R., M.M. Mehran, C.G. Tim, et al (1995). Switched Reluctance Motor Control via Fuzzy Adaptive Systems. IEEE control system magazine, pp. 8-15.

Theocharis, J., Petridis (1994). Neural Network Observer for Induction Motor Control. IEEE control system magazine, pp.26-37.

Seo, Y.R., C.H.Chung (1994). Application of a fuzzy controller with a self-learning structure. Proceedings of the Asian Control Conference, pp.483-486.

Pedrycz, W., A.Rocha (1993). Knowledge-based neural network. IEEE Trans. Fuzzy Systems 1, pp.254-266.

HIGH - QUALITY MAGNESIA AND MAGNESIA - CHROME MATERIALS BASED ON THE FIRED MAGNESIA MATERIALS IMPORTED FROM KOREA AND CHINA.

Professor Wladyslaw Bieda
M. Sc. Boguslaw Bieda

Polish Academy of Science
ul.Slawkowska 17
PL-31-916 Krakow, Poland

Dept. of Management
Academy of Mining and Metallurgy
ul.Gramatyka 10
PL-30-067 Krakow, Poland

Abstract:This paper describes the production of the high-quality magnesia and magnesia-chrome materials based on the fired magnesia materials imported from Korea and China. Practical experiments permitted to develop oryginally technology of the production of the: three, four and five components refractory materials.We illustrate this technology with detailed characteristics gived in Table 1 and Table 2. *Copyright © 1998 IFAC*

1. INTRODUCTION

Refractory materials industry in East Europa used small quantities of magnesia materials from Korea and China to produce the magnesia and magnesia - chrome materials. Only refractory materials industry in Poland used big quantities of fired magnesia materials from Far East which are composed of high contents of $Si\ O_2$ at the cost of Mg O. Refractory Materials Factory in Poland is based on fired magnesia materials imported from Korea and China.

2. DESCRIPTION OF TECHNOLOGY

Practical experiments with magnesia refractory materials permitted allowed to bring in practice the oryginally technology in order to manufacture the refractory materials for modern steelmaking and ironmaking processing (like oxygen converters, tubs, etc.).

This paper descibes a general approach to the producing of :
three components refractory materials: MgO - Ca O - $Si\ O_2$,
four components refractory materials: MgO - Ca O - $Si\ O_2$ - $Fe_2\ O_3$,
five components refractory materials: MgO - Ca O - $Si\ O_2$ - $Fe_2\ O_3$ - $Al_2\ O_3$.

Magnesia chrome bricks contained two kinds of fired magnesia and chrome ore which in technology processing enables us to get refractory materials. The chemical composition of these bricks contained from 44.0 to 68.6 % of MgO, from 1.0 to 3 % of $Si\ O_2$ and from 16 to 24.6 % of $Cr_2\ O_3$. The low porosity (15.4 - 18.9 %), high cold crusing strenght (44.0 - 62.2 MPa) and the high refractoriness under load ($T_{0.6}$ 1660 - 1700 ^{o}C) prefered these bricks to vacuum degasing plant (DH), ladle refractory lining in VOD / VAD processing, oxygen - blow converters refractory lining and in furnace lining in copper industry.

But in steel ladleselectric furnaces and oxygen - blow converters is used unfired magnesia - carbon bricks. The chemical composition of these bricks contained from 89.0 to 94.0 % of Mg O, from 1.8 t to 4.5 % of $Si\ O_2$ and from 7.0 to 9.0 % of carbon.

These materials are characterized by high compactness, low bulk density (2.70-2.80 g/cm^3), low porosity (8.0-12.0%) and high cold crushing (25.0-35.0 Kpa). Magnesia - carbon materials have positive influence on the live of steel ladles because in large

quantities of ladle heats have been not observed overgrowth of slag in the ladles. Laboratory and industial test of magnesia - chrome and magnesia carbon bricks have been prooved the high quality of these materials due to the optimal content of SiO_2 (from 2.0 to 3.0 and 4.0%).
Crystallographic structure, mineralogical composition and texture have been investigated by microscopy. These elements gived the resistance to chemical agents of metallurgical process in high temperatures.

The detailed characteristics of the magnesia - chrome bricks are listed in Table 1. In refractory materials industry in Poland are producted many kinds of these materials.

Table 1. Magnesia - chrome bricks.

Properties		Measure Unit	Product							
			MC10	MC9	MCV	MCT	MXT0	MXT60	MX40	CM30
Chemical Composi-tion	MgO	%	68.6	54.8	54.2	57.6	75.3	60.8	44.6	34.3
	Cr_2O_3	%	15.6	22.1	22.1	22.2	9.5	17.5	24.6	33.8
	SiO_2	%	1.5	1.0	1.0	1.3	1.6	1.2	2.9	3.3
	Fe_2O_3	%	7.9	14.5	13.5	10.9	8.3	13.0	13.3	17.5
	CaO	%	0.9	0.8	0.9	0.9	0.6	0.8	0.9	1.0
	Al_2O_3	%	5.5	7.6	8.2	6.7	4.2	6.2	13.4	9.9
Bulk Density		g/cm^3	3.10	3.25	3.25	3.31	3.12	3.12	3.17	3.20
Apparent Porosity		%	18.8	16.2	16.4	16.7	16.4	17.9	19	17.7
Cold Crushing Strength		MPa	38	55	62	53	50	44	30	35
Refractoriness Unter Load	$T_{0.6}$	°C	1660	1700	1700	1700	1700	1700	1660	1660
Thermal Conductivity	700°C	W/m. °C	3.7	3.7	3.7	3.7				
	1100°C	W/m. °C	2.8	2.8	2.8	2.8				
Specific Head (Average)	20-40	kJ/kg. °C	1.008	1.088	1.088	1.088				
	20-1000	kJ/kg. °C	1.172	1.172	1.172	1.172				

On the contrary Table 2 gives the parameters of the unfired magnesia - carbon bricks. These unfired materials produced on the base the high quality flake graphite which has not only high refractory quality but simultanously limits in magnesia plastics the wettability of the magnesia part under influence of slag chemical action in high temperatures.

Table 2. Unfired magnesia - carbon brikcs

Properties		Measure unit	Product				
			MWK51	MWK6	MWK8	MWK81	MWK101S
Chemical composition	MgO	%	91	92	94	94	90
	SiO_2	%	2.0	2.5	1.4	1.0	4.0
	CaO	%	1.8	2.0	1.9	2.0	2.0
Carbon contents		%	5.5	7.0	8.0	9.0	11
Residual carbon		%	4.0	4.0	5.5	7.0	9
Bulk density		g/cm^3	2.85	2.70	2.75	2.80	2.70
Apparent porosity		%	14	12	10	8	5
Cold crushing strength		MPa	28	25	30	35	30

In this way slag chemical aggressive action have been considerably reduced, what perfectly prolongs refractory lining worktime. The microscopic image the fired refractory materials microstructure in the reflected light shows the perfect cristals and compact texture of the magnesia plastics with the growths of the chrome cristals in the spinel phase part.

3. CONCLUSION

The microscopic investigation have been proved the high physical and chemical as well as thermo - mechanical properties of those refractory materials described above.

AN APPLICATION OF IMAGE PROCESSING METHOD FOR POSITION CONTROL OF PUSHER CAR IN COKE PLANT

Il-Seop Choi Jong-Hag Jeon

Instrumentation and Control Research Team
POSCO Technical Research Lab
1,Koedong-dong,Pohang-shi,790-785,Korea
pc546803smail.posco.co.kr pc542888smail.posco.co.kr

Abstract: With the aim of saving energy and labor, automatization and systematization for moving machines in coke plant have been done. An essential technology for automatization of coke moving machines is the position control which enables a pusher car to position exactly in front of a target coke oven and to push coke safely. Because of very high temperature inside a coke oven, there is a relative position error due to deformation of a coke oven, which can't be compensated by an existing system. This paper presents an implementation of the system which make it possible to measure the relative position error by using the image processing method and control by driving the main travel motor of a pusher car. The developed system was installed and tested for no.2 coke plant in Kwang-Yang Steel Works,POSCO. *Copyright © 1998 IFAC*

Keywords: moving machine, coke plant, position control, image processing, pusher car, automatization

1. INTRODUCTION

A coke process is to produce coke from coal which is heated without air supply. COG(Coke Oven Gas) and liquid tar are also obtained as a by-products. Coke is mainly used to deoxidize iron ore in blast furnace and also used in small coke form in the sintering process. COG is utillized as fuel for plants. There are many facilities to produde coke in coke plant. The main facilities in coke plant are coke ovens and moving machines. Each oven consists of a coking chamber, a heating flue chamber, and a regenerative chamber. Coke moving machines are mainly composed of a coal charging car, a coke pusher car, a transfer car and a locomotive. (Jun S. and N, 1993)

Many operations in coke process are conducted under the unfavorable working condition. There are many tasks to be done in the outdoors which has gas leakage, tar and dust. So the development of automatic operation system is strongly required in addition to improve working condition in coke oven plant. POSCO promotes gradually the automatization of coke moving machines to satisfy this requirements.

A position control system can be considered as an essential technology for automatization of coke moving machines. The position control system in a pusher car enables a pusher car to position exactly in front of a target coke oven in order to push coke safely. A pushing operation has to be done when the center line of a coke oven exactly coincide with the center line of a pusher car ram. An existing position control system in POSCO is based on an infrared sensor and a distance integration method using motor encoder. But both devices are located in the lower part of a coke oven , it is possible to control only the absolute position error. Actually, because of very high temperature inside a coke oven, there is the relative position error due to deformation of a coke oven, which can't be compensated by an existing system.

This paper presents an useful measuring method of the center position error including the relative position error. This method is based on image processing. The center position error means the position difference between the center line of a coke oven and the center line of a pusher car ram. The measured position error is controlled by interfacing with an existing PLC for position control. The developed system have been installed and tested for coke plant in Kwang-Yang Steel Works,POSCO. By applying this system to conventional system, it can be possible to control a pusher car within 5 mm position error and get better reliability and safety. (Michio T. and Hiroshi, 1988), (Masao M. and Yuji, 1989)

2. CENTER POSITION CONTROL SYSTEM

2.1 *Control System Overview*

The overview of the developed control system is shown in Fig.1. The moment a coke pusher car stops in front of a target coke oven in order to conduct a pushing operation, the developed system acquires the image of the surface of a coke oven and ram of a pusher car from a vision system. It applys edge detection algorithm to acquired image and detects the edges of coke oven door frame and ram of a pusher car. Because the door frame projects and have different gray level compared with other part of a coke oven, it is not difficult to detect the edges. From the information of the edges, it is possible to compute a position difference between the center line of coke oven door and the center line of ram. The center position error is defined as the position difference. The measured position error is displayed in MMI for operator and sent to PLC of a pusher car to control. (Jeon and CHOI, 1996)

2.2 *Measurement of the Center Position Error*

The concept of position error measurement is shown in Fig.2. A couple of CCD camera which is about 6m from a coke oven acquire the image of buckstay and ram head of a pusher car. The camera A scan the left hand side of ram head and buckstay. The camera B scan the right hand side of ram head and buckstay. Edge detection algorithm detect the line 1, line 2 for buckstay and line A for ram head. The length D1 means the distance between the left hand side of buckstay and ram head. The D2 is the distance for the right hand side. The center position error(CPE) is computed by subtracting D2 from D1. If the CPE is zero, it means a pusher car exactly positions in front of a target coke oven. If the CPE is positive, it means a pusher car positions at the right hand side of center line of a target coke oven and has to be moved to the left hand side as much of the CPE in order to push safely. If the CPE is negative , it has to be moved to the right hand side as much of the CPE.

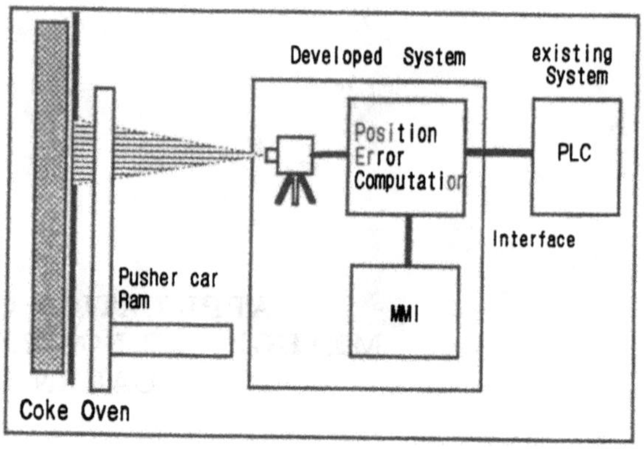

Fig. 1. Position Control System Overview

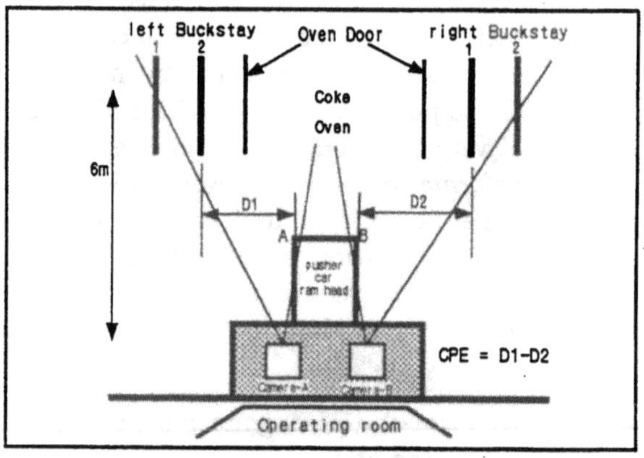

Fig. 2. Concept of Position Error Measurement

2.3 *Hardware Configuration*

The position error measurement system has follwing H/W configuration. It consists of a vision system, PLC interface and a computer system for position error computation and MMI. A vision system has a couple of CCD cameras and was installed outside of operating room of a pusher car. It is about 6m from a coke oven. A computer system including DT3851-4 image board was installed inside of operating room. It is a industrial 486 computer system and uses windows 95 operating system. The PLC interface was established by serial link between the computer system and an existing PLC.

2.4 *Software Configuration*

The software configuration for real time operation is shown in Fig.4. It consists of the position error

computation, calibration, MMI and data link with PLC.

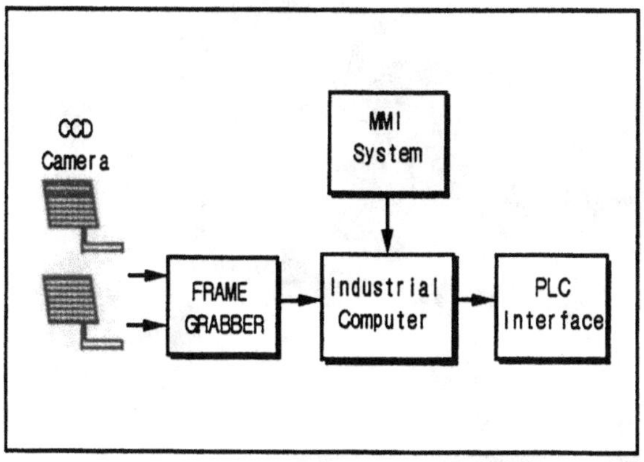

Fig. 3. Hardware configuration for Measurement System

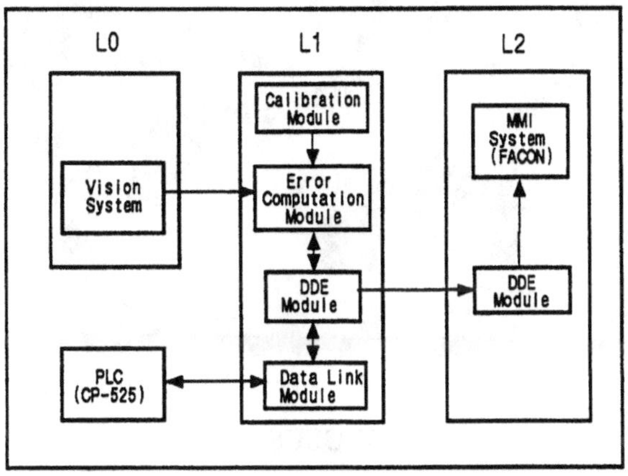

Fig. 4. Software Configuration for Measurement System

2.4.1. *Position Error Computation*

This module computes the center position error using image information acquired from a vision system. It acquires the images of a pusher car ram and coke oven door frame from a couple of camera. After acquisition of image, the region of image is selected to reduce processing time. For selected region, it gets grey level and applies threshold value for all pixels of the region. The gray levels passed threshold value are stored to temporary buffer. It evaluates everage for each column in temporary buffer and saves to another buffer. Then the edge line is drawed from a point on which has maximum change of the gray level for buffer. Two lines of edge are found for one camera image. The distance is computed by subtracting x coordination of these two lines. It is the distance between the buckstay and ram head for the left hand side. The distance for the right hand side is computed by repetition of this process for the right hand side of image. The center position error is computed by subtracting the second computed distance from the first computed distance. This error value is expressed in pixels. In order to change this value to real distance,it is multiplied by calibration data which include the distance for a pixel. This module determine if this error value is applied to control or not. If the error is smaller than the predefined value, it is only displayed in MMI. The position error is deliverd to the PLC to control when it exceeds the predefined value. (Phillips, 1994)

2.4.2. *Calibration*

In order to cope with the situation of camera parameter tuning, the function of camera calibration is required because the distance per a pixel is changed. To do this, a plate for calibration was installed in front of a pusher car at the home position. The plate of calibraton is composed of two grid of which the distance was previously measured. When a pusher car positions at the home position, calibration sequence is executed by operator pushing the button for calibration. The calibration module acquires the image of calibration plate and detects the edges of grid. The distance between two edges can be measured in pixels. Then the distance per a pixel, calibration data can be computed by dividing the known distance by computed pixels. Computed calibration data is saved in file and referenced when the center position error is computed.

2.4.3. *Man Machine Interface*

To guide operator to the computed center position error, MMI system is used. It offers the center positon error monitoring, working ststus of pusher car and historic management of the center position error for individual oven. The center position error monitoring is shown in Fig.5. It represents the center position error, oven number for working and current position of moving machines. Fig.6 represents historic data of the center positon error for worked oven number. From this function, the management of the position error for individual oven is possible.

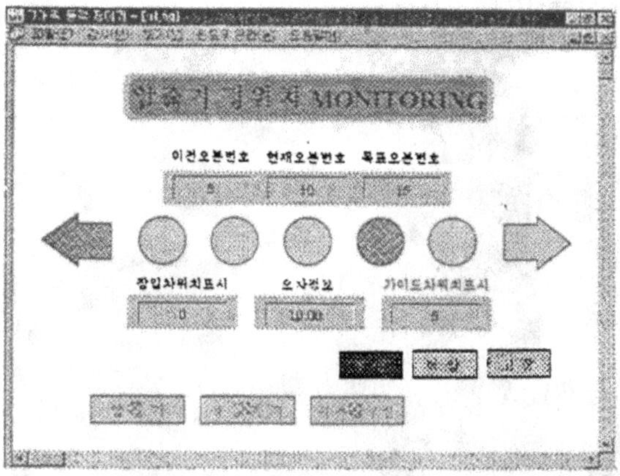

Fig. 5. The view of center position error monitoring

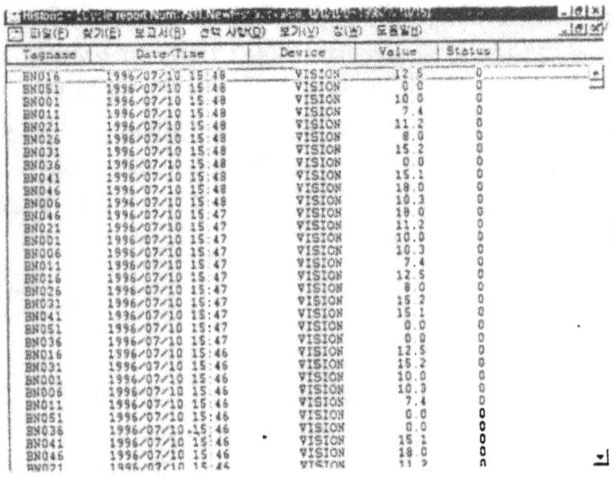

Fig. 6. The view of historic data management

2.4.4. *Data Link with PLC*

The function of this module is to perform data exchange with the PLC of the travel motor. The PLC offers the position error measurement system information concerned with working status of a pusher car. This is done when the PLC performs the sending sequence. Information deliverd to PLC is the center position error to control. This is done when the PLC performs the fetch job. The data exchange is realized by serial link using 3964R communication protocol.

3. EXPERIMENTAL RESULTS

The developed system was installed and tested for no.2 coke plant in Kwang-Yang Steel Works,POSCO. When a pusher car positions at a target oven number to push coke, it acquires the image of a coke oven. Fig.7 represents the image of a coke oven which is seen at operating room. The processing results about both sides of camera image are shown in Fig.8 and Fig.9. From the detected line, the distance between buckstay and ram head is measured easily. The difference of the distance computed in Fig.8 and Fig.9 means the center position error. When the position error larger than the predefined value, it is deliverd to motor driving PLC to control.

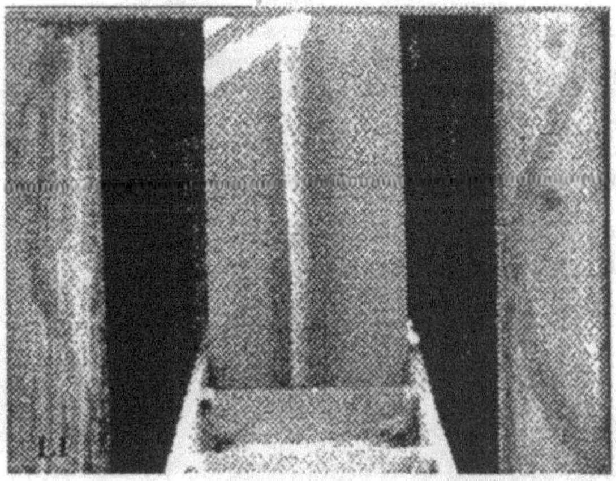

Fig. 7. The image of a coke oven seen at operating room

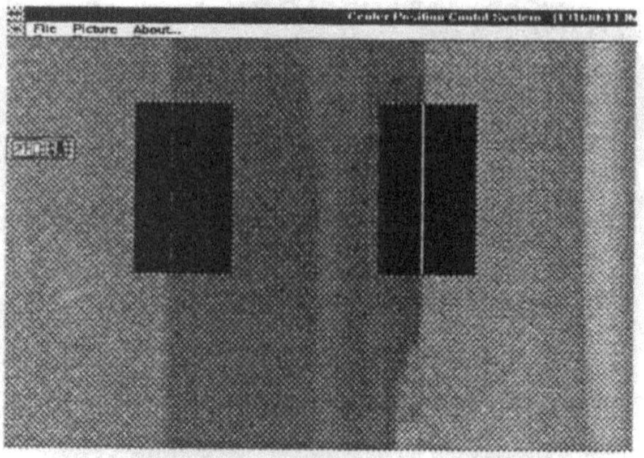

Fig. 8. The image processing result of camera A

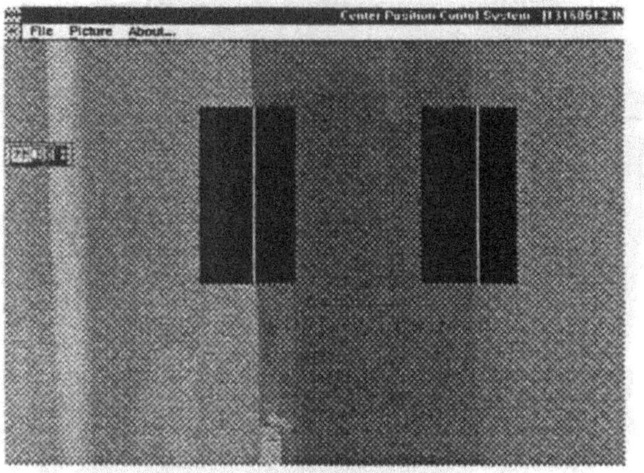

Fig. 9. The image processing result of camera B

4. CONCLUSION

In order to save energy and labor, automatization for moving machines in coke plant have been done. As a essential technology of automatization for a pusher car, a new position control system is proposed. This system has good points in that it can cope with the relative positon error which is come from deformation of a coke oven due to very high temperature inside oven. The measuring method of the position error including the relative position error is baseed on image processing. By applying this system to an existing position control system, a pusher car can be controlled within 5 mm position error.

5. REFERENCES

Jeon, J.H. and I.S. CHOI (1996). A development of center position control system in coke moving machines. *POSCO Technical Report.*

Jun S., Keihachirou T., Youichi N. and Youji N (1993). Automatic coking control system. *Second IEEE Conference on Control Application* pp. 13–16.

Masao M., Hiroshi U., Yoji N. and Yuji (1989). Development of automatic operation system for coke oven machines at yawata works of nippon steel corporation. *NIPPON STEEL TECHNICAL REPORT.*

Michio T., Minoru S., Akio Y. and Hiroshi (1988). Automatic operaton system for coke oven machiner. *Ironmaking Conference Proceeding* pp. 109–115.

Phillips, Dwayne (1994). *Image Processing in C.* R&D Publications,Inc.. Kansas.

A MATHEMATICAL MODEL FOR GAS-WIPING PHENOMENA IN HOT-DIP GALVANIZING PROCESS

T.Kametani, K.Andachi, T.Nakagawa,
H.Shigemoto, T.Yoshioka, M.Hirata

Mizushima Works, Kawasaki Steel Corporation,
Kawasakidori 1-chome, Mizushima Kurashiki 712, Japan
phone : 086-447-3609
E-mail : kametani@miz. kawasaki-steel. co. jp

Abstract:In the gas wiping for continuous hot-dip galvanizing, a model analysis representing the relationship between its zinc coating weight and operating factors was carried out. By using this model taking the characteristics of two-dimensional free jet into consideration, new knowledge such as the fact that zinc coating weight is not influenced by its nozzle-slit gap in a jet range of narrow distance from nozzle to strip, etc., were obtained. Also with this model verified in Mizushima CGL, it was confirmed that the calculated value of zinc coating weight was in good agreement with its measured values. Furthermore, with these knowledge applied to the improvement of its operation using a zinc coating weight control, an optimization of nozzle-sit gap, etc., good results were achieved. *Copyright © 1998 IFAC*

Keywords : distance, gap, gas, mathematical model, nozzles, pressure control, steel industry

1. INTRODUCTION

Recently, a lot of high anticorrosion alloy hot-dip galvanized steel plates(GA) have been used as an exterior steel of automobiles. Therefore, the GA ratio accounting for the amount of production in the continuous hot-dip galvanizing line in our Mizushima Works is very high as well. Gas wiping is an important process in which the surface quality of this GA product is produced. Consequently, the relationship between its coating weight and operating factors, especially, the influence of nozzle-slit gap on its coating weight, and the relationship between the wiping nozzle and the strip in the range of a near distance are important for improving the wiping operations such as coating weight control, highspeed thin galvanizing, etc..

This paper reports that, with the model analysis of wiping phenomena carried out. this was verified in Mizushima CGL, and together with this, a good result was achieved by applying the above result to the coating weight control and the improvement of operation.

2.MATHEMATICAL MODEL FOR GAS WIPING

In the gas wiping process, continuously annealed steel sheets are lifted up after the immersion in the molten zinc pot as shown in Figure 1, and its molten zinc liquid film adhered to the front- and back surfaces of the steel sheet is adjusted by a high pressure gas jet from its wiping nozzles arranged in the height of about 300~500 mm over the pot to a specified zinc coat weight (30~70g/m2=4.5~9.0μ m).

With Navier-Stokes equation applied with the boundary layer theory to the zinc liquid film adhered to the steel sheet lifted up from the zinc pot and the mass conservation method simultaneously arranged[1], and with the two dimensional jet theory[2] and isoentropic flow in the near field region and the far field region applied to the wiping jet, the relationship between the coating weight and the operating factor is determined. It is suggested that zinc coating weight is related to influencing factors as shown in equations (1) and (2) and the dependency on the distance from the nozzle to the strip varies in the near field region and the far field region by the characteristics of two dimensional jet flow jetting from a slit with a large aspect ratio, and the coating weight is not influenced by the nozzle-slit gap in the near field region. Also because of an influence of zinc temperature at wiping point, the above equations are constructed taking the change in wiping gas temperature in adiabatic expansion, zinc bath temperature, and heat transfer coefficient of wiping gas into consideration.

1) Near Field Region ($D/B \leqq C$)

$$W = k1 \times \rho \times \left(\frac{\kappa - 1}{2 \times \eta \times \kappa \times Pa} \right)^{0.5} \times D^{0.5} \times \left(\frac{\mu \times V}{(P/Pa)^{\frac{\kappa-1}{\kappa}} - 1} \right)^{0.5} \quad \text{----- (1)}$$

2) Far Field Region ($D/B > C$)

$$W = k2 \times \rho \times \left(\frac{\kappa - 1}{2 \times \eta \times \kappa \times Pa} \right)^{0.5} \times \frac{D}{B^{0.5}} \times \left(\frac{\mu \times V}{(P/Pa)^{\frac{\kappa-1}{\kappa}} - 1} \right)^{0.5} \quad \text{----- (2)}$$

<List of symbols>

B ; nozzle slit gap, D ; nozzle-strip distance,
H ; height of nozzle, P ; nozzle plenum pressure,
Pa; atomospheric pressure, V ; strip speed,
W; coating weight, η ; nozzle efficiency,
κ ; ratio of specific heat of gas,
μ ; molten metal viscosity,
ρ ; molten metal density

3.VERIFICATION RESULTS OF MATHEMATICAL MODEL

The analytical model was verified using experimental and operating data collected in Mizushima CGL. Zinc coating weight was evaluated using the measured value obtained from an on line coat weighting instrument and the value obtained from wet chemical analysis, and with the distance from the nozzle to the strip measured with a range finder installed immediately over the wiping nozzle, front and back surfaces of the strip were evaluated for each.

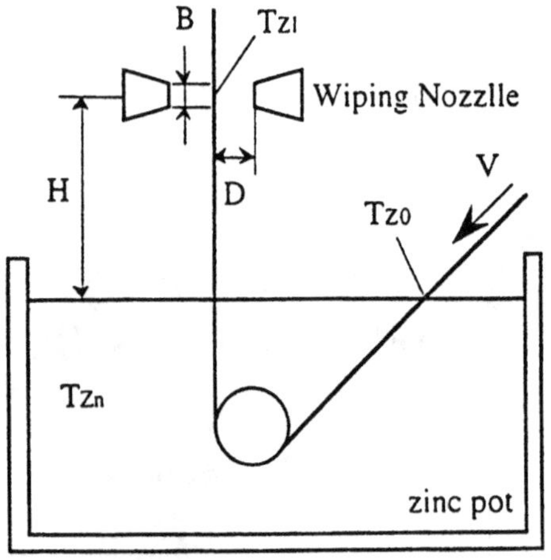

Fig.1-1 Schematic of hot dip galvanizing process

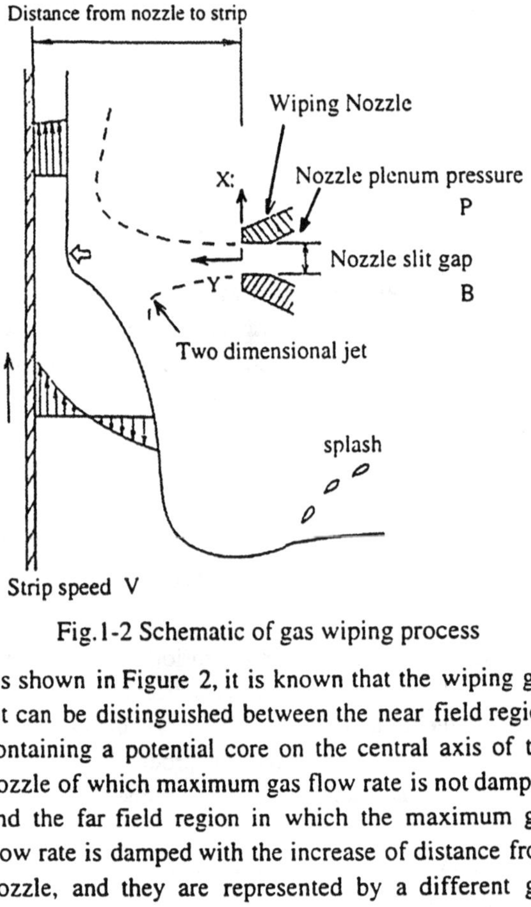

Fig.1-2 Schematic of gas wiping process

As shown in Figure 2, it is known that the wiping gas jet can be distinguished between the near field region containing a potential core on the central axis of the nozzle of which maximum gas flow rate is not damped and the far field region in which the maximum gas flow rate is damped with the increase of distance from nozzle, and they are represented by a different gas

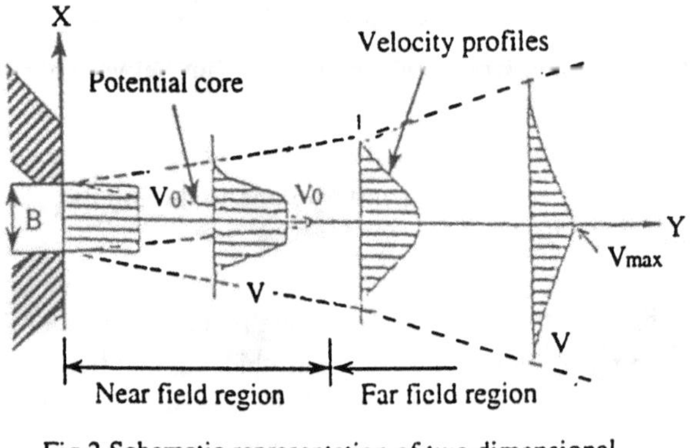

Fig.2 Schematic representation of two dimensional subsonic turbulent free jet

flow rate distribution for each. However, its boundary is different from its nozzle shapes, etc. In the nozzle of Mizushima CGL, its boundary is about D/B = 8 ~ 10 as shown in Figure 3.

Figure 4 shows the relationship between zinc coating weight and nozzle pressure. The zinc coating weight is proportional to one second power of nozzle pressure term and in good agreement with its model equation.

Also as shown in Figure 5, the zinc coating weight is proportional to one second power of strip speed.

Figure 6 shows the relationship between zinc coating weight and distance from nozzle to strip. The zinc coating weight is proportional to one second power of the distance from nozzle to strip in the near field region, and proportional to first power of that in the far field region. Consequently, it revealed that there is a different dependency on the distance from the nozzle to the strip.

Figure 7 shows the comparison of the measured value of zinc coating weight obtained from the operating data in the near field region carried out varying the nozzle-slit gap with the calculated value from the model equation. In the near field region, it was confirmed that the zinc coating weight is not influenced by the nozzle-slit gap. Also the analytical model equation is capable of predicting the zinc

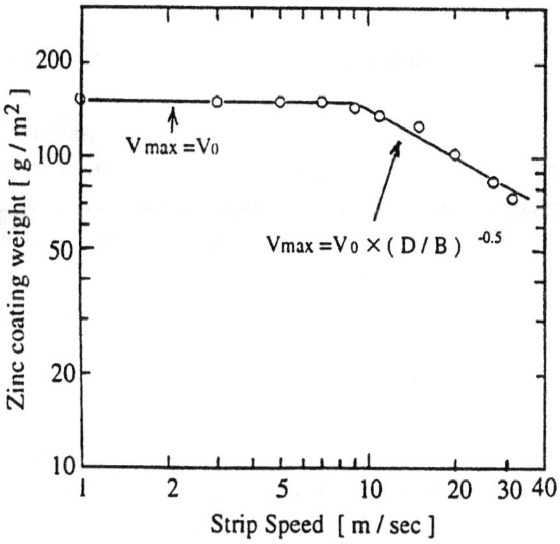

Fig.3 Damping characteristic of maximum wiping gas flow rate

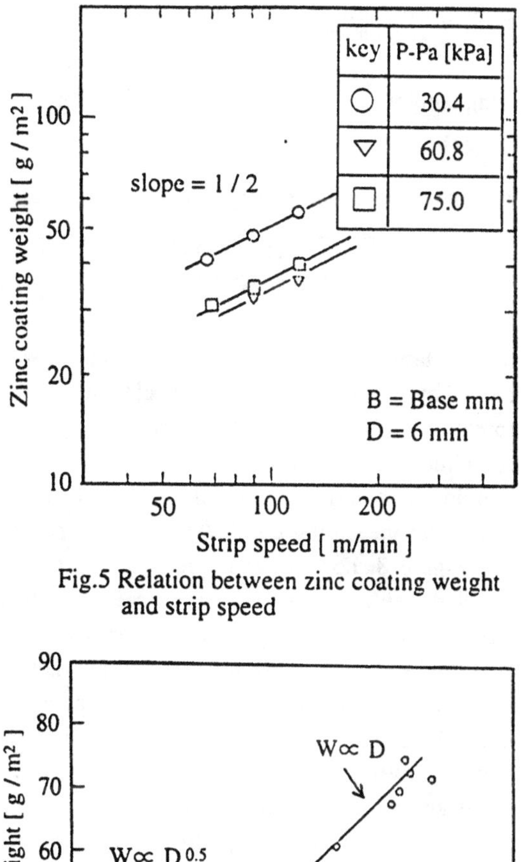

Fig.5 Relation between zinc coating weight and strip speed

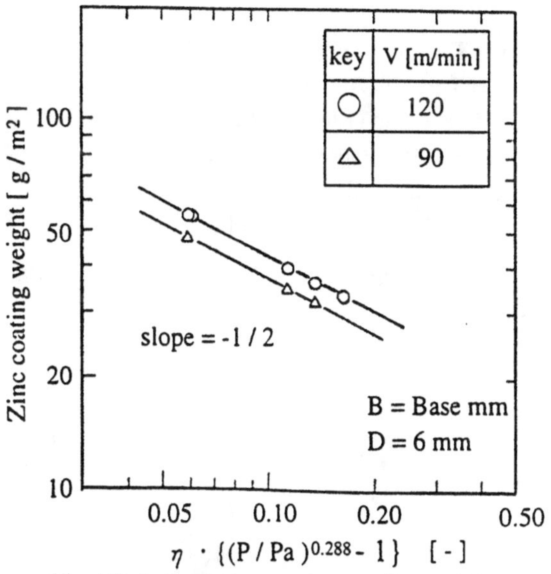

Fig.4 Relation between zinc coating weight and nozzle plenum pressure

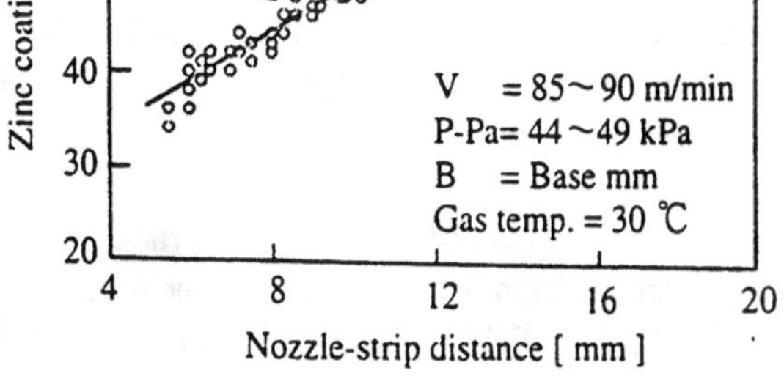

Fig.6 Relation between zinc coating weight and nozzle-strip distance

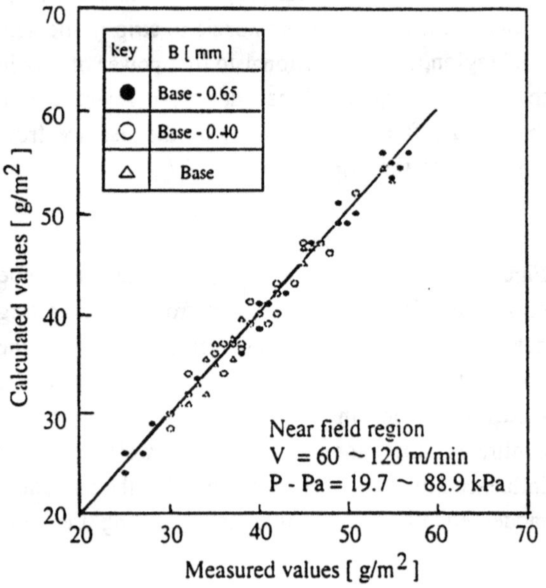

Fig.7 Comparison of calculated and measured coating weight on near field region

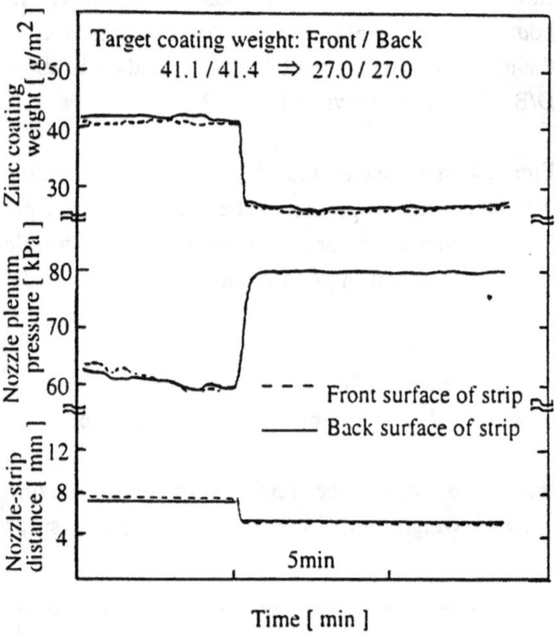

Fig.8 Automatically coating weight controlled results

coating weight in the actual operation of Mizushima CGL with the accuracy of $\pm 2g/m^2$.

4.EXAMPLES OF RESULTS APPLIED TO ACTUAL MACHINES

The automation of zinc coating weight control was achieved by introducing this analytical model to the automatic control of zinc coating weight in Mizushima CGL, and together with this, by improving its control system, and by introducing a range finder measuring the distance from nozzle to strip. Figure 8 shows an example of automatically controlled results in the case that a targeted coating weight is same in front and back surfaces of strip (equal thickness galvanizing). At the point where a targeted coating weight is changed, its response is raised taking into consideration the lower limit of distance between nozzle and strip and the lower- and upper limits of nozzle pressure, and changing the distance from nozzle to strip and its nozzle pressure. Also in the case that a targeted coating weight is different in the front and back surfaces of strip(different thickness galvanizing),automatic controls are carried out independently for each of surfaces.

It is considered that, based on this analytical model, zinc coating weight is not affected by the nozzle-slit gap in the near field region, and a thin galvanizing can be effectively achieved by narrowing the nozzle-slit gap, and suppressing its zinc splash and the lowering of zinc coating temperature. Based on these concepts, with the nozzle-slit gap optimized, and together with the automation of zinc coating weight control and the improvement of control accuracy promoted, effects have been achieved for improving the operation.

5.CONCLUSION

The model which predicts the coating weight for gas wiping in hot-dip galvanizing process was devised and was applied in Mizushima CGL.

REFERENCES

[1] J.A.Thornton and H.F.Graff (1976) ; Met.Trans.,B,7b, , 106

[2]N.Rajaratnam (1981) ; " Turbulent Jets ", japanese translation Morikita pub.

[3]K.Andachi, et al . (1993); ISIJ Meeting Materials and Processes,vol.6,No.5,430

General Purpose Control System for Steel Making Process Control Using VME Bus Based System and OS-9 WINDOWS

°Je-Young You, Sangchul Won and *Seok-Ju Bang

Dept. of Electr. Engr., POSTECH, KOREA, 279-2894; *fax*: 279-2903; E-mail: gabriel@postech.ac.kr, *POSCO

Abstract: In this paper, a general purpose control system was constructed to provide accumulation of the technology in integrated control system design under open architecture environment. The VMEbus system with OS-9 was chosen to simulate a process control system due to the popularity in open market. Interactive screen graphics including system function diagrams, menus, controlled system status and update information are standard in this model. The simulation program was ported on a 32bit MVME147S-1 system and successfully tested. *Copyright © 1998 IFAC*

Keywords: Machine-Machine Interface(MMI), Real-Time Operating System(RTOS), OS-9, OS-9 WINDOWS, VMEbus, Programmable Logic Controller(PLC), Direct Digital Controller(DDC), Distributed Control System(DCS)

1. INTRODUCTION

Most of the control systems in steel industry are designed, manufactured and integrated uniquelly by suppliers and delievered by turn-key base. Due to suppliers' closeness of technology, many control engineers in steel industry frequently encountered difficulties whenever they try to modify or expand their control system. The difficulties of the closeness are as follows. First, It is impossible to add the additional node to the existing system. Second, dne to the supply of the unique protocol in each control system, it is impossible that one control system communicates with another control systems. Third, software prting is impossible by ignorance of hardware specification. So open architecture of control system is essential.
Also the existing control systems has many problems for the loop control. In general, the sequential control system using PLC and the distributed control system using microprocessors are used for the control of the steel making process. The sequential controllers have fast response time, supplies sufficient I/O modules, and guarantee the reliability. However the closed loop control and MMI environment is not supported generally. On the other hand, microprocessor based controllers support the closed loop control scheme and many application program such as MMI environment, system management, and database management tools. But bacause it does not support high response time and sufficient reliability, it is not adequate to be executed in high speed.
A general purpose control system using the advantages of the sequential and closed loop control systems are composed in this paper. The distinctive features of this control system include the open architectured system bus such as VMEbus, sufficient I/O module products independent on CPU modules, the open architectured network protocol such as TCP/IP, the general programming language such as C and C++, and RTOS for the application of the control algorithms. In accordance with these design phylosophy, the VMEbus system has been generally adopted as the system bus in the industrial fields recently. Due to its open architecture, any VME I/O module can be designed easily by necessity. So I/O and CPU modules for VMEbus are very popular in the market.
The real-time control capability in the steel making process is very essential for precision control. Therefore we adopted OS-9 as RTOS and OS-9 WINDOWS as the window program executed in RTOS.
The remainder of this paper is organized as follows. Section 2 describes the general requirements for the industrial control system. The open architectured control system based on VMEbus is presented in section 3. The description of control system software developed under OS-9 WINDOWS is presented in section 4. Finally,

section 5 presents the constructed control system and MMI console screen are shown.

2. THE REQUIREMENTS FOR CONTROL SYSTEM

The control system has the characteristics as follows.

Open architectural system bus: The system bus is the most closed part of the control systems. Because each system is constructed on the characteristic backplane bus, the modification of the system hardware and the additional development of the system software are pratically impossible. Therefore open architure standard system bus is necessary.

Intelligent I/O system modules: Because PLC carries out mostly passive function, all functions of PLC are executed by the only command of the main processor. So the work of intelligent data acquisition and monitoring function, which is the periodic or triggered function, are impossible. Therefore intelligent I/O system modules are essential for the construction of the closed loop control scheme.

Operating system: RTOS for the satisfaction of the real-time control performance has been adopted in some DCS and PLC. The general characteristics of RTOS are determinism, responsiveniess, user control, reliability, and failsoft. Determinism is the tendency to execution of the process in the determined time. RTOS has the minimal time of determinism compared with the general RTOS. Responsiveness is the period for process conversion. These two conditions are the most important properties of the real-time conditions. User control, which is the useful property of RTOS, is the function for system control by privileged users. The main processes of RTOS pre-allocate the essential resources(i.e. memory field, program and data load) by reliability.
But because most of the adopted RTOS prohibit user access, the modification and addition of the application softwares are very difficult. Also window environment constructed on RTOS is essential for MMI. So the necessary conditions of operating system are a real-time property and open I/O control scheme.

Programming language: Because the advanced control systems need complex function and high technology, general purpose programming languages are essential. These languages, which are C and C++, etc., are independent of a hardware and easy to modify programs.

MMI tool: Because MMI tools must express some graphic displays, this is slow generally and give rise to the decrease of total system speed. Therefore the real-time window system for RTOS is adopted for the real-time processing. Real-time window system preserves the characteristics of RTOS and show the graphical representation.

Multi purpose and high speed network: The network among sub-systems is the main problem for the real-time control. Generally the network doesn't proceed the real-time process. So the generally solution for this problem is the high speed and general prupose network.

3. HARDWARE IMPLEMENTATION

3.1 VMEbus

The classical hardware configuration is constructed using the specific system bus and I/O modules. Because of the closed different protocols, the network between other systems is impossible. So we constructed the system with the open architecture system bus and general intelligent I/O system modules. The standard open architecture system bus makes it possible to compose the distributed system with the different system modules.

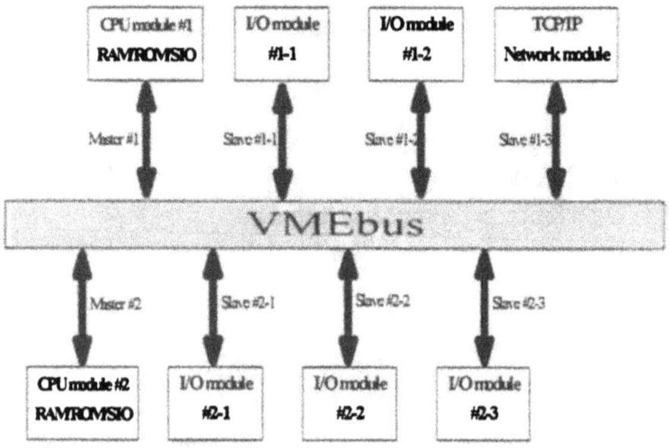

Figure 1. General VMEbus control system configuration

One of the mainly used system buses is VMEbus, which has been used as the steel making process control system recently. VMEbus is the typical asynchronous, open architectured, multiprocessor, and multiprocessing system bus. Therefore MOTOROLA 68x, INTEL, SPARC, DSP, and RISC CPUs are all used. And because VMEbus has many kinds of OS and VME I/O modules, the control system does not depend on the product of special maker. This means that VMEbus does not have any problem for VME CPU and I/O modules.
Figure 1 shows the general configuration of VMEbus based control system, where some CPU modules exist on a VMEbus and each CPU modules manage some I/O modules exclusively. In this system, each CPU module executes the role as a master and each I/O modules executes the role as a slave.

3.2 Control system configuration

VMEbus based controller is used as control system hardware in this paper. This system is composed of CPU module, I/O(A/D, D/A, DIO) modules, TCP/IP communication module, graphic module, and signal

detectors. Because this control system can be connected with other VMEbus based controllers using commuincation network, each VMEbus control system can execute the execlusively shared control job.
The system components are as followed. Motorola 68030 CPU is used as main processor, which manages data I/O and communication. I/O modules are used for acquiring the switch and sensor data. The communication network is used for data communication among other VMEbus based systems. These are TCP/IP ethernet and serial comminucation. Ethernet lines support the communication between the VMEbus controllers. Serial lines are connected to the communication between I/O modules and measurement equipments. A graphic module is used for the MMI representation of the I/O data. This module adopted DSP CPU for the graphic data processing.

4. SOFTWARE IMPLEMENTATION

Application program modules of this control system were composed of C language. OS-9 is used as RTOS of this control system. And OS-9 WINDOWS is adopted as the MMI tool, which is executed on the exclusively graphic processing module. Because OS-9 WINDOWS was developed for the execution of OS-9 real-time function on the graphic environment, it has also real-time performance.

4.1 Real-time software tool

OS-9 has the performance of the multi-tasking, the shape of module, and the classes for the real-time execution. As these characteristics, when the hardware or software problem is occurred, the full system is executed stably by the partial repairment or replacement of some modules.
The text mode monitor and control functions are only possible in the classical OS-9 environment on VMEbus. But the I/O data visualization function is possible on OS-9 WINDOWS environment. The controller based on this idea can do real-time control using the simple interface between I/O module and plant. OS-9 WINDOWS uses APLIB as the application library, which is composed by the object oriented programming method. So the structure of OS-9 and OS-9 WINDOWS are similar. All APLIB components, which are I/O data, internal process signal, and user defined signal, are managed by each message handler. MMI environment is implemented by using APLIB.

4.2 Software configuration for process control

The softwares for process control classify to three parts, which are data acquisition, control process, and MMI representation group. Figure 2 shows the block diagram for process control.

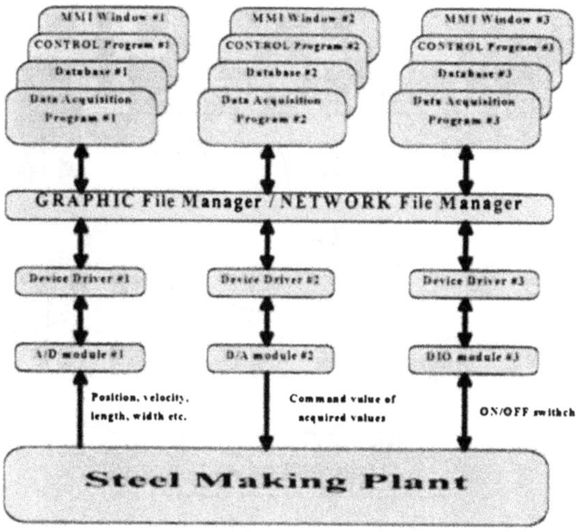

Figure 2. Total software block diagram

Data acquisition group : This group is implemented by using C language. And it is composed with data initialization routine, data acquisition routine, and database. Data initialization routine initializes the database for storing of digital and analog data. And this routine connects the control hardware and software from database setting. Data acquisition routine get switch, voltage, and current data from I/O modules. These acquired data are stored to the database.

Control process group : The role of this group is the command transmission to the control I/O module. After acquiring data from data acquisition group, the control command is calculated. And these commands are transmitted through the control modules. Therefore this group executes the essential control jobs.

MMI representation group : This group represents the control and monitor data using the graphic tool. This is divided into two parts, which are control and monitor graph.
Control graph sets the position, speed, length, and width of the steel products. Monitor screen compares the setting values with the real values. And each sensor data can be referred by this graph.

5. SIMULATION

This section represents the control system using OS-9 and OS-9 WINDOWS as RTOS and real-time MMI tool based on VMEbus. Because this control system has adopted the open architecture, this is composed manually and modified easily by coustomers. Figure 3 shows the flow chart of the MMI and control processes. Figure 4 represents the execution graph.

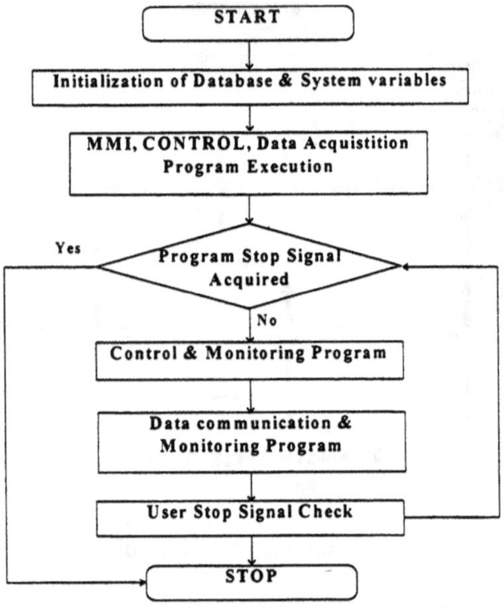

Figure 3. Flow chart of MMI and control process

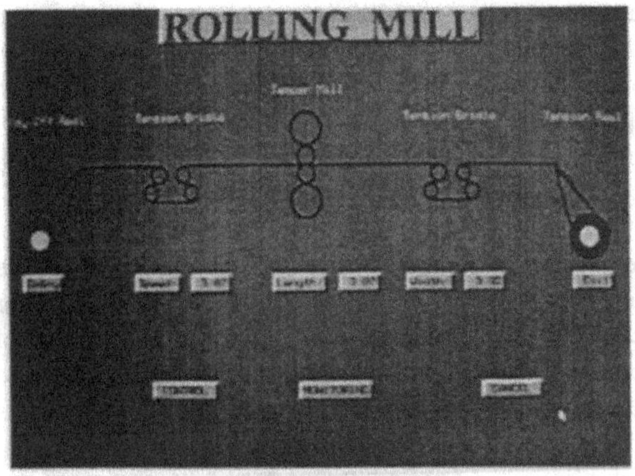

(a) Main MMI screen

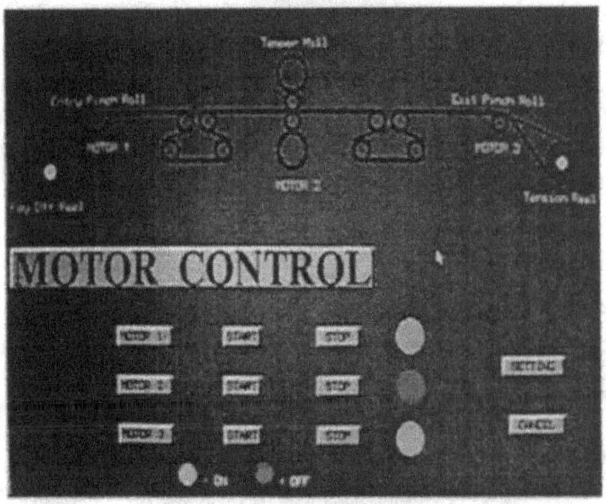

(b) Switch control command MMI screen

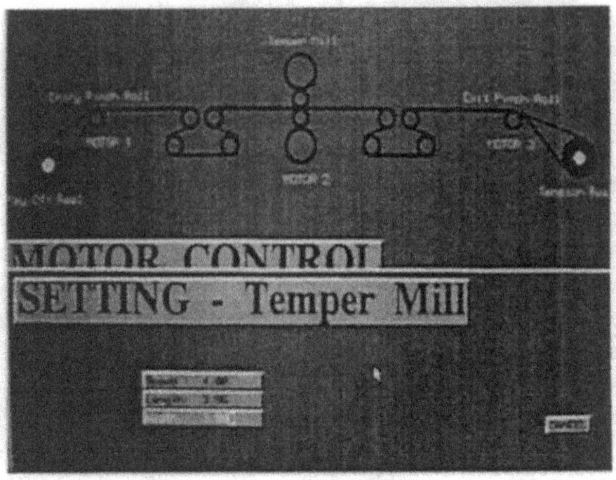

(c) Analog control value setting MMI screen

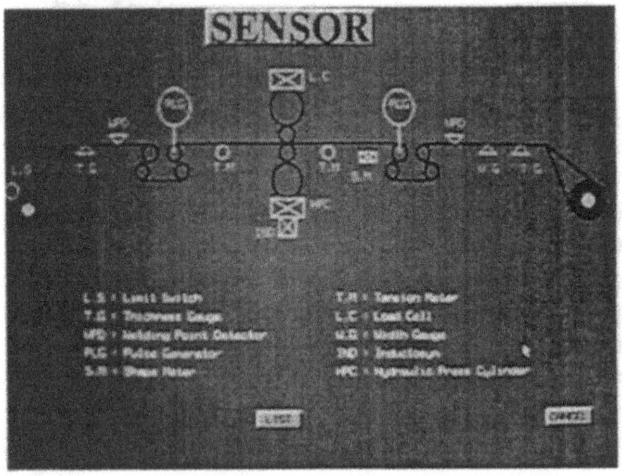

(d) Sensor position and explanation screen

Figure 4. MMI screen

6. CONCLUSIONS

Because the control system managed by the closed system bus has occurred many problems, we implemented the control system using the open architedtured system bus. And we constructed the open architectured system using OS-9 and OS-9 WINDOWS as the real-time software.

REFERENCES

[1] "Computer Systems For Automation And Control" by Gustaf Olsson & Gianguido Piani. Prentice Hall International LTD. 1992.

[2] "Data Communications Computer Networks and Open Systems" Fred Halsall, 3rd Ed., Addison Wesley

[3] "OS-9 Insights" by Peter C. Dibble, Second Edition, Microware system co., 1992

[4] "OS-9 WINDOWS/Level-2 Programming Guide" 1993. Computer System GmbH

THE COMPARISON OF PID WITH INTELLIGENT CONTROL ALGORITHM FOR MOLD LEVEL CONTROL IN A CONTINUOUS CASTING PROCESS

Joo-Man Kim*, Jin S. Lee, Duk-Man Lee****

**Maintenance Technology Department, Pohang Works, Pohang Iron & Steel CO., Ltd. Pohang P.O. Box 35, Dongchon-dong, Nam-gu, Pohang, Kyungbuk, 790-360, Korea*
***Department of Electrical and Electronic Engineering, Pohang University of science and Technology, San31, Hyoja-dong, Nam-gu, Pohang, Kyungbuk, 790-784, Korea*

Abstract : This paper describes the design and implementation of an intelligent control technology for mold level control in continuous casting process. The continuous casting process is difficult to control due to various changes of the process parameter and unexpected disturbances. To solve this control problem the proposed controller adopted a fuzzy control algorithm with feedback linearization technique. The simulation results show that the proposed intelligent control algorithm is superior performance to the conventional PID control. *Copyright © 1998 IFAC*

Keywords : PID control, Fuzzy control, Feedback linearization, Intelligent controller

1. INTRODUCTION

In the steel making process, continuous casting continuously converts the molten steel from the steel making furnace directly into slabs, blooms or billets without passing through the ingot making and slabbing mill. Especially, the slab is formed by transferring the molten steel from the tundish through a nozzle into the mold or water-cooled jacket.

The quality of continuous casting slab is normally affected by mold level(molten steel level in the mold) fluctuation. If the mold level fluctuates severely, the powder which is added to the steel to improve the flow properties is mixed with molten steel and becomes the defect of solidified steel.
Therefore, it is very important to keep the mold level constants at desired value. However, the precise control of mold level is difficult in itself because it is with high nonlinearity in its dynamics, under various changes of the process parameter and disturbances. The main sources of disturbances are the variation of flow rate constant, casting speed, sensor limitation and noise. The variation of flow rate constant mainly comes from hardware wearing whereas the variation of casting speed results from productivity and quality of slab.

Under these conditions, it is extrimely difficult to control accurately the mold level with the conventional PID algorithm only.
In order to achieve high performance regulation even under various disturbances and nonlinearities, this paper presents the intelligent control algorithm which adopted a fuzzy control and feedback linearization.
To compare the effectiveness of proposed algorithm, several simulation test are performed. The simulation results show that the proposed algorithm is more

robust and more effective than that of PID controller.

This paper is organized as follows. In section 2, the mathematical model of the plant is derived. In section 3, the design of proposed intelligent controller is presented. Section 4 shows the simulation results and section 5 concludes the paper.

2. PLANT MODELING

The Fig.1 shows the schematic flow of continuous casting process and its control system.

There are two sliding gate valves with a position controller. One controls the level of molten steel in the tundish and the other controls the level of molten steel in the mold. The weight of molten steel in the tundish is controlled by the first ladle sliding gate and the molten steel is passed from the tundish into the mold under control of the second sliding gate. The mold acts as a cooling jacket with circulating water removing the heat. A surface solidified slab passes from the bottom of the mold through a cooling water into the slab handling facilities.

The rate of withdrawal from the mold is controlled by a pinch roll which determines the casting speed. To regulate the mold level, the sliding gate position is controlled to keep a balance between the flow rate of molten steel from the tundish to the mold and the flow rate of molten steel out of the mold.

In the development of the mathematical model, it was assumed that the molten steel is incompressible. The dynamics of mold level is discribed as

$$V_i = dL/dt = (Q_i - Q_o) / A \qquad (1)$$

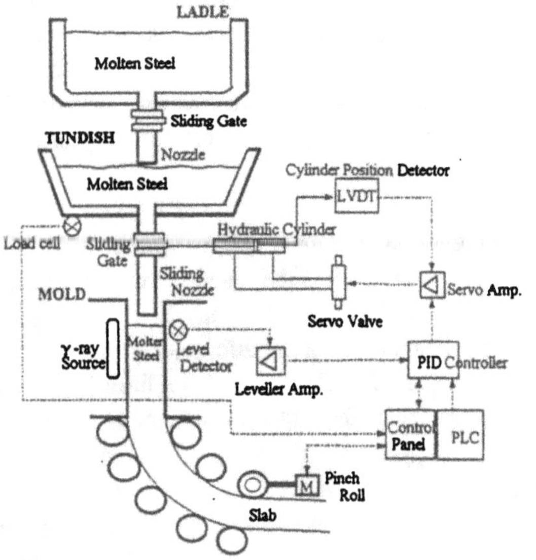

Fig. 1. Schematic flow of continuous casting

where, V_i is the changing rate of the mold level, L is mold level, Q_i is inlet molten steel flow rate to the mold, Q_o is outlet molten steel flow rate from the mold, and A is the cross section area of the mold.

Since, the flow rate at some specific point is equal to the flow velocity multipled by the cross section area, the outlet flow rate of the mold is

$$Q_o = V_o A \qquad (2)$$

where, V_o is the outlet flow velocity from the mold.

Actually the inlet flow rate of the molten steel is controlled by the opening of the sliding gate. In a mechanical stucture, the sliding gate consists of the two holed plates ; The one is the fixed part and the other is movable part which is connected to hydraulic cylinder.

The operation of the sliding gate is shown in Fig. 2. The operating region of the sliding gate ranging from 0 to 40mm is the dead zone in which region the molten steel does not come out of the turndish. This means that the molten steel can flow when the moving distance of sliding gate is over 40mm, and the flow rate can be maximum when it is 120mm.

The inlet molten steel flow rate to the mold througth the sliding gate is

$$Q_i = V_t S_u \qquad (3)$$

where, V_t is the inlet flow velocity of molten steel from the turndish into the mold, S_u is the opening area of sliding gate.

It was assumed that this inlet flow velocity is equal to that of free falling if the friction force between the molten steel and the inner wall of the sliding nozzle is ignored and if there is no weight change of the molten steel in the tundish. Therefore, the following equation is derived from Bernoulli equation.

$$V_t = \sqrt{2gh} \qquad (4)$$

where g is the acceleration of gravity, h is the falling height of the molten steel from the turndish to the mold.

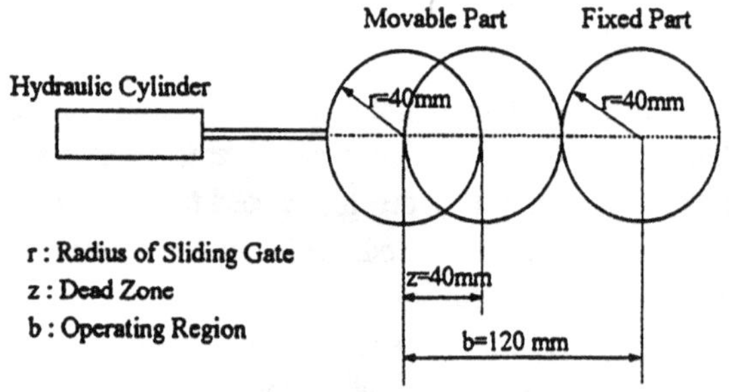

Fig. 2. Operation of sliding gate

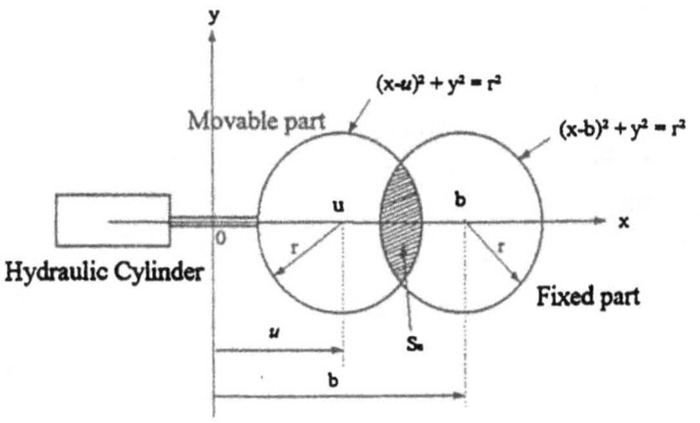

Fig. 3. Opening area of the sliding gate

The opening area of sliding gate can be calculated from the shade area in Fig. 3.
Thus S_u can be written as

$$S_u = 0, \quad u \le z$$

$$S_u = 4\int_{\frac{u+b}{2}}^{u+r} \sqrt{r^2-(x-u)^2}\,dx, \qquad u > z$$

$$= \pi r^2 - 2r^2 \arcsin\left(\frac{b-u}{2r}\right) - \left(\frac{b-u}{2}\right)\sqrt{4r^2-(b-u)^2} \qquad (5)$$

where u is control input(movable part position of sliding gate) which decides the opening area of the sliding gate.

Finally, by combining the above equation from (1) to (5), the mathematical model of plant is given as

$$\dot{L} = \frac{\sqrt{2gh}}{A}\left[\pi r^2 - 2r^2 \arcsin\left(\frac{b-u}{2r}\right) - \left(\frac{b-u}{2}\right)\sqrt{4r^2-(b-u)^2}\right] - V_o \qquad (6)$$

where all parameters are known from the actual plant geometry.

3. CONTROLLER DESIGN

From the equation (1), if the outlet flow rate Q_o is exactly known, the ideal inlet molten steel flow rate Q_i can be given as

$$Q_i = AK_p e + AK_I \int e\,dt + Q_o \qquad (7)$$

where K_p and K_i are respectively the proportional and integral gain of controller, error e is difference of measuring level L from the desired mold level L_d. Substituting equation (7) into equation (1) yields

$$\dot{L} = AK_p e + AK_I \int e\,dt \qquad (8)$$

Since $\dot{e} = -\dot{L}$, from equation (8)

$$\ddot{e} + K_p \dot{e} + K_I e = 0 \qquad (9)$$

which implies that if K_p and K_I are positive $e \rightarrow 0$ as $t \rightarrow \infty$. This control technique is called feedback linearization.
But, it is difficult to know Q_o value accurately because of noise or uncertain information. Therefore, the estimated $\hat{Q}_o$ is used instead of Q_o. The estimated $\hat{Q}_o$ can be calculated by the desired value of mold width and measuring value of casting speed.
The realizable form of the controller is

$$\hat{Q}_i = AK_p e + AK_I \int e\,dt + \hat{Q}_o \qquad (10)$$

In addition to thease, an additional fuzzy controller for the compensation of nonlinearity, the various noise existing the system certainly and the estimated error of Q_o is added into equation.

Thus, the output of the proposed controller is shown as follows

$$Q_i = Q_c + AK_p e + AK_I \int e\,dt + \hat{Q}_o \qquad (11)$$

In the equation (11), Q_c is the output of fuzzy controller which uses the singleton fuzzifier, center of area defuzzifier and product inference rule. Table 1 shows the fuzzy rule base, where NB is negative big, NM is negative middle, NS is negative small, ZO is zero, PB is positive big, PM is positive middle, PS is positive small.
Fig. 4 shows the configuration of proposed control system.

4. SIMULATION

The applied modeling plant is No. 2 continuous casting at Pohang Works.
The control system for simulation is configured respectively with proposed intelligent controller and conventional PID controller and each performance of the control system in case of changing the operation

Table 1 Rule base of fuzzy controller

e \ Δe	NB	NM	NS	ZO	PS	PM	PB
NB				NB	NM		
NM				NM			
NS				NS	ZO		PM
ZO	NB	NM	NS	ZO	PS	PM	PB
PS	NM		ZO	PS			
PM				PM			
PB			PM	PB			

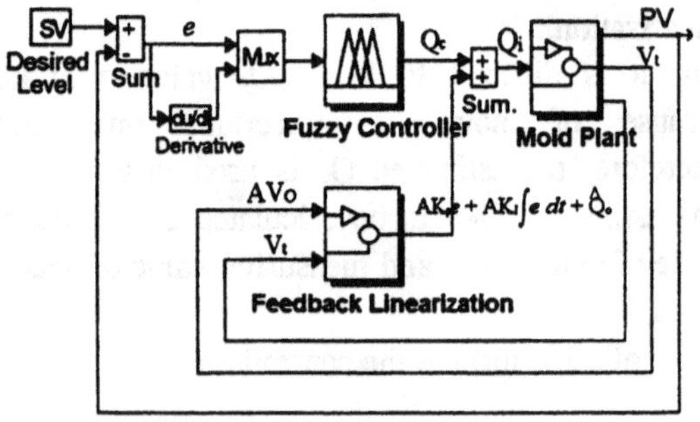

Fig. 4. Block diagram of intelligent control system

condition or noise is compared.

The Fig. 5 shows the step response when the desired level(set point value) changes 150mm to 170mm. The response time of the intelligent controller is faster than that of PID controller and no overshoot is occurred.

Fig. 6 shows the simulation result when step input is applied and casting speed is varied with a pattern of Fig. 7. In this figure, the mold level of intelligent controller is very smooth but PID controller changes a lot. Unlike the actual speed pattern, it is assumed that the changing pattern of the casting speed is step shape.

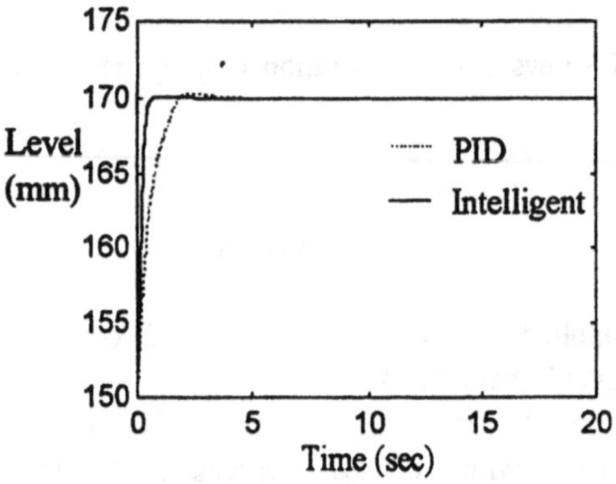

Fig. 5. Output response in case of step input

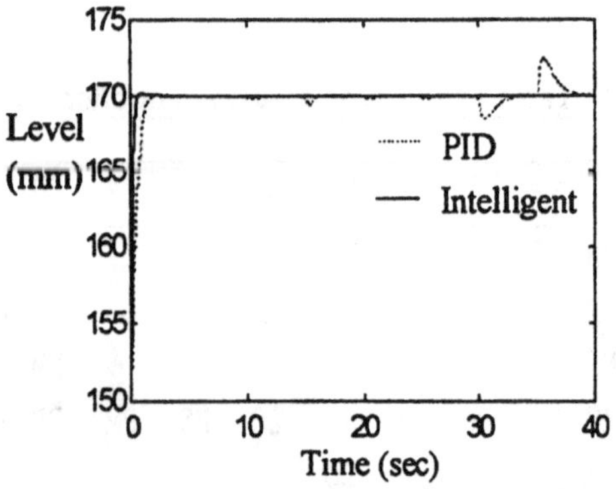

Fig. 6. Output response in case of step input and variation of casting speed

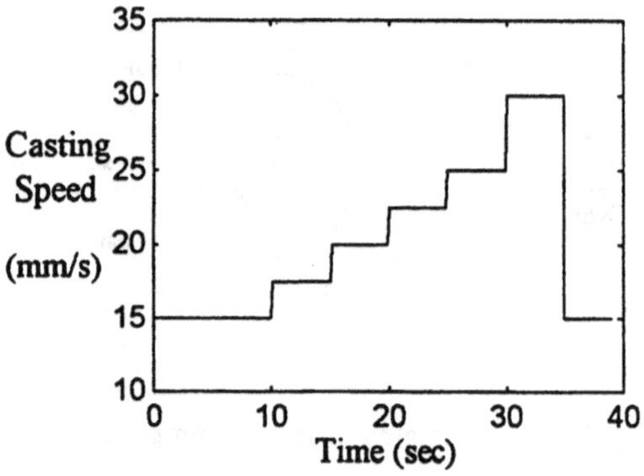

Fig. 7. Changing pattern of the casting speed

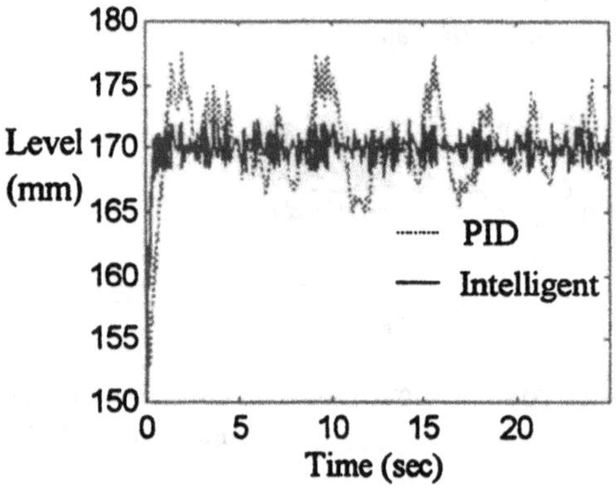

Fig. 8. Output response in case of step input, variation of casting speed and sensor noise applied

Fig. 8 shows the control result with noise in the level measuring system. The regulation error is within ± 3 mm in the case of the intelligent controller but ± 10 mm in the case of the PID controller. The applied noise's amplitude is maximum 4mm and frequency is 120 Hz.

When the time delay terms is applied in the model with variation of casting speed, the simulation result is following Fig. 9. The output of intelligent

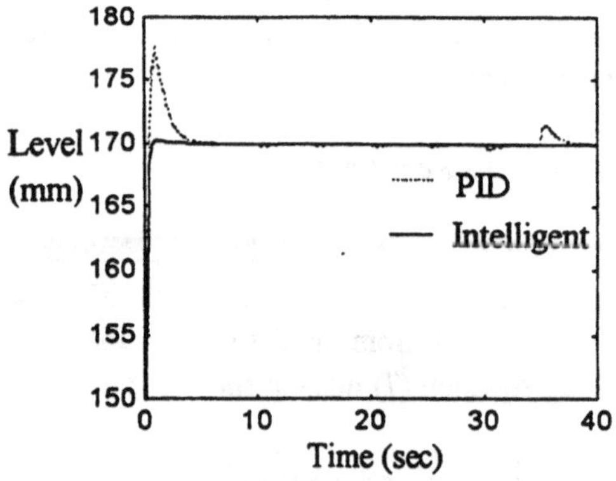

Fig. 9. Output responce in case of step input, variation of casting speed and time delay term added

controller shows better performance than that of PID controller.

5. CONCLUSION

According to the simulation results, it is known that the intelligent controller with fuzzy control and feedback linearizing control strategy is better than that of conventional PID controller. This means that the rising time is fast, the flexibility for the variation of operating condition is excellent, and the robustness against noise or time delay terms is also good.

Therefore, besides the mold level control of the continuous casting, the proposed technique is can be applied with similar process which has nonlinearity and unexpected noise or disturbance.

REFERENCES

Carli A. De, P. Liguori and A. Marroni (1994), "A Fuzzye-PI Control Strategy," *Control Engineering Practice,* **Vol. 2,** No. 1, pp.147-153.

Hesketh T., D. J. Clements and R. Williams (1993), "Adaptive Mold Level Control for continuous Steel slab Casting," *Automatica,* **Vol. 29,** No.4, pp.851-864.

Hur Y. G. (1994), " Realization of Fuzzy Control Algorithm for Advanced DCS and Application to Continuous Casting process of Iron & Steel Plants," *M. E. Thesis of KAIST,* Korea.

Kuribayashi T., K. Fukuda, A. Koyama, T. Takawa and H. Kato (1986), "Process Control Technology of Continuous Caster," *SMI,* **Vol. 38,** No.4, pp.31-38.

Kurokawa T., T. Kondo, T. Mita, K. Liu and M. Sampei (1993), " Development of Mold Level Control in Continuous Casting by H^{∞} Control Theory," *Second IEEE Conference on Control applications,* p865-871, Vancouver.

Lee H. Y. (1989), "A Study on the Mold Level Control of Continuous Casting System," *Project Report of RIST,* Pohang, Korea.

Manayathara T. J., T.C. Tsao, J. Bentsman and D. Ross (1996), "Rejection of Unknown Periodic Load Disturbances in Continuous Steel Casting Processing Using learning Repetitive Control Approach," *IEEE Transactions on Control Systems Tech.,* **Vol. 4,** No. 3, pp. 259-265.

Slotine J. E. and W. Li (1991), *Applied Nonlinear Control,* Chapter 6, Prentice-Hall, Newjersey.

Tomono T., T. Tsuruta and H. kagari (1987), "Development of Technology for Mold Bath Level Control System of Continuous Caster," *Iron & Steel,* **No. 2,** pp.327-332.

Wang L. X. (1994), *Adaptive Fuzzy Systems and Control ; Design and Stability Analysis,"* Prentice-Hall, Newjersey.

Zhao Z. Y., M. Tomizuka and S. Isaka (1993), "Fuzzy Gain Scheduling of PID Controllers," *IEEE Transactions on Systems, Man, and Cybernetics,* **Vol.23,** No.5, pp.1392-1398.

ON THE CONTROL OF SILICON RATIO IN FERROSILICON PRODUCTION

Helgi Thor Ingason

Icelandic Alloys Ltd.

Gudmundur R. Jonsson

Dep. of Mech. and Ind. Engineering
University of Iceland

Abstract: A Generalized Predictive Control, *GPC*, has been implemented at Icelandic Alloys Ltd. two ferrosilicon furnaces, for controlling the silicon ratio in tapped metal. The control is based on an *ARX* model of the furnace. The model is used to predict the silicon ratio one and two time steps ahead. The use of *GPC* has resulted in decreased variations in the chemical analysis of tapped metal and better results are also achieved in terms of keeping the silicon ratio near the desired value. *Copyright © 1998 IFAC*

Keywords: Predictive Control, Parameter estimation, Modelling, Implementation.

1. INTRODUCTION

Ferrosilicon (*FeSi75*) is a two-phase mixture of the chemical compound $FeSi_2$ and the element silicon. *FeSi75* contains about 76 *wt%* silicon, 22 *wt%* iron and 2 *wt%* aluminum, calcium and other trace elements. Icelandic Alloys Ltd. produce 70000 tons annually *FeSi75* in two 48 MVA submerged arc furnaces. The balance between silicon and iron in the tapped metal is regulated by adjusting the input of raw-materials to the furnace. Raw materials are charged batchwise to the top of the furnace. Each batch consists of a fixed amount of quartz (SiO_2) and a variable quantity of coal/coke (*C*) and iron oxide (Fe_2O_3). The quantity of coal/coke in each batch is adjusted to regulate the carbon balance of the raw material charge in the furnace and this control is beyond the scope of this paper.

The silicon ratio is regulated by increasing or decreasing the quantity of iron oxide in each batch. Such a change can be made once per day. This paper describes the Generalized Predictive Control (*GPC*) which has been implemented on both furnaces at Icelandic Alloys Ltd. at Grundartangi, Iceland. A short description of the underlying theoretical principles is given with a discussion about the practical implementation of the control. Finally, an analysis of actual furnace data is presented, giving indication about the performance of the control.

2. THEORY

2.1 The chemical balance of the charge

A simple model of the relation between silicon ratio (*Si%*) in tapped metal and the quantity of iron oxide per batch (*kgFe*) is as follows, (Halfdanarson, 1995):

$$Si\% = 0.98/(1+1.5 \cdot kgFe/(700 \cdot n)) \qquad (1)$$

The variable n represents the relative utilization of the quartz that is input to the furnace. The constant 0.98 expresses that *Si* and *Fe* are 98 *wt%* of the tapped metal, and the constant 700 represents that there are 700 kg quartz in each batch of raw materials. The constant 1.5 is based on experience at Icelandic Alloys Ltd. where 89% utilization of quartz is typically achieved with 120 kg of iron oxide per batch, when the *Si%* is 76%. The chemical analysis of the tapped metal varies from day to day. These changes may be attributed partly to variations in the internal environment of the furnace, e.g. temperature, and partly to fluctuations in the chemical composition of the raw materials. It is thus necessary to vary the quantity of iron oxide input in order to control the silicon ratio and keep it as close as possible to the desired value, 76% in the case of Icelandic Alloys Ltd.

2.2 ARX model of the furnace

The Generalized Predictive Control uses a mathematical model of the furnace for predicting the silicon ratio. The model is of the type Auto Regressive, auXiliary input (*ARX*) and can be written as, see e.g. (Ljung, 1987):

$$A(q)y(t) = B(q)x(t-nk) + e(t) \qquad (2)$$

AR refers to the autoregressive part $A(q)y(t)$ and *X* refers to the extra input $B(q)x_{t-nk}$. q^{-1} is a delay operator, $y(t)q^{-1} = y(t-1)$ and nk is the time delay, i.e. the number of delays from input to output. $e(t)$ is the prediction error, i.e. the difference between the actual output value and the predicted output value. Finally, $A(q)$ and $B(q)$ are polynomials in the delay operator:

$$A(q) = 1 + a_1 q^{-1} + \ldots + a_{na} q^{-na}$$

$$B(q) = b_0 + b_1 q^{-1} + \ldots + b_{nb} q^{-nb+1} \qquad (3)$$

with na and nb as the orders of the respective polynomials.

The *ARX* model, upon which the *GPC* is based, is of the form (see section 3.1):

$$Si_t + Si_{t-1} a_1 = b_1 Fe_{t-1} + d + e_t \qquad (4)$$

i.e. the *Si%* value today (Si_t) is based on *Si%* and *kgFe* yesterday and a constant d.

2.3 Generalized Predictive Control (GPC)

The predicted value of *Si%* is compared to the set-point or desired value of *Si%* and the control calculates what *kgFe* must be in order for these to be equal. In practice, too high variations from day to day in the iron oxide input to the furnace are unacceptable. The control therefore includes limitations as to how much it allows *kgFe* to change from one day to another. Following is a short description of *GPC* algorithm, see e.g. (Clarke et al., 1987) and (Palsson et al., 1994) for more details.

The cost function to be minimized was chosen to be

$$J = E\left[\sum_{j=N_1}^{N_2} \left(y_{t+j} - w_{t+j}\right)^2 + \sum_{j=1}^{N_u} \lambda_j \left(\Delta u_{t+j-1}\right)^2\right] \qquad (5)$$

with N_1 as the minimum costing horizon, N_2 as the maximum costing horizon, N_u is the control horizon and w_{t+j} as the future reference output. Eq. (5) shows that besides penalizing deviations in the output from the reference value, the penalty is on changes in the input, u_t, i.e. Δu_t. The parameter λ is typically found by trial and error.

The cost function may also be expressed in vector form as

$$J = E\left[(\mathbf{y}_t - \mathbf{w}_t)^T (\mathbf{y}_t - \mathbf{w}_t) + \Delta\mathbf{u}_t^T \Lambda_t \mathbf{D} u_t\right] \qquad (6)$$

with

$$\mathbf{y}_t = \left[y_{t+N_1}, \cdots, y_{1+N_2}\right]^T,$$

$$\mathbf{w}_t = \left[y_{t+N_1}, \cdots, w_{1+N_2}\right]^T,$$

$$\Delta\mathbf{u}_t = \left[\Delta u_{t+N_1}, \cdots, \Delta u_{1+N_2}\right]^T$$

and

$$\Lambda_t = diag\{\lambda_1, \ldots, \lambda_{Nu}\}$$

The output may be written as a sum of two independent parts, i.e. the predicted output and an error term or

$$\mathbf{y}_t = \hat{\mathbf{y}}_t + \tilde{\mathbf{y}}_t \qquad (7)$$

with $\hat{\mathbf{y}}_t$ containing the predictions

$$\hat{\mathbf{y}}_t = \left[\hat{y}_{t+1|t}, \cdots, \hat{y}_{t+N|t}\right]^T$$

and $\tilde{\mathbf{y}}_t$ containing the prediction errors, i.e.

$$\tilde{\mathbf{y}}_t = \left[\tilde{y}_{t+N_1|t}, \cdots, \tilde{y}_{t+N_2|t}\right]^T$$

Inserting this into Eq. (6) yields

$$J = E\left[(\hat{\mathbf{y}}_t - \mathbf{w}_t)^T(\hat{\mathbf{y}}_t - \mathbf{w}_t) + \Delta\mathbf{u}_t^T \Lambda_t \Delta\mathbf{u}_t\right] + E\left[\tilde{\mathbf{y}}_t^T \tilde{\mathbf{y}}_t\right] \quad (8)$$

The *GPC* assumes that the predictions can be expressed as a linear combination of present and future controls. Following the lines in (Palsson et al., 1994), the predicted output is written as

$$\hat{y}_{t+j|t} = \sum_{i=1}^{j} h_{i,t+j} u_{t+j-i} + v_{j,t} \quad (9)$$

where the former term contains unknown values until the control values are known, but the latter one contains known values at time t. Eq. (9) can be written on vector form as

$$\hat{\mathbf{y}}_t = \mathbf{H}_t \mathbf{u}_t + \mathbf{v}_t \quad (10)$$

where

$$\hat{\mathbf{y}}_t = \left[\hat{y}_{t+N_1|t}, \cdots, \hat{y}_{t+N_2|t}\right]^T,$$

$$\mathbf{u}_t = \left[u_t, \cdots, u_{t+N_U-1}\right]^T,$$

$$\mathbf{v}_t = \left[v_{N_1,t}, \ldots, v_{N_2,t}\right]^T$$

and

$$\mathbf{H}_t = \begin{bmatrix} h_{1,t+N_1} & 0 & \cdots & 0 \\ h_{2,t+N_1+1} & h_{1,t+N_1} & \cdots & 0 \\ \vdots & \vdots & \vdots & \vdots \\ h_{N_U,I} & h_{N_U-1,I} & \cdots & h_{1,I} \\ h_{N_U+1,I-1} & h_{N_U,I-1} & \cdots & h_{2,I-1}+h_{1,I-1} \\ & & & \vdots \\ \vdots & \vdots & & \\ h_{N_2-N_1+1,t+N_2} & h_{N_2-N_1,t+N_2} & \cdots & \sum_{i=1}^{N_2-N_1-N_U+2} h_{i,t+N_2} \end{bmatrix}$$

where $I = t + N_1 + N_U - 1$. Note that the matrix $\mathbf{H}_t$ is shown time dependent to indicate that this version of the algorithm can cope with deterministic time varying system parameters, see (Palsson et al., 1994). Inserting Eq. (10) into Eq. (9) yields

$$J = E\left[(\mathbf{H}_t \mathbf{u}_t + \mathbf{v}_t - \mathbf{w}_t)^T(\mathbf{H}_t \mathbf{u}_t + \mathbf{v}_t - \mathbf{w}_t) + \Delta\mathbf{u}_t^T \Lambda_t \Delta\mathbf{u}_t\right] + E\left[\tilde{\mathbf{y}}_t^T \tilde{\mathbf{y}}_t\right] \quad (11)$$

In order to prevent offset, the control signal may be filtered and may be written as, see (Palsson et al., 1994)

$$\Delta\mathbf{u}_t = \mathbf{F}\mathbf{u}_t + \mathbf{g}_t \quad (12)$$

If $\Delta u_t = u_t - u_{t-1}$, then $\mathbf{g}_t = \left[-u_{t-1}, 0, \cdots, 0\right]^T$ and

$$\mathbf{F} = \begin{bmatrix} 1 & 0 & \cdots & 0 & 0 \\ -1 & 1 & \cdots & 0 & 0 \\ 0 & -1 & \cdots & 0 & 0 \\ \vdots & \vdots & & \vdots & \vdots \\ 0 & 0 & \cdots & -1 & 1 \end{bmatrix}$$

Inserting Eq. (12) into Eq. (11) gives

$$J = E\left[(\mathbf{H}_t \mathbf{u}_t + \mathbf{v}_t - \mathbf{w}_t)^T(\mathbf{H}_t \mathbf{u}_t + \mathbf{v}_t - \mathbf{w}_t) + (\mathbf{F}\mathbf{u}_t + \mathbf{g}_t)^T \Lambda_t \mathbf{F}(\mathbf{u}_t + \mathbf{g}_t)\right] + E\left[\tilde{\mathbf{y}}_t^T \tilde{\mathbf{y}}_t\right] \quad (13)$$

The *GPC* is then obtained by differentiating the cost function with respect to $\mathbf{u}_t$, i.e.

$$\frac{\delta J}{\delta \mathbf{u}_t} = 0$$

yielding the solution:

$$\mathbf{u}_t = \left(\mathbf{H}_t^T \mathbf{H}_t + \mathbf{F}_t^T \Lambda_t \mathbf{F}_t\right)^{-1} \left[\mathbf{H}_t^T(\mathbf{w}_t - \mathbf{v}_t) - \mathbf{F}^T \Lambda_t \mathbf{g}_t\right] \quad (14)$$

Since only the first element in $\mathbf{u}_t$ is used, the control signal becomes

$$u_t = [1, 0, \ldots, 0]\left(\mathbf{H}_t^T \mathbf{H}_t + \mathbf{F}_t^T \Delta \mathbf{F}_t\right)^{-1} \left[\mathbf{H}_t^T(\mathbf{w}_t - \mathbf{v}_t) - \mathbf{F}^T \Lambda_t \mathbf{g}_t\right] \quad (15)$$

The are several advantages in using *GPC* in the present case instead of e.g. Minimum Variance Control.For instance, *GPC* is more robust than *MVC* with regard to changes in system time delays and to changes in the parameters of the system model.

The following guidelines were used in chosing the design parameters, i.e. N_1, N_2 and N_u:

N_1: Naturally, $N_1 \leq k$ = system time delay. If k is unknown, then $N_1 = 1$.

N_2: Usually, $N_1 \geq nb$ with nb the degree of the B polynomial in the *ARX* model.

N_u: Often it is sufficient to put $N_u = 1$ but for complicated systems, N_u is chosen at least as large as the time constants corresponding to illconditioned poles of the system.

It was found, see the following section, that an adequate *ARX* model is given by

$$Si_t + a_1 Si_{t-1} = b_1 Fe_{t-1} + d + e_t \tag{16}$$

where d is a constant. With $N_1 = 1$ and $N_2 = N_u = 2$, the control becomes, cf. Eq. (14):

$$\mathbf{u}_t = \begin{bmatrix} Fe_t & Fe_{t+1} \end{bmatrix}^T$$

$$\mathbf{H}_t = \begin{bmatrix} b_1 & 0 \\ -a_1 b_1 & b_1 \end{bmatrix}^T \tag{17}$$

$$\mathbf{v}_t = \begin{bmatrix} -a_1 Si_t + d \\ a_1^2 Si_t - a_1 d + d \end{bmatrix}$$

Furthermore,

$$\mathbf{w}_t = \begin{bmatrix} w & w \end{bmatrix}^T, \ \mathbf{F}_t = \begin{bmatrix} 1 & 0 \\ -1 & 1 \end{bmatrix}, \tag{18}$$

$$\mathbf{g}_t = \begin{bmatrix} -Fe_{t-1} & 0 \end{bmatrix}^T, \ \Lambda = \lambda I$$

3. METHODOLOGY

3.1 Implementing GPC

A thorough analysis, involving e.g. cross validation, was carried out in order to find a suitable system model. The result was the following first order *ARX* model (parameter standard deviation in parenthesis):

$$\begin{aligned} Si_t - 0.44 Si_{t-1} &= -0.0028 Fe_{t-1} + 46.1 + e_t \\ (0.07) \qquad & \qquad (0.001) \qquad\quad (5.6) \end{aligned} \tag{19}$$

The parameter estimation was based on daily data from furnace 1 in the year 1995. The numbers in parenthesis are the standard deviations of the corresponding coefficients.

It was found through simulations that a reasonable value for λ in Eq. (18) is 0.0007, see Fig. 1 which shows how the variance of output *Si%* as a function of the variance of input *kgFe* for various values of λ. If $\lambda = 0$, i.e. no penalty on the input, the variance of the output is about 0.147 but the variance of the input is much higher compared to when $\lambda = 0.0007$. A variance of 20 *kgFe* was found acceptable by Icelandic Alloys Ltd.

3.2 User interface

A sample is taken from each tap for chemical analysis. All samples from each of the three shifts are crushed and mixed together to form one representative sample for that shift. The results from chemical analysis are available at 10.00 in the morning, six days a week (excluding Sundays).

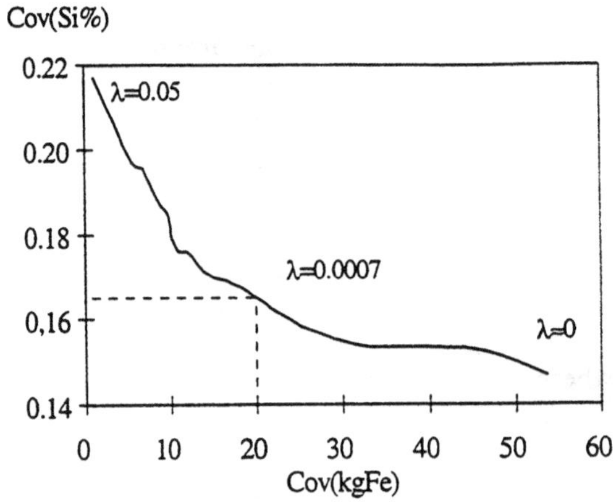

Fig. 1 Variance of *Si%* as a function of the variance of *kgFe* for different values of λ.

The results are then input to an Excel based user interface, specially developed for this purpose. The program uses the average of the three analysis for its calculation. After the program calculation, results are presented as indicated in Fig. 2.

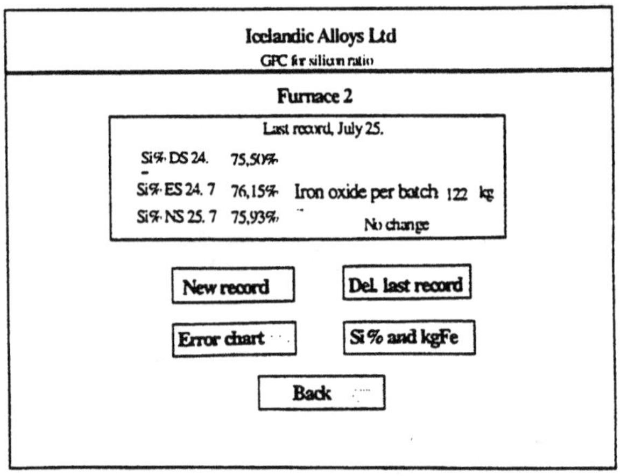

Fig. 2. Result window for the *GPC* program (Gudmundsson, 1996).

The operator can delete the last record, if it is discovered that an error has been made in the input. The operator can view the *Si%* (average values) and *kgFe* for the last 60 days. Finally, the operator can plot the prediction error, e_t, for the *ARX* model. The error chart gives the operator a change to monitor the process as a part of the Statistical Process Control system at Icelandic Alloys Ltd.

4. RESULTS

4.1 Time series

The silicon ratio for the period 1.1.95 - 31.7.96 is shown in Fig. 3. The dotted line shows the actual values but the solid line shows a running average, based on the last 7 values.

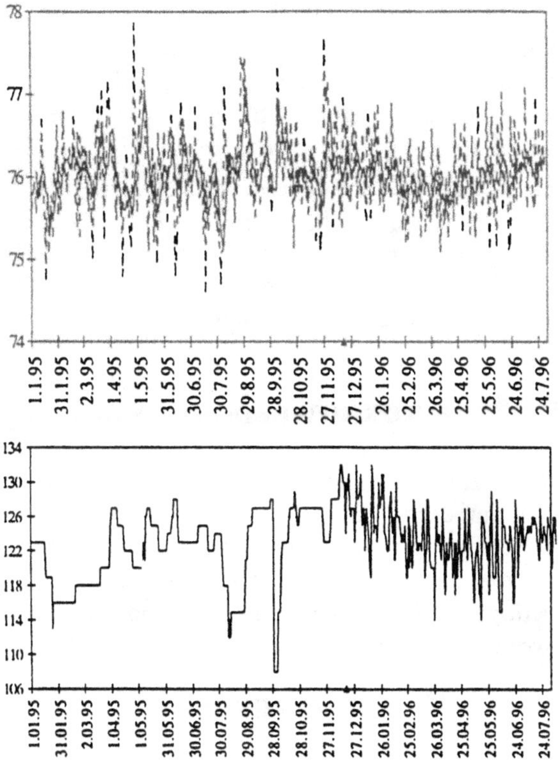

Fig. 3 a) The silicon ratio and b) the iron oxide per raw material batch during the period 1.1.95-31.7.96.

The *GPC* was implemented 18.12.95 which is indicated by a small solid triangle in Fig. 4.

4.2 Analysis

In order to evaluate the performance of the *GPC*, a simple analysis is done. The period 1.1.95-31.7.96 is divided into periods of 14 days and the average and standard deviation of the silicon ratio is calculated for each period. Fig. 5 shows plots of the results.

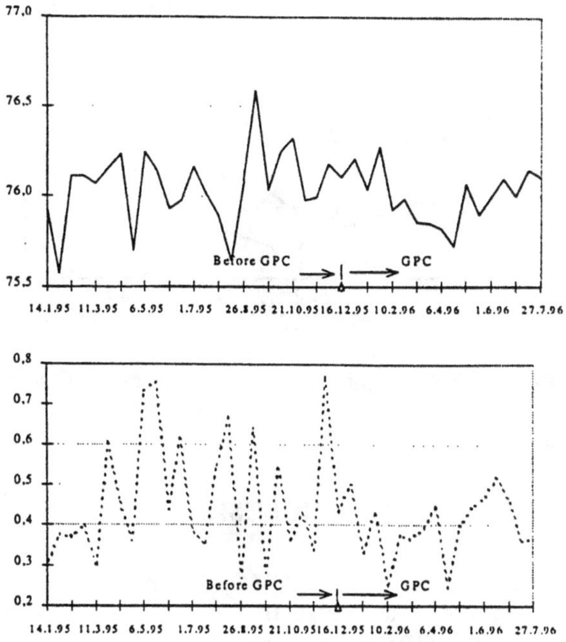

Fig. 4 a) Average values and b) standard deviation, for the silicon ratio, based on the period 1.1.95-31.7.96. The period is subdivided into 14 days periods and the parameters are calculated for each 14 day period.

The distribution of changes in the iron oxide input from day to day is analyzed by making a frequency plot of two equally long periods, before and after the implementation of *GPC*. This is shown in Fig. 5.

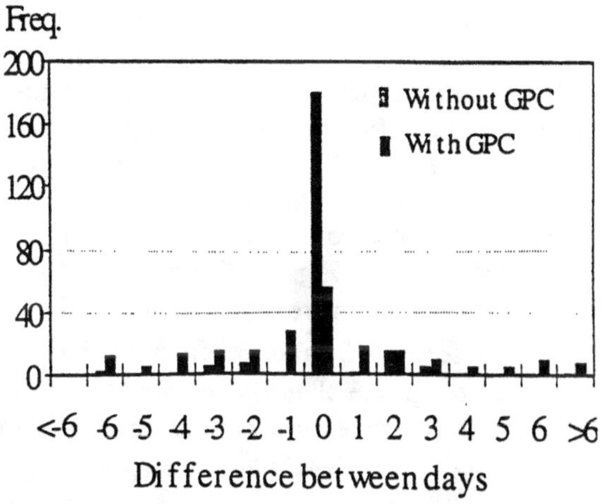

Fig. 5 A frequency plot of the distribution of difference in *kgFe* between two adjacent days. Total umber of data points is 221.

Furthermore, the use of *GPC* has significantly reduced the correlation in the *Si%* as can be seen in Fig. 6 where the sample autocorrelation function of the *Si%* from both periods is shown.

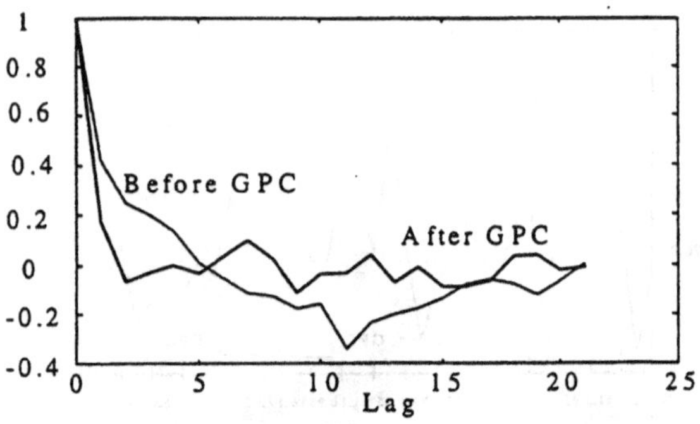

Fig. 6 The autocorrelation functions of Si% before and after implementing *GPC*.

5. CONCLUSIONS

The time series presented in Fig. 3 indicate that there is a change in the process behavior when the *GPC* is implemented. Fig. 3a clearly shows that *kgFe* is much more dynamic after than before. This is explained by the fact that the control before *GPC* was much more rigid, it allowed *Si%* to vary freely within defined limits, and reacted only if those limits were crossed. Furthermore, changes in *kgFe*, from day to day, that exceeded 3 kg were not allowed. Fig. 3a indicates that the *Si%* is kept near 76%, and oscillations in the signal are less after the implementation of *GPC*. These observations are confirmed by the statistical analysis presented in Fig. 4 where oscillations in the 14 days average values for *Si%* seem to decrease somewhat with the implementation of the *GPC*. The impact of *GPC* on *Si%* is best seen in the standard deviation presented in Fig. 4b. The standard deviation drops significantly and is near 0.4 after the *GPC* is implemented. This is the expected variance of the signal (0.16), as discussed in the Theory section. The increased dynamics of *kgFe*, observed in Fig. 3b is confirmed in Fig. 5. Changes in *kgFe*, between two days, were none in the majority of cases before *GPC*, and never more than ±3 kg. This diagram changes drastically after *GPC* and the changes are now smeared out to the limits of ±6 kg. Finally, the autocorrelation functions presented in Fig. 6, indicate that correlation in *Si%* has dropped significantly after the implementation of *GPC*.

The *GPC* assumes that chemical analysis is done every day. This is not the case at Icelandic Alloys Ltd. where no chemical analysis is made on Sundays. In addition, chemical analysis is done irregularly around holidays, e.g. Christmas and Easter, but no more than 2 days pass between analysis of shift samples. This means that once a week the orders of the *GPC*, regarding *kgFe* for that day, are not followed. This must have a negative effect on the control performance. Other problems have been encountered. Both Figs. 3a and 4a show that *Si%* decreases in February and March 1996 but increases again in April. This is explained by slow variations in the internal furnace environment. The model parameters were based on furnace data for 1995. These constants are not final and must be re-calculated regularly as the furnace performance may differ from time to time, e.g. because of different raw materials. By re-calculating the coefficient *d* it is possible to some extent account for those slow variations in the furnace. Given the average of the *kgFe* for a 60 days period of time, the coefficient is calculated by $d = (1+a_1)\cdot 76 - b_1 kgFe_{60d}$. The coefficient *d* has been updated irregularly, using this equation, when a trend has been discovered in the data. In the future, the coefficient along with other model parameters will be calculated and updated on-line.

6. ACKNOWLEDGEMENTS

The authors wish to thank Mr. Atli B. Gudmundsson for his work on designing the *GPC* and help regarding the statistical analysis of the performance of the control. Icelandic Alloys Ltd. is thanked for supporting this work.

REFERENCES

Clarke, D.W., C. Montadi, and P.S. Tuffs (1987). Generalized Predictive Control - Part II. The Basic Algorithm, *Automatica*, **Vol. 23**, No. 2. pp 137 - 148.

Gudmundsson, A.B (1995). Control charts and statistical quality control at Icelandic Alloy Ltd. (in Icelandic), BSc thesis, Department of Mechanical and Industrial Engineering, University of Iceland.

Gudmundsson, A.B (1996). An Excel Based User Interface for *GPC* - Revised Version, Excel program, Icelandic Alloys Ltd.

Halfdanarson, J.H. (1995). The relation between Si%, quarts utilization and kgFe, Internal report, Icelandic Alloys Ltd.

Ljung, L. (1987). *System Identification - Theory for the User*, Prentice-Hall, Englewood Cliffs, N.J.

Palsson, O.P, H. Madsen and H.T. Sogaard (1994): Generalized Predictive Control for Non-stationary Systems, *Automatica*, **Vol. 30**, No. 12. pp 1991 - 1997.

ROBUST OPTIMAL DESIGN FOR SHAPE AND GAUGE COMPOSITE CONTROL OF COLD STRIP MILL

Anke Xue and Youxian Sun

Institute of Industrial Process Control
Zhejiang University, Hangzhou, 310027, P. R. China
E-mail: akxue@iipc.zju.edu.cn

Abstract: A new idea and approach of robust optimal control for shape and gauge on cold strip mill are proposed in this paper. The paper discusses three major topics: first, we develop a composite regulation model with parameter uncertainties for shape and gauge on cold strip mill, then a robust optimal control system to guarantee the strip quality is proposed, last, we give simulation results of the designed system by the proposed method.

Keywords: Shape, gauge, cold strip mill, robust control, optimal control.

1. INTRODUCTION

Shape and gauge are two critical performance indexes to evaluate the product quality of strip steel. Up to now, the technique of gauge control is very successful in applications. However, shape control is still difficult due to the complex rolling process and the lack of on-line measuring devices as well as effective control methods (Anders, et. al., 1991). Furthermore, shape and gauge composite control is a significant aim for operators in strip rolling.

Experimental investigation and practical measurement show that the aid of backup roll and work roll bending is of benefit to regulate shape and gage of strip when the vertical and lateral rigidities of cold strip mill satisfy some design conditions. According to the basic conditions of shape and gauge composite regulation, we have developed a model with the follow-up of the backup roll bending for $\phi 300\ mm$ 4-high reversible cold strip mill (Xue, 1996). Because the rigidities are affected by various kinds of factors, it will be impossible to guarantee the given conditions in mill making and rolling process. On the other hand, with both large uncertainties and rapid response, it is difficult to get accurate models of rolling process to satisfy the control requirements. Therefore, the developed model will be that with parameter uncertainty.

Although many results of robust feedback control for uncertain systems have been presented, most of previous results only guarantee the designed systems robust stability (Bien et al., 1992; Ezzine, 1987). Developing efficient and practical robust optimal control algorithms for shape and gauge composite regulation of cold strip mill is of essential benefit. In this paper, we will examine a linear quadratic regulator (LQR) based control design which is both robust to parameter uncertainties and optimal to a given performance index. We shall refer to this class of controllers as robust LQR (RLQR). Sufficient conditions of guaranteeing RLQR with the gain margin (GM) of $(1/2, \infty)$ and phase margin (PM) of 60 degrees are obtained based on the derived robust modified return difference equality with uncertainties. Then, for the parameter uncertainties, some robust optimal bounds of keeping RLQR robust stability and guaranteed margins are proposed. Furthermore, the corresponding optimization techniques to improve the bounds are presented. The laboratory test results show that the system designed by the proposed approach guarantees not only the flat shape but also the gauge accuracy.

2. PARAMETER UNCERTAIN MODEL FOR SHAPE AND GAUGE

A $\varphi 300mm$ 4-high reversible cold strip mill system is illustrated in Fig. 1. We obtain two basic

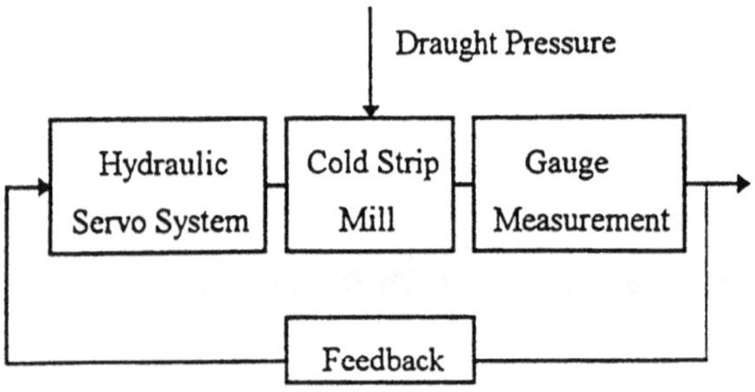

Fig. 1. Cold Strip Mill Control System

conditions for the mill system to realize shape and gauge composite regulation

$$\Delta h = \frac{\Delta P}{M_P} - \frac{\Delta(2S_2)}{M_{S2}} \quad (1)$$

$$\Delta(\Delta h_b) = \frac{\Delta P}{K_P} - \frac{\Delta(2S_2)}{K_{S2}} \quad (2)$$

where Δh, Δh_b, ΔP and $\Delta(2S_2)$ are gauge, shape, draught pressure and backup roll bending force excursion respectively, M_P and M_{S2} are vertical rigidities of cold strip mill and backup roll bending respectively, K_P and K_{S2} are lateral rigidities respectively.

Mechanism analysis and experiences of cold rolling process show that the ratio of vertical rigidities should be equal to that of lateral rigidities in order to guarantee the good quality of shape and gauge, i.e., the following condition should be held

$$\frac{K_{S2}}{K_P} = \frac{M_{S2}}{M_P} \quad (3)$$

When (3) is satisfied, shape and gauge composite control can be performed only by regulating backup roll bending. Both backup roll and work roll bending should be regulated simultaneously as (3) is not satisfied. In this case, the backup rolls are regulated to ensure the gauge accuracy, and the work rolls to improve the shape performance. Hence, the mechanism condition of (3) should be effectively guaranteed in cold strip mill design and making for the purpose of realizing shape and gauge composite control perfectly. However, it is very difficult in practice. On the other hand, the control system including two devices of regulating both backup roll and work roll bending is too complex to be used in plants. In order to solve these difficulties, this paper presents a new method-composite control method based on developed hybrid modeling algorithm.

As stated above, $\Delta h = 0$ and $\Delta(\Delta h_b) = 0$ are guaranteed conditions of getting good quality of shape and gauge for the cold strip mill system. From (2), we get a condition for $\Delta(\Delta h_b) = 0$

$$\Delta(2S_2) = \frac{K_{S2}}{K_P} . \Delta P \quad (4)$$

In view of the basic relation of the cold rolling mill

$$h = S_0 + \frac{P}{M_P} - \frac{2S_2}{M_{S2}} \quad (5)$$

where h is gauge, S_0 is initial clearance between backup rolls.

Considering the characteristic of hydraulic servo system and the structures of on-line measurements for shape and gauge, we can derive a mechanism model for shape and gauge composite control. Theoretically, the reversible cold strip rolling process is affected by various kinds of factors. Many of these factors are random, nonlinear, unmeasurable, and closely coupled. Therefore, a typical model of the process requires about on the order of 100 equations. As a result, they are cumbersome to manipulate in performing the input/output analysis required to design an appropriate control structure. Design of an appropriate control system, in contrast, requires an understanding of the dominant dynamic features and coupling between the manipulable inputs and outputs. Our work focuses on the development of a low order model which reveals the essential dynamics of the process. The low order model is useful in revealing the basic limitations posed by the process physics and input/output coupling. With both large uncertainties and rapid response, it is difficult to get accurate models of the process. Robust modeling techniques are of essential benefit. The modeling of the process involves analysis of the elastic deformation of backup rolls and work rolls, and the characteristic of raw strips, et. al.. The composite model with parameter uncertainties for regulating shape and gauge of the $\varphi 300mm$ 4-high reversible cold strip mill is expressed in the form:

$$\dot{x}(t) = (A_0 + \Delta A)x(t) + Bu(t) + Cf(t) \quad (6)$$

where

$$A_0 = \begin{pmatrix} -3.33 & -229.64 \\ 0 & -10.0 \end{pmatrix}, \quad B = \begin{pmatrix} 0 \\ 80 \end{pmatrix}, \quad C = \begin{pmatrix} 0.0476 \\ 18.9 \end{pmatrix}$$

$x = (x_1, x_2)^T = (\Delta h, \Delta P_l)^T$, $f = \Delta P$, P_l is hydraulic pressure, ΔA is an uncertain matrix which will be in different forms with the different assumptions in the modeling of the process.

The model accuracy is evaluated by comparing the in-situ measuring values. The relevant comparison and analyses for experiment results show that the model (6) is approximately consistent with the cold strip rolling process (Xue, 1996).

3. ROBUST OPTIMAL DESIGN OF UNCERTAIN LINEAR SYSTEM

For the uncertain multivariable linear system (6), we consider an optimal control problem with H_2

performance index

$$J=\int_0^{\infty}(x^T(t)Qx(t)+u^T(t)Ru(t))dt \quad (7)$$

where $Q \geq 0,\ R > 0$.

We want to derive a LQR controller which is robust to this model with parameter uncertainty. In order to tackle this problem, a robust modified return difference equality, in terms of the LQR design parameters, is derived. We begin by stating the LQR Riccati equation for the nominal system (A_0, B) and the index (7):

$$PA_0+A_0^TP+Q-PBR^{-1}B^TP=0 \quad (8)$$

Theorem 1 For RLQR problem, if P satisfies (8), then the following equality holds

$$(I+K\Phi(-s)B)^TR(I+K\Phi(s)B)$$
$$=R+B^T\Phi(-s)(-PA-A^TP+PBR^{-1}B^TP)\Phi(s)B \quad (9)$$

where

$$A=A_0+\Delta A,\ \Phi(s)=(sI-A)^{-1},\ K=R^{-1}B^TP.$$

The proof of Theorem 1 is a direct result of algebraic and matrix manipulation. (9) is called the robust modified return difference equality (RMRDE) for RLQR. From the RMRDE (9), it is clear that if

$$-PA-A^TP+PBR^{-1}B^TP\geq 0 \quad (10)$$

then

$$-PA-A^TP+PBR^{-1}B^TP=F^TF \quad (11)$$

for some matrix F. If we define $T_R(s)=R^{1/2}K\Phi(s)BR^{-1/2}$, then, it is clear that

$$\sigma_i(I+T_R(s))=\sqrt{R+\sigma_i^2(F\Phi(s)B)} \quad (12)$$

when (10) is satisfied, then

$$\sigma_i(I+T_R(s))\geq 1 \quad (13)$$

Therefore, the performance robustness is guaranteed. Inequality (13) means that the singular values of the sensitivity function are less than unity, i.e., $\sigma_i(I+T_R(s))^{-1}\leq 1$. This is a guaranteed property of LQR designs in the absence of uncertain parameters. We will now state a theorem which provides sufficient conditions of RLQR designs with the guaranteed GM of $(1/2,\infty)$ and the PM of 60 degrees as the standard LQ optimal regulators.

Definition 1 For uncertain system (6) with the performance index (7), if exist $P>0$ satisfying the Riccati equation (8) and a linear state feedback

$$u(t)=-Kx(t)=-R^{-1}B^TPx(t) \quad (14)$$

so that the closed-loop system

$$\dot{x}(t)=(A_0+\Delta A-BK)x(t) \quad (15)$$

is robust stability with the guaranteed stability margins GM of $(1/2,\infty)$ and PM of 60 degrees, then the uncertain system RLQR design is said to be robust optimal.

Theorem 2 The RLQR design for uncertain systems is robust optimal, if the uncertainty ΔA satisfies

$$\Delta A^T\Delta A\leq Q-P^2 \quad (16)$$

Proof: From the inequality (10), it is easy to get

$$P\Delta A+\Delta A^TP\leq Q \quad (17)$$

Notice that the following identity holds for any $m\times n$ real matrices X and Y

$$X^TY+Y^TX\leq X^TX+Y^TY \quad (18)$$

Hence

$$P\Delta A+\Delta A^TP\leq \Delta A^T\Delta A+P^2 \quad (19)$$

If

$$\Delta A^T\Delta A+P^2\leq Q \quad (20)$$

then, condition (16) is satisfied. It is obvious that (13) will be guaranteed for all i when (16) holds. In view of the well-known frequency condition of optimum systems, we complete the proof of the theorem.

Corollary 1 A robust optimal bound for the system (6) with uncertainty ΔA is

$$\|\Delta A\|_2\leq\sqrt{\underline{\sigma}(Q-P^2)} \quad (21)$$

The inequality (16) is a sufficient condition to guarantee RLQR with the GM of $(1/2,\infty)$ and PM of 60 degrees, i, e., RLQR robust optimal. The robust optimal bound derived by using the sufficient condition (16) satisfies any forms of unstructured and structured uncertainties. Next, we will give some robust optimal bounds for three types of parameter uncertainties using the proposed results.

Theorem 3 If the parameter uncertainty ΔA is of the form

$$\Delta A=(1/\rho)^{1/2}BD(r) \quad (22)$$

then the robust optimal bound for the system (6) with RLQR design is

$$D^T(r)D(r)\leq Q-1/\rho PBB^TP \quad (23)$$

where ρ is a positive constant, B is a known constant real matrix, $D(\cdot)$ is a continuous matrix function, the uncertainty $r\in\varphi\subset R^p$, where φ is a compact set.

Theorem 4 If the parameter uncertainty ΔA is of the form

$$\Delta A = DF(t)E \tag{24}$$

then the robust optimal bound for the system (6) with RLQR design is

$$\varepsilon^2 I \le (E^T E)^{-1}(Q - PDD^T P) \tag{25}$$

where D and E are known constant real matrices, $E^T E > 0$, $F(t)$ is an unknown matrix function satisfying $F^T(t)F(t) \le \varepsilon^2 I, \varepsilon > 0$ is a given constant.

From Theorem 2, the solutions of Theorem 3 and 4 are clear.

Theorem 5 Considering the parameter uncertainty

$$\Delta A = \sum_{i=1}^{l} k_i A_i \tag{26}$$

where k_i are real uncertain parameters and A_i are given real constant matrices, then the uncertain closed-loop system (15) with RLQR design is robust optimal if

$$\sum_{i=1}^{l} |k_i|^2 \le \frac{\bar{\sigma}(Q - P^2)}{\sum_{i=1}^{l} \bar{\sigma}(A_i)} \tag{27}$$

Proof: In view of Theorem 2, an equivalent express of (16) is

$$\bar{\lambda}(\Delta A^T \Delta A) \le \bar{\lambda}(Q - P^2) \tag{28}$$

From (28), we have

$$\bar{\sigma}^2(\Delta A) \le \bar{\sigma}(Q - P^2) \tag{29}$$

Substitute in (29) the actual value of ΔA matrix (26) to get

$$\bar{\sigma}^2(\sum_{i=1}^{l} k_i A_i) \le \bar{\sigma}(Q - P^2) \tag{30}$$

thus

$$(\sum_{i=1}^{l} k_i \bar{\sigma}(A_i))^2 \le \bar{\sigma}(Q - P^2) \tag{31}$$

In view of Cauchy-Schwarz inequality, we have

$$(\sum_{i=1}^{l} k_i \bar{\sigma}(A_i))^2 \le \sum_{i=1}^{l} k_i^2 . \sum_{i=1}^{l} \bar{\sigma}^2(A_i) \tag{32}$$

Hence, (31) holds if

$$\sum_{i=1}^{l} k_i^2 . \sum_{i=1}^{l} \bar{\sigma}^2(A_i) \le \bar{\sigma}(Q - P^2) \tag{33}$$

the proof is completed.

Moreover, from Theorem 5, a robustification procedure for RLQR design leads to the following optimization problem

$$\sum_{i=1}^{l} k_i^2 = \max_{Q,P} \frac{\bar{\sigma}(Q - P^2)}{\sum_{i=1}^{l} \bar{\sigma}(A_i)} \tag{34}$$

Thus, based on (34), we may improve the robust optimal bounds by optimizing (34) with respect to the choice of the weighting matrix Q and solving Riccati equation (8).

It is clear that those proposed robust optimal bounds are of benefit to deal with the uncertainty with complex forms in cold rolling process shown in Section 2.

4. CONTROLLER DESIGN AND SIMULATION

For the purpose of guaranteeing the shape and gauge accuracy in cold rolling process, a robust optimal controller based on the developed composite regulation model with parameter uncertainties is designed. First, according to the rolling principles of $\varphi 300mm$ 4-high cold strip mill and the guaranteed conditions of shape and gauge composite regulation, a subsystem with the follow-up of backup roll bending and rolling force feedforward is designed. Then, based on the designed subsystem and the proposed model in Section 2, a robust optimal controller of composite regulation for shape and gauge of strip is developed. For details, the reader may refer to References (Xue, 1996).

By practical experience and experiments, an appropriate performance index has been selected. The design parameter used were

$$R = I, \quad Q = \begin{pmatrix} 3.02 & 0.01 \\ 0.01 & 0.43 \end{pmatrix} \tag{35}$$

To illustrate the use of the proposed results on robust optimal bounds, we suppose the uncertainty of the system (6) in the form

$$\Delta A = \begin{pmatrix} \alpha_1 & 0 \\ 0 & \alpha_2 \end{pmatrix} \tag{36}$$

where α_1 and α_2 are uncertain parameters. The resulting robust optimal bounds of α_1 and α_2 are

$$|\alpha_1| \le 1.74, \qquad |\alpha_2| \le 0.66 \tag{37}$$

For any uncertainties satisfying (37), the resulting RLQR feedback law is

$$u(t) = \begin{pmatrix} 1.69 & -3.05 \end{pmatrix} x(t) \tag{38}$$

The laboratory test results of the designed robust optimal system are shown in Fig. 2 and Fig. 3. Fig. 2 shows the output transient of state x_1 when the

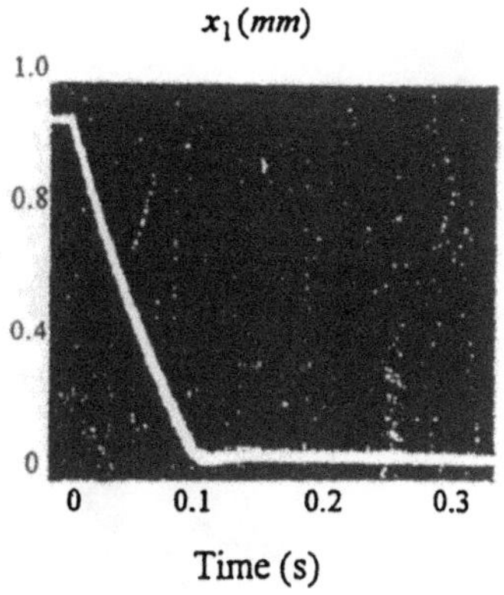

Fig. 2. Transient response of x_1

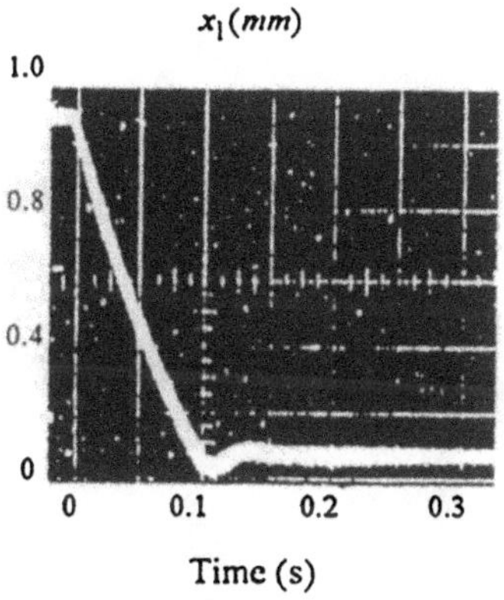

Fig. 3. Transient response of x_1

uncertain parameter α_1 and α_2 satisfy condition (37). When (37) is not satisfied for α_1 and α_2, the corresponding result is shown in Fig. 3. In this case, the optimality of the system is by no means guaranteed. It indicates the performance unrobustness.

5. CONCLUSIONS

A new way of composite regulation for shape and gauge by backup roll and work roll bending is presented. The modeling method for the cold strip rolling process with parameter uncertainties is very available and simple for practice. For shape and gauge composite regulation, the combination of robust and optimal control seems to be a better choice than a separate use of those controls. It is also clear that the proposed robust optimal bounds for RLQR are more significant and practicable in control system design. The optimization algorithm on the bounds is helpful to reduce the conservatism of the results.

Future work should focus on considering the problems of uncertain dynamic delay for such a specific system.

Acknowledgment: The Project Supported by Zhejiang Provincial Natural Science Foundation of China.

REFERENCES

Anders, G. Garlstedt and Olov Keijser (1991). Modern approach to flatness measurement and control in cold rolling. Iron and steel Engineer (68), 34-37.

Bien, Z. and Jin-Hoon Kim (1992). A robust stability bound of linear systems with structured uncertainty. IEEE Transactions on Automatic Control (AC 37), 1549-1551.

Ezzine, J. (1995). Robust stability bounds for sampled -data systems. Int. J. Systems Sci. (26), 1951-1966.

Xue, Anke (1996). Study on optimal control system for gauge and shape composite regulation of strip mill. Electric Drive (in Chinese) (26), 25-35.

ON THE DESIGN OF SELF-TUNING CONTROLLER USING GENERALIZED PREDICTIVE CONTROL

Kang Sup Yoon* and Man Hyung Lee**

* *Research Institute of Mechanical Technology, Pusan National University, Pusan, 609-735, Korea*
** *School of Mechanical Engineering, Engineering Research Center for Net-Shape and Die Manufacturing, and Research Institute of Mechanical Technology, Pusan National University, Pusan, 609-735, Korea*

Abstract: Generalized Predictive Control(GPC) has been reported as a useful self-tuning control technique for systems with unknown time-delay and parameters, and thus has won popularity among many practicing engineers. Despite its success, GPC does not guarantee its nominal stability. In this paper, GPC is rederived in the frequency domain instead of in the time domain to guarantee its nominal stability. Derivation of GPC in the frequency domain involves spectral factorization and Diophantine equations. Frequency domain GPC control system is stable because the characteristic polynomial are strictly Schur. Recursive least square algorithm is used to identify unknown parameters. To see the effectiveness of the proposed controller, the controller is simulated for a numerical problem which changes in dead-time, in order and in parameters. *Copyright © 1998 IFAC*

KeyWords: Generalized Predictive Control, Frequency Domain Generalized Predictive Control, Spectral Factorization, Diophantine equations

1. INTRODUCTION

GPC(Clarke, 1987a; Clarke, 1987b) is known as a useful algorithm for systems with unknown time delay and parameter. It is known that GPC is easy to understand as well as to implement, while it has the defect that it can not guarantee stability.
Despite studies on stability(Clarke, 1989; Robinson, 1991) most of them were limited to special cases, such as mean level and dead beat. Constrained Receding Horizon Predictive Control (CRHPC) suggested by Clarke et al.(1991), using Lagrange multiplier, can be hardly regarded as a systematic study about stability. But Stable Generalized Predictive Control(SGPC) suggested by Kouvaritakis et al.(1992) makes a system stable at first by using Bezout identity and then applies GPC to the system, so that it solves the stability problem of GPC, and makes it possible to study about the robustness of GPC systematically.

In this paper, to guarantee the stability of GPC, the standard GPC control law which is developed in the time domain is rederived in the frequency domain. Derivation of GPC in the frequency domain involves spectral factorization and Diophantine equations. After leading out j-step predictor in the frequency domain, frequency domain GPC which has the same 2 degree-of-freedom(DOF) as the time domain GPC is induced. Then, to prove that the frequency domain GPC can always ensure the nominal stability, the proposed controller is simulated for a unstable non-minimum phase plant. Also to see the effectiveness of the frequency domain GPC, the frequency domain GPC compared with the time domain GPC through numerical simulation.

2. PLANT MODEL

The plant is assumed to be described by the following discrete time Controlled Auto-Regressive Integrated Moving Average(CARIMA) model

$$A(z^{-1})y(k) = B(z^{-1})u(k) + C(z^{-1})\frac{\xi(k)}{\Delta} \tag{1}$$

where $y(k)$ is the plant output, $u(k)$ is the control input., z^{-1} is the backward shift operator, Δ is the backward difference operator($\Delta = 1-z^{-1}$). $\xi(k)$ is assumed to be the Gaussian white noise whose mean is 0 and variance is Q_d.

$$A(z^{-1}) = 1 + a_1 z^{-1} + \cdots + a_{na} z^{-na} ,$$
$$B(z^{-1}) = b_1 z^{-1} + \cdots + b_{nb} z^{-nb} ,$$
$$C(z^{-1}) = c_0 + c_1 z^{-1} + \cdots + c_{nc} z^{-nc} .$$

Eq.(1) can be rewritten in the following transfer function form

$$y(k) = W(z^{-1})\Delta u(k) + W_d(z^{-1})\xi(k) \tag{2}$$

where $W = \dfrac{B(z^{-1})}{\Delta A(z^{-1})}, \quad W_d = \dfrac{C(z^{-1})}{\Delta A(z^{-1})}.$ (3)

Polynomials $A(z^{-1})$ and $B(z^{-1})$ are not necessarily coprime so long as the plant does not contain any unstable hidden modes. The poles of system(i.e. poles of $W(z^{-1})$) can be either stable or unstable. Numerator polynomial $B(z^{-1})$ can be Schur polynomial that zeros exist within unit circle($|z|=1$) in z-plane or non-minimum phase polynomial. Numerator polynomials $C(z^{-1})$ is assumed to have no zeros on the unit circle in z-plane.

3. FREQUENCY DOMAIN j-STEP PREDICTOR

To lead out the GPC law in frequency domain, firstly frequency domain j-step predictor is to be required. To have j-step predictor defined in the frequency domain, Wiener filter formulation given by Grimble(1985) is referred.

In time k, given the set of output and control input as $\Omega=\{y(k_1),\ u(k_2);\ k_1\le k,\ k_2\le k+j-d\}$, let's put the j-step predictor in the following linear estimator form

$$\hat{y}(k+j|k)=H_{pj}(z^{-1})y(k)+z^jH_0(z^{-1})u(k)\,. \tag{4}$$

Then the j-step ahead predictor which minimizes the estimation error

$$\tilde{y}(k+j|k)=y(k+j)-\hat{y}(k+j|k) \tag{5}$$

under the cost function

$$J^P=E[\,\tilde{y}^2(k+j|k)\mid\Omega\,] \tag{6}$$

is given as

$$H_{pj}(z^{-1})=\frac{H_{nj}(z^{-1})}{H_d(z^{-1})}\,, \tag{7}$$

where $E[\cdot]$ is the expectation operator, denominator polynomial $H_d(z^{-1})$ is the Schur spectral factor which satisfies the following equation

$$H_dH_d^*=CQ_dC^*\,. \tag{8}$$

The numerator polynomials $H_{nj}(z^{-1})$ is computed from the solution H_{nj}(with F_j of smallest degree) of the following Diophantine equation

$$\tilde{A}F_{pj}+H_{nj}H_d^*z^{-g_p}=CQ_dC^*z^{-g_p+j} \tag{9}$$

where, $\tilde{A}(z^{-1})=\Delta A(z^{-1})$, g_p is the smallest positive integer which ensures the above Diophantine equation to be in the power of z^{-1}, $g_p=nc+j$. And, superscript '*' means the adjoint of a polynomial (e.g. $A^*(z^{-1})=A(z)$).
And, H_0 is

$$H_0=(1-z^{-j}H_{pj})\frac{B}{A}\,. \tag{10}$$

Given that j-step predictor $\hat{y}(k+j|k)$ is described by (7)-(10), it is proved whether the cost function (6) can be minimized by the following process.
First, from Eq. (1),

$$y(k+j)=z^j\frac{B}{A}u(k)+z^j\frac{C}{\Delta A}\xi(k)\,. \tag{11}$$

Substituing (4) and (11) into (5), the estimation error becomes

$$\tilde{y}(k+j|k)=\left\{\frac{B}{A}(1-z^{-j}H_{pj})-H_0\right\}z^ju(k)+\frac{C}{\Delta A}(1-z^{-j}H_{pj})z^j\xi(k)\,. \tag{12}$$

Since the estimation has to be unbiased (i.e. $E\{\tilde{y}(k+j|k)\}=0$), H_0 is the same as (10) and the estimation error $\tilde{y}(k+j|k)$ can be expressed as

$$\tilde{y}(k+j|k)=\frac{C}{\tilde{A}}(1-z^{-j}H_{pj})z^j\xi(k)\,. \tag{13}$$

Using a discrete form of Parseval's theorem, cost function (6) can be expressed as (Grimble 1988)

$$J^P=\frac{1}{2\pi i}\oint_{|z|=1}(1-z^{-j}H_{pj})z^j\frac{C}{\tilde{A}}\Phi_{\xi\xi}\frac{C^*}{\tilde{A}^*}z^{-j}(1-z^{-j}H_{pj})^*\frac{dz}{z} \tag{14}$$

where, $\Phi_{\xi\xi}$ is the power-density spectra of white noise $\xi(k)$. Variance of $\xi(k)$ is Q_d, so that the cost function (14) can be described as the following complete square form

$$J^P=\frac{1}{2\pi i}\oint_{|z|=1}\left\{z^j\frac{CQ_dC^*}{\tilde{A}H_d}-H_{pj}\frac{H_d^*}{\tilde{A}}\right\}\left\{z^j\frac{CQ_dC^*}{\tilde{A}H_d}-H_{pj}\frac{H_d^*}{\tilde{A}}\right\}^*\frac{dz}{z}\,. \tag{15}$$

By using Diophantine equation (9), square term in (15) will be rewritten as

$$z^j\frac{CQ_dC^*}{\tilde{A}H_d}-H_{pj}\frac{H_d^*}{\tilde{A}}=z^{g_p}\frac{F_{pj}}{H_d}\,. \tag{16}$$

Therefore the cost function (15) can be described as

$$J^P=\frac{1}{2\pi i}\oint_{|z|=1}\frac{F_{pj}F_{pj}^*}{H_dH_d^*}\frac{dz}{z}\,. \tag{17}$$

From the cost function (17), the term $F_{pj}F_{pj}^*/H_dH_d^*$ is independent to the predictor, so that the above cost function shows minimum cost.

The proposed optimal linear predictor $H_{pj}(z^{-1})$ is asymptotically stable since $H_d(z^{-1})$ is strictly Schur. Also, optimal j-step predictor can be written as

$$\hat{y}(k+j|k)=H_{pj}\frac{C}{\Delta A}\xi(k)+\frac{B}{A}u(k+j) \tag{18}$$

$$=H_{pj}d(k)+W\Delta u(k+j) \tag{19}$$

where $d(k)=\dfrac{C}{\Delta A}\xi(k)$ is disturbance. (20)

4. FREQUENCY DOMAIN GENERALIZED PREDICTIVE CONTROL

In this section, let's try to rederive GPC in the frequency domain, which is developed in the time domain. The frequency domain derivation of GPC law minimizes the following cost function(Clarke, 1987a)

$$J^G=E\left[\sum_{j=N_1}^{N_2}\tilde{e}^2(k+j|k)+\lambda\sum_{j=1}^{N_u}\Delta u^2(k+j-1)\right] \tag{21}$$

where, N_1 is minimum costing horizon, N_2 is maximum costing horizon, N_u is control horizon, λ is scalar control weighting. And, $\hat{e}(k+j|k)$ is the future predicted tracking error as

$$\hat{e}(k+j|k) = r(k+j) - \hat{y}(k+j|k) \qquad (22)$$

where $r(k+j)$ is the future reference input.

The form of the time domain GPC controller can be seen through 2 DOF controller which is composed of prefilter and feedback loop controller, so the form of frequency domain GPC controller is also made to be composed of 2 DOF controller, and the block diagram for its control system can be described as Fig. 1.

Therefore, future control input increment signal from the frequency domain GPC can be defined as

$$\Delta u(k+j) = K_1 r(k+j) - K_0 \hat{y}(k+j|k)) \qquad (23)$$

where, $K_0 = K_{0n}/K_{0d}$ is the feedback loop controller, $K_1 = K_{1n}/K_{1d}$ is the reference following controller. The reference input, $r(k)$, is assumed to be generated by white noise $\zeta(k)$. Then $r(k)$ can be described as,

$$r(k) = W_r(z^{-1})\zeta(k) \qquad (24)$$

where, $W_r(z^{-1}) = E(z^{-1})/A_r(z^{-1})$ is assumed to be asymptotically stable and casual. The polynomial $E(z^{-1})$ is assumed to have no zeros on the unit circle in z plane. $\zeta(t)$ is assumed to be the Gaussian white noise whose mean is 0 and variance is Q_r. And, white noise sources $\xi(t)$ and $\zeta(t)$ are assumed to be mutually statistically independent.

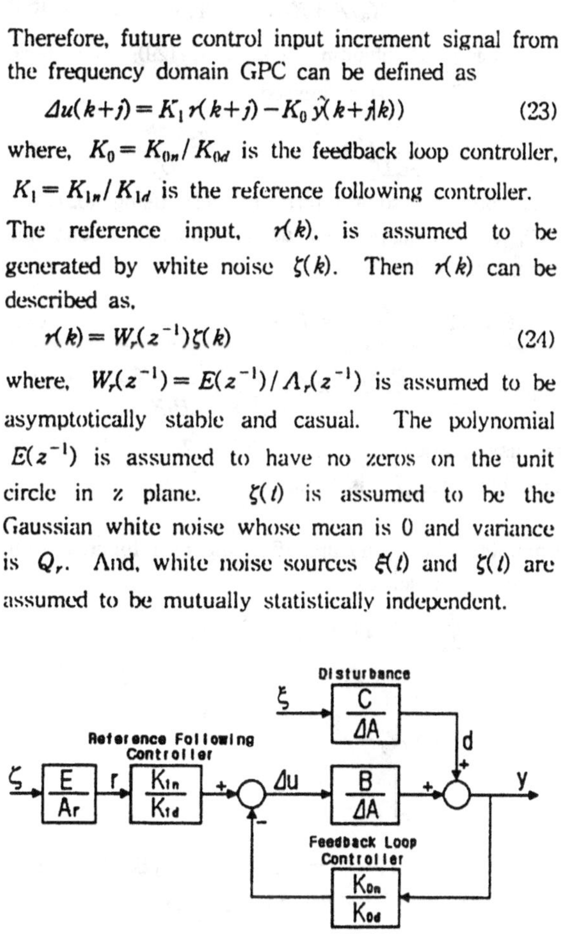

Fig. 1 Frequency domain GPC system

Under the assumption that the delay-time of plant is unknown, set $N_1 = 1$, and suppose that $N_u = N_2$ and future reference input signal $r(k+j)$ is given.
Optimal feedback loop controller and optimal reference input following controller which can minimize the cost function of GPC (21) will be respectively(Grimble, 1984; Grimble, 1994)

$$K_0 = \frac{G}{H} \qquad (25)$$

$$K_1 = \frac{Yz^{N_2-1}D_f}{D_r H}. \qquad (26)$$

Optimal feedback loop controller of (25) can be computed from the solutions G, H (with F of smallest degree) of the following two coupled Diophantine equations

$$D_c^* G z^{-g_g} + F\tilde{A} = B^* D_f z^{-g_g} \qquad (27)$$

$$D_c^* H z^{-g_g} - FB = \lambda \tilde{A}^* D_f z^{-g_g}. \qquad (28)$$

And optimal reference following controller of (26) can be computed from the solutions Y(with Z of smallest degree) of the following Diophantine equations

$$D_c^* Y z^{-g_r} + ZA_r = B^* D_r z^{-g_r - N_2 + 1} \qquad (29)$$

where g_g and g_r are the smallest positive integer which ensures the above Diophantine equations to be in the power of z^{-1}, $g_g = \max(nb, na+1)$, $g_r = \max(nb-1, na+1, nb-N_2+1)$. And D_f, D_c, D_r are strictly Schur spectral polynomials, which are

$$D_f D_f^* = \sum_{j=1}^{N_2} H_{nj} H_{nj}^* \qquad (30)$$

$$D_c D_c^* = \lambda \tilde{A}\tilde{A}^* + BB^* \qquad (31)$$

$$D_r D_r^* = \sum_{j=1}^{N_2} E Q_r E^*. \qquad (32)$$

Therefore, optimal control law is

$$\Delta u(k+j) = \frac{D_f Y z^{N_2-1}}{D_r H} r(k+j) - \frac{G}{H}\hat{y}(k+j). \qquad (33)$$

In case of using above optimal controller, characteristics polynomial of the closed loop system will be

$$\tilde{A}K_{0d} + BK_{0n} = 0 \quad \text{or} \quad \tilde{A}H + BG = 0. \qquad (34)$$

On the other hand, eliminating F from two Diophantine equation (27) and (28), it will be

$$\tilde{A}H + BG = D_f D_c \qquad (35)$$

where D_f and D_c are strictly Schur spectral polynomials. So the closed loop system is stable.

When frequency domain generalized predictive controller is given as (25)~(33), minimization of the cost function(21) will be proved as the following process.
First, the cost function (21) can be rewritten in the following form.

$$J^G = \frac{1}{2\pi i}\oint_{|z|=1}\left\{\sum_{j=1}^{N_2}\Phi_{e_j e_j} + \lambda\sum_{j=1}^{N_2}\Phi_{\Delta u_j \Delta u_j}\right\}\frac{dz}{z} \qquad (36)$$

where $\Phi_{e_j e_j}$ and $\Phi_{\Delta u_j \Delta u_j}$ are power-density spectra of the future predicted tracking error $\hat{e}(k+j|k)$ and the future control input increment $\Delta u(k+j-1)$, respectively.
Substituting (19) into (23), the future control input increment $\Delta u(k+j-1)$ becomes

$$\Delta u(k+j-1) = K_1 S r(k+j-1) - H_{nj} K_0 S d(k-1) \qquad (37)$$

where, $S = 1/(1+WK_0)$ is sensitivity function.

Future predicted tracking error $\hat{e}(k+j|k)$ can be described as the following form by substituting (19) and (37) into (22)

$$\hat{e}(k+j|k)=(1-K_1SW)r(k+j)-H_{pj}(1-K_0SW)d(k) . \quad (38)$$

Substituting (37) and (38) into (36), the cost function becomes

$$J^G=\frac{1}{2\pi j}\oint_{|z|=1}\{T_1(z^{-1})+T_2(z^{-1})\}\frac{dz}{z} \quad (39)$$

where,

$$T_1= \sum_j H_{pj}(1-K_0SW)\Phi_{dd}(1-K_0SW)^*H_{pj}^* + \sum_j \lambda H_{pj}SK_0\Phi_{dd}K_0^*S^*H_{pj}^* \quad (39a)$$

$$T_2= \sum_j(1-K_1SW)\Phi_{rr}(1-K_1SW)^* + \sum_j \lambda K_1S\Phi_{rr}S^*K_1^* \quad (39b)$$

where Φ_{rr} and Φ_{dd} are power-density spectra of future reference input $r(k+j)$ and disturbance $d(k)$, respectively.

Let the generalized spectral factors Y_f, Y_r, Y_c be defined as

$$Y_fY_f^*=\sum_j H_{pj}\Phi_{dd}H_{pj}^* \quad (40)$$

$$Y_rY_r^*=\sum_j \Phi_{rr} \quad (41)$$

$$Y_cY_c^*= WW^*+\lambda. \quad (42)$$

Then, $T_1(z^{-1})$ and $T_2(z^{-1})$, given in (39), can be described as the following complete square form

$$T_1= \left(K_0SY_cY_f-\frac{Y_fW^*}{Y_c^*}\right)\left(K_0SY_cY_f-\frac{Y_fW^*}{Y_c^*}\right)^* + Y_fY_f^*-\frac{W^*Y_fY_f^*W}{Y_cY_c^*} \quad (43)$$

$$T_2= \left(K_1SY_cY_r-\frac{W^*Y_r}{Y_c^*}\right)\left(K_1SY_cY_r-\frac{W^*Y_r}{Y_c^*}\right)^* + Y_rY_r^*-\frac{W^*Y_rY_r^*W}{Y_cY_c^*} . \quad (44)$$

Next, (40) by using (23) and (20), (41) by using (24), (42) by using (23) will be rearranged as the following form, respectively

$$Y_fY_f^*=\sum_j H_{pj}\Phi_{dd}H_{pj}^*=\sum_j\frac{H_{pj}CC^*H_{pj}^*}{\bar{A}\bar{A}^*} \quad (45)$$

$$Y_rY_r^*=\sum_j \Phi_{rr}=\sum_j\frac{EQ_rE^*}{A_rA_r^*} \quad (46)$$

$$Y_cY_c^*=\frac{\lambda\bar{A}^*\bar{A}+B^*B}{\bar{A}^*\bar{A}} . \quad (47)$$

Therefore, the generalized spectral factor Y_f, Y_r, Y_c can be defined as

$$Y_f=\frac{D_f}{\bar{A}},\quad Y_r=\frac{D_r}{A_r},\quad Y_c=\frac{D_c}{\bar{A}} \quad (48)$$

where D_f, D_c and D_r are strictly Schur spectral polynomials and those are same with (30), (31), (32).

The term $K_0SY_cY_f$ which appears in (43) can be described as

$$K_0SY_cY_f=\frac{D_fD_cK_{0n}}{\bar{A}(\bar{A}K_{0d}+z^{-1}BK_{0n})} \quad (49)$$

and using Diophantine equation (27), the term W^*Y_f/Y_c^* can be described as

$$\frac{W^*Y_f}{Y_c^*}=\frac{zB^*D_f}{\bar{A}D_c^*}=\frac{G}{\bar{A}}+\frac{F}{D_c^*z^{-g_c}} . \quad (50)$$

Therefore, the square term in (43), using (49), (50) and (35), can be expressed as a stable and unstable term, separately

$$K_0SY_cY_f-\frac{W^*Y_f}{Y_c^*}=T_1^+ + T_1^- \quad (51)$$

where, $T_1^+=\frac{HK_{0n}-GK_{0d}}{\bar{A}K_{0d}+BK_{0n}}$ is asymptotically stable, $T_1^-=F/D_c^*z^{-g_c}$ is unstable since D_c is Schur.

The term $K_1SY_cY_r$ which appears in (44) can be described as

$$K_1SY_cY_r=\frac{D_cK_{0d}K_{1n}D_r}{(K_{0d}\bar{A}+K_{0n}B)K_{1d}A_r} , \quad (52)$$

and using Diophantine equation (29), the term W^*Y_r/Y_c^* can be expressed as

$$\frac{W^*Y_r}{Y_c^*}=\frac{B^*D_r}{A_rD_c^*}=\frac{Yz^{N_2-1}}{A_r}+\frac{Zz^{g_r+N_2-1}}{D_c^*} . \quad (53)$$

Therefore, the square term in (44), using (52) and (53), can be expressed as a stable and unstable term, separately

$$K_1SY_cY_r-\frac{W^*Y_r}{Y_c^*}=T_2^+ - T_2^- \quad (54)$$

where $T_2^+=\frac{D_cK_{0d}K_{1n}D_r-Yz^{N_2-1}K_{1d}(K_{0d}\bar{A}+K_{0n}B)}{(K_{0d}\bar{A}+K_{0n}B)K_{1d}A_r}$

is asymptotically stable, $T_2^-=Zz^{g_r+N_2-1}/D_c^*$ is unstable since D_c is Schur.

From (43), (44), (51) and (54), the cost function (39) can be expressed as

$$J^G=\frac{1}{2\pi j}\oint_{|z|=1}\{(T_1^+-T_1^-)(T_1^+-T_1^-)^*+T_1^0 + (T_2^+-T_2^-)(T_2^+-T_2^-)^*+T_2^0\}\frac{dz}{z} \quad (55)$$

where, $T_1^0=Y_fY_f^*-\frac{W^*Y_fY_f^*W}{Y_cY_c^*}$

$$T_2^0=Y_rY_r^*-\frac{W^*Y_rY_r^*W}{Y_cY_c^*} .$$

The terms $T_i^-T_i^{+*}$ $(i=1,2)$ are analytic for $|z|\le 1$, so that by the residue theorem the sum of the residues are zero (i.e. $\oint T_i^-T_i^{+*}\frac{dz}{z}=0$, $i=1,2$), and $\oint_{|z|=1}T_i^{-*}T_i^+\frac{dz}{z}=-\oint T_i^-T_i^{+*}\frac{dz}{z}=0$ $(i=1,2)$.

Therefore, the cost function (55) becomes

$$J^G=\frac{1}{2\pi j}\oint_{|z|=1}\{T_1^+T_1^{+*}+T_1^-T_1^{-*} + T_2^+T_2^{+*}+T_2^-T_2^{-*}+T_1^0+T_2^0\}\frac{dz}{z} . \quad (56)$$

Since $T_1^-T_1^{-*}$, $T_2^-T_2^{-*}$, T_1^0, and T_2^0 are independent of the controller transfer function, the cost function is minimized when $T_1^+T_1^{+*}$ and $T_2^+T_2^{+*}$ are zero (i.e. T_1^+ and T_2^+ is zero).

The optimal feedback loop controller is computed by setting T_1^+ to zero($T_1^+=0$).

$$K_0=\frac{K_{0n}}{K_{0d}}=\frac{G}{H}. \tag{57}$$

The optimal reference input following controller is computed by setting T_2^+ to zero ($T_2^+=0$),

$$K_1=\frac{K_{1n}}{K_{1d}}=\frac{D_fYz^{N_2-1}}{D_rH}. \tag{58}$$

When it has optimal controller like (57) and (58), the minimal cost function is given by

$$J_{\min}^G=\frac{1}{2\pi j}\oint_{|z|=1}\left\{\frac{FF^*}{D_cD_c^*}+\frac{ZZ^*}{D_cD_c^*}+\frac{\lambda D_fD_f^*}{D_cD_c^*}+\frac{\lambda D_rD_r^*\tilde{A}\tilde{A}^*}{D_cD_c^*A_rA_r^*}\right\}\frac{dz}{z}. \tag{59}$$

4.1 Self-tuning frequency domain GPC

To expand the above frequency domain GPC controller into unknown system parameter, self-turning control scheme is adopted which can get controller parameters by estimated parameters after estimating system parameters.

For the parameter estimator, recursive least-squares estimation algorithm with exponential forgetting is used (Astrom 1989),

$$\begin{aligned} K(k)&=P(k-1)\phi(k)[\mu I+\phi^T(k)P(k-1)\phi(k)]^{-1}\\ P(k)&=[P(k-1)-K(k)\phi^T(k)P(k-1)]/\mu \\ \hat{\theta}(k)&=\hat{\theta}(k-1)+K(k)[y(k)-\phi^T(k)\hat{\theta}(k-1)] \end{aligned} \tag{60}$$

where, $\hat{\theta}(k)=[a_1,\cdots,a_{na},b_1,\cdots,b_{nb}]^T$,

$$\phi(k)=[-y(k-1),\cdots,-y(k-na),u(k-1),\cdots,u(k-nb)]^T.$$

Computing procedure of self-tuning frequency domain GPC algorithm can be summarized as

① selection of prediction horizon(N_2) and control weighting(λ),

② estimation of system parameters $a_1,\cdots,a_{na}$, $b_1,\cdots,b_{nb}$, using parameter estimator,

③ solving of prediction equation (9), using estimated $A(z^{-1})$ and $B(z^{-1})$,

④ computing of stable spectral factor D_f (30), D_c (31), and D_r (32),

⑤ computing of Diophantine equations (27) and (28), and (29),

⑥ determining optimal controller from (25) and (26),

⑦ first control increment is computed by (33), added to previous control value, and resulting control value is applied to the plant input,

⑧ repetition of ②-⑦ step.

Proposed frequency domain GPC has similar design papameter($N_1, N_2, N_u=N_2, \lambda$) as time domain GPC. And it is receding horizon controller which is structured in 2 DOF controller like time domain GPC.

The stability of time domain GPC is determined according to design parameters, while in frequency domain GPC characteristics polynomial of closed loop system is always strictly Schur, so that stability of frequency domain GPC is always guaranteed. However, it does not directly mean that the receding horizon frequency domain GPC is stable. But, if transfer function of plant or disturbance is time invariant and prediction interval is fixed, controller will be fixed and characteristics polynomial of closed loop system is strictly Schur. So we can see that frequency domain GPC will stabilize the closed loop system.

5. SIMULATION

The objective of these simulations is to show that the proposed frequency domain GPC can always guarantee stability of closed loop system and to show how proposed self-tuning frequency domain GPC can cope with a plant which changes in delay-time, in order and in parameters compared with a time domain GPC.

Firstly, to compare the stability of time domain GPC with that of frequency domain GPC, consider the following unstable non-minimum phase plant

$$\begin{aligned}&(1-4.0023z^{-1}+3.4903z^{-2})y(k)\\ &=(-0.7621z^{-1}+1.2501z^{-2})u(k).\end{aligned} \tag{61}$$

The prediction horizon(N) is chosen to be 10 (i.e. $N_1=1$, $N_2=10$). The control horizon(N_u) is also chosen to be 10. The control input weighting(λ) is chosen to be 0.01, 0.1, 1, 10, 100. Simulation results of time domain GPC and frequency domain GPC are shown in Fig. 2. In each case unit step response is expressed by changing a design parameter(i. e. control input weighting into 0.01, 0.1, 1, 10, 100. Fig. 2 shows that the stability of time domain GPC depends on a design parameter(especially control input weighting). But it also indicates that the frequency domain GPC always guarantee stable closed loop system regardless of a design parameter. In the above example, the time domain GPC can guarantee stability when $0.02<\lambda<14.65$.

Secondly, to compare the performance of proposed frequency domain GPC with that of the time domain GPC about the plant, which is changing in delay-time, in order and in parameters, consider the plant given in Table 1.(Clarke 1987a)

During the first 10 samples the control input was fixed at 10 for parameter estimation. To estimate the system parameter, $A(z^{-1})$ and $B(z^{-1})$ are assumed to be the 2nd and 5th degree polynomial respectively. The prediction horizon is chosen to be 10 (i.e. $N_1=1$, $N_2=10$). The control horizon(N_u) is also chosen to be 10. The control input weighting

(λ) is chosen to be 0.01. Simulation results of time domain GPC and frequency domain GPC are shown in Fig. 3. Fig. 3 show that both time domain GPC and frequency domain GPC can produce positive response characters.

Table 1 Transfer functions of the simulated models.

Number	Samples	Model
1	1 -100	$\frac{1}{1+10s+40s^2}$
2	101-200	$\frac{e^{-1.5s}}{1+10s+40s^2}$
3	201-300	$\frac{e^{-1.5s}}{1+10s}$
4	301-400	$\frac{1}{1+10s}$
5	401-500	$\frac{1}{10s(1+2.5s)}$

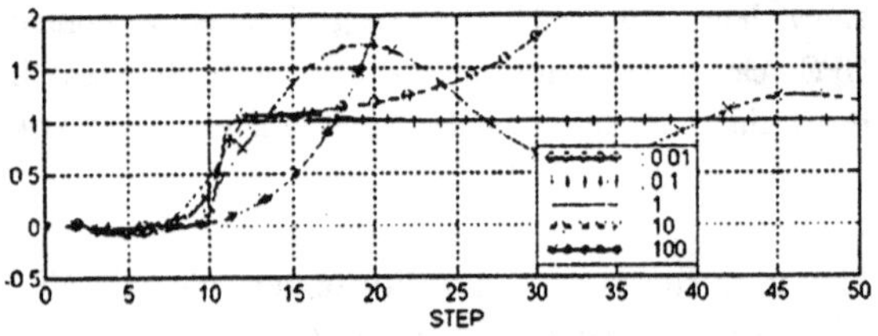

(a) Output of time domain GPC.

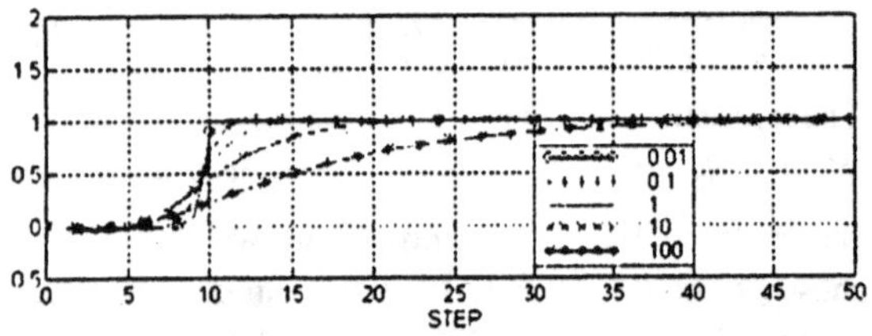

(b) Output of frequency domain GPC.

Fig. 2 Responses of time domian GPC and frequency domain GPC for non minimum phase plant.

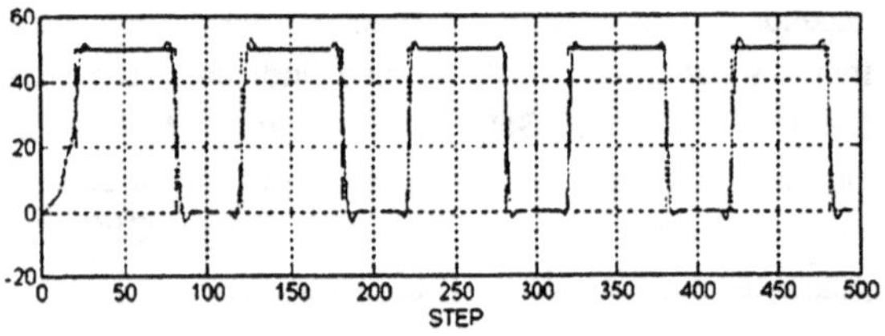

(a) output of time domain GPC.

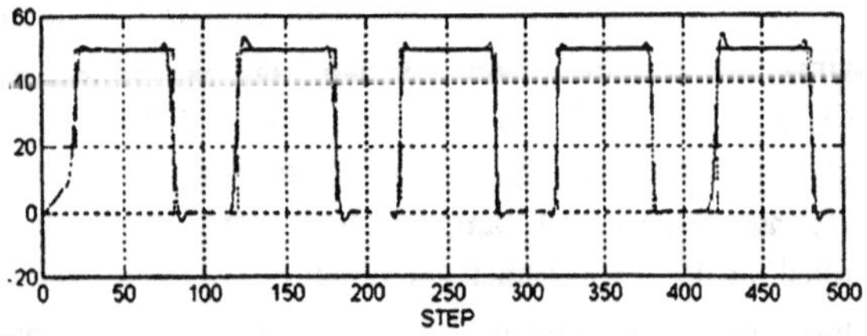

(b) output of frequency domain GPC.

Fig. 3 Responses of time domian GPC and frequency domain GPC. for the plant given in Table 1.

6. CONCLUSION

To guarantee nominal stability of GPC, the standard GPC control law which is developed in time domain is rederived in the frequency domain. By designing GPC in frequency domain with the proposed method, stable control system can be made regardless of a designed parameter. Through simulation of unstable non-minimum phase plant, the proposed frequency domain GPC is proved to be always stable, while the stability of the time domain GPC depends on a design parameter(especially control input weighting). Besides, in terms of performance, the frequency domain GPC can be seen to have almost similar performance with the time domain GPC.

And, study of robustness of the proposed frequency domain GPC is to be required.

REFERENCES

Astrom, K. J. and B. Wittenmark (1989). *Adaptive Control*, Addison-Wesley.

Clarke, D. W., C. Mohtadi, and P. S. Tuffs (1987a). "Generalized Predictive Control -Part I. The Basic Algorithm," *Automatica*, Vol. 23, pp. 137-148.

Clarke, D. W., C. Mohtadi, and P. S. Tuffs (1987b). "Generalized Predictive Control--Part II. Extensions and Interpretations," *Automatica*, Vol. 23, pp. 149-160.

Clarke, D. W., and C. Mohtadi (1989). "Properties of Generalized Predictive Control," *Automatica*, Vol. 25, No. 6, pp. 859-875.

Clarke, D. W., and R. Scattolini (1991). "Constrained Receding-Horizon Predictive Control," *IEE-Proceedings* Part D, Vol. 138, No. 4, pp. 347-354.

Grimble, M. J. (1985). "Polynomial Systems Approach to Optimal Linear Filtering and Prediction," *International Journal of Control*, Vol. 41, No. 6, pp.1545-1564.

Grimble, M. J. and A. Johnson (1988). *Optimal Control and Stochastic Estimation: Theory and Application, Volume 1, 2*, John Wiley & Sons, Chichester, London.

Grimble, M. J. (1994). Robust Industrial Control - Optimal Design Approach for Polynomial Systems, Prentice Hall.

Kouvaritakis, B., J. A. Rossitar, and A. O. T. Chang (1992). "Stable Generalized Predictive Control: an Algorithm with guaranteed stability," *IEE-Proceedings*, Part D, Vol. 139, No. 4, pp. 349-362.

Robinson, B. D. and D. W. Clarke (1991). "Robustness Effects of a Prefilter in Generalized Predictive Control," *IEE-Proceedings-D*, Vol. 138, No. 1, pp. 2-8.

IDENTIFICATION AND ADAPTIVE CONTROL OF DYNAMIC SYSTEMS USING SELF-ORGANIZED DISTRIBUTED NETWORKS

Jong-Soo Choi*, Hyongsuk Kim, and Young-Joo Moon***

**Automation Research Division, RIST, KOREA*
joschoi, yjmoon@risnet.rist.re.kr
***Dept. of Control & Instrumentation Eng., Chonbuk National University, KOREA*
hskim@moak.chonbuk.ac.kr

Abstract: An adaptive control technique, using system identification based on Self-Organized Distributed Networks (SODNs), is presented for a class of discrete nonlinear dynamic systems having unknown dynamics. The SODN belongs to the category of distributed local learning networks and is composed of two main networks called the learning network and the distribution network. The learning network consists of subnets each responsible for a subproblem. The distribution network is responsible for input space decomposition. The learning of the SODN is fast and precise because of the local learning mechanism. In this paper, methods for identification and indirect adaptive control of nonlinear systems using the SODN are presented. Through extensive simulation, the SODN is shown to be effective both for identification and adaptive control of nonlinear dynamic systems. *Copyright © 1998 IFAC*

Key Words: Adaptive control, Identification, Dynamic systems, SODN.

1. INTRODUCTION

Artificial neural networks have been proposed for use in a wide range of control applications (Miller *et al.*, 1990; Narendra and Parthasarathy, 1991; Sastry *et al.*, 1994). The most representative neural networks used for control problem is multilayer neural networks (MNNs) with the well-known error backpropagation learning algorithm (Rumelhart, *et al.*, 1986). Provided enough training data, neural networks can learn to generalized the mapping. While there are a lot of possible applications using MNN, people have also realized that some severe drawbacks remain hard to solve. One problem is that the learning easily gets trapped in a *local minima*. If the number of input variables is large or target function is complicated, backpropagation learning can be very difficult. An impractically large number of learning iterations may be required and the result is often not precise. In some applications such as system identification, precision could be very important. A little improvement in precision in the MNN may require a significant increase in learning time as well as the network size. In many application areas such as control problem, a single function approximator or neural network may not be sufficient to handle different modes of system behavior. An obvious approach is to use several neural networks, one for each mode of behavior, with each neural network being trained separately.

The separated or distributed local learning approaches such as RBF networks (Chen *et al.*, 1993, Sanner *et al.*, 1992), CMAC (Albus, 1975), and modular neural architecture (Jacobs, *et al.*, 1991, 1993) have been studied by researchers to solve control problems. The distributed local learning approaches are not affected by the interference problems that affect global neural network like the MNN. In (Jacobs et al., 1993) multi-network or modular neural architecture that learn to perform control tasks using piecewise linear controller is described. The architecture's networks compete to learn the training patterns. As a result, a plants' parameter space is adaptively partitioned into a number of region, and a different network learns a control law in each region.

Self-Organized Distributed Networks (SODNs) proposed by (Kim et al., 1994) are similar to the modular neural architecture. The SODN is composed of two main networks called the learning network and the distribution network. The learning network consists of subnets each responsible for a subproblem. The distribution network is responsible for input space decomposition. The learning of the SODN is fast and precise because of the local learning mechanism.

In this paper, an indirect adaptive control technique, using system identification based on SODN, is presented for a class of discrete nonlinear dynamic system having unknown dynamics. Through extensive simulation, the SODN is shown to be effective both for identification and adaptive control of nonlinear dynamic system.

2. SELF-ORGANIZED DISTRIBUTED NETWORKS (SODN)

2.1 Structure and learning of the SODN

The SODN belongs to the category of distributed local learning networks. A SODN consists of two networks called the distribution network and the learning network as shown in Fig. 1. The learning networks consist of a group of subnets. Each learning subnetwork consists of small sized multilayer perceptron with *limiting network.* The limiting networks (LNs) have the output range of each subnetwork to be limited to a dynamic range of target values in each subregion. The distribution network consists of self-organized units and its learning is based on the winner-take-all rule (Kohonen 1982).

The output of each self-organized unit is linked to the output of a learning subnetwork through multiplication as shown in the Fig. 1. While an input vector is presented to the distribution network, only the neuron with the closest weight vector to the input vector is activated. Weights from input units to the activated unit are updated toward the input vector. Such weight vector of each self-organized unit is called network center of each subnetwork. The activated unit invokes its counterpart in the learning network. Thus, the distribution network divides a problem into subproblems and each learning subnetwork learns a subproblem.

While any input vector is assigned to the subnetwork whose network center is the closest to the input vector, the input space can be divided rigidly according to distance from the network centers. Each subnetwork learns only data within its region exclusively. Some discontinuity can be appeared at the boundaries of each subregion. Noise shown at the input of distribution networks in the Fig. 1 is for improving the continuity at boundaries. With the noise, some data placed at border of one subnetwork can be moved to other subnetwork. For learning efficiency, network generation/extinction mechanism is, also, employed in the SODN learning. The subnetwork connected with dotted lines in the Fig. 1 is the deleted one with this mechanism.

Let the input vector of the SODN be $\boldsymbol{x}$, the output vector of learning network be $\boldsymbol{y}(\boldsymbol{x})$ and output vector of the distribution network be $\boldsymbol{u}$, the total network output $o_t(\boldsymbol{x})$ is

$$o_t(\boldsymbol{x}) = \boldsymbol{u}^t(\boldsymbol{x})\, \boldsymbol{y}(\boldsymbol{x}) \tag{1}$$

where the jth element of $\boldsymbol{u}$ is

$$u_j = \begin{cases} 1 & \text{if the jth subnet is selected} \\ 0 & \text{otherwise.} \end{cases}$$

The total network output is the output of a learning subnet selected by the distribution network. If $t(\boldsymbol{x})$ is a target value, the output error will be

$$error(\boldsymbol{x}) = t(\boldsymbol{x}) - o_t(\boldsymbol{x}). \tag{2}$$

According to the backpropagation rule, the error $\delta(\boldsymbol{x})$ to be backpropagated will be

$$\delta(\boldsymbol{x}) = error(\boldsymbol{x})\, f'(\, \boldsymbol{u}^t(\boldsymbol{x})\boldsymbol{y}(\boldsymbol{x})). \tag{3}$$

If the function f is ramp function, (3) can be rewritten as

$$\delta(\boldsymbol{x}) = t(\boldsymbol{x}) - o_t(\boldsymbol{x}). \tag{4}$$

Thus the error backpropagated to the jth learning network is

$$\delta_j(\boldsymbol{x}) = u_j\delta(\boldsymbol{x}) = u_j(t(\boldsymbol{x}) - o_t(\boldsymbol{x})). \tag{5}$$

The equation indicates that only the selected learning subnet is updated with the backpropagated error $\delta(\boldsymbol{x})$.

2.2 Improvement of continuity of output function

In SODN, the output function is built through combining many subfunctions. Matching the outputs of subfunctions at boundaries is important. There are two strategies of improving the continuity in the proposed network. One is by adding random noise to input of the distribution networks as explained in section 2.1. The other one is by using LN. With the limiting network, output of each subnetwork is limited to the dynamic range of target values, which improves the continuity. The LN in Fig. 1 is for providing such extrapolation constraints. Each LN is connected to each output terminal of the MNN and is composed of a scaling weight s_w and a linear perceptron with a bias. The scaling weight is the dynamic range of target values, the activation function of the perceptron is a ramp function and its bias is the minimum value of the targets. Let kth output of a learning subnetwork be y_k' and the output of the LN be y_k, then y_k is

$$\begin{aligned} y_k &= f(y_k' s_w + b) \\ &= y_k' s_w + b \end{aligned} \quad (6)$$

where the neural unit is a linear combiner with ramp activation function. Since s_w is the difference between the maximum value t_{max} and the minimum value t_{min}, and bias b is the minimum value t_{min}, (6) becomes

$$y_k = y_k'(t_{max} - t_{min}) + t_{min} \quad (7)$$

where the range of y_k' in (7) is $0.0 \leq y_k' \leq 1.0$. Therefore, the range of y_k is limited to

$$t_{min} \leq y_k' \leq t_{max} \quad (8)$$

The output of the subnetwork is always within this range. With this LN, extrapolation with abrupt increment/decrement will be avoided. The value of scaling weight and the bias is updated by not through learning but through testing the samples data at every iteration in each subregion.

2.3 Subnetwork generation and extinction

A function may be more complicated in some areas in the input space than other areas. If the complicated area is divided into smaller regions and a subnet is assigned to each small region, learning can be much easier. In addition to self-organizing the networks, a strategy, generation of new subnets and extinction of redundant networks, is employed in SODN learning for the efficient use of networks. Backpropagated error to the distribution network is exploited as such network generation/extinction energy.

In the SODN structure, the output of each subnetwork is multiplied by that of its counterpart in the distribution network and become the input of the output unit out_k^o. Let the *j*th output of the *i*th subnetwork be y_{ji}, the number of subnetworks be M and the input to the *k*th output unit be out_k^o. Then, the out_k^o is

$$out_k^o = \sum_{i=1}^{M} y_{ki}\, u_i \quad (9)$$

If the activation function of the output unit is f, the k th output o_k will be

$$\begin{aligned} o_k &= f(out_k^o) \\ &= f(\sum_{i=1}^{M} y_{ki}\, u_i) \end{aligned} \quad (10)$$

In (10), y is the output of the learning network and f is the activation function of the output unit. Thus, the output layer is equivalent to a single layer network whose input, weight and activation function are y, u and f respectively. Learning can be done with the ordinary backpropagation rule for the selected subnetwork. Note that the u which is the output of distribution is considered as a kind of weight. If some error is occurred at the output terminal, changing of weight u is required. The magnitude of the required weight changing for each subnetwork can be used as a measure of learned status of the subnetwork. Let the weight located between the *i*th output unit to *j*th self-organized unit be u_{ji}. Then, its change Δu_{ji} is

$$\begin{aligned} \Delta u_{ji} &\propto \delta_j\, u_i \\ &= (t_j - o_j) u_i. \end{aligned} \quad (11)$$

Summation of squared value of Δu_{ji} from all output terminal is used as the generation/extinction energy for *i*th subnetwork. Let such energy be e_i, then e_i is

$$e_i = \sum_{j=0}^{N} (\Delta u_{ji})^2. \quad (12)$$

The energy is accumulated at every iteration. If summation of energies of all subnetworks reaches to some threshold value E_{th}, a new subnetwork is generated around the subnetwork with the largest energy. And also the subnetwork with the smallest energy is deleted. After each generation/extinction operation, energies of all subnetworks are initialized to zero. For this purpose, an energy storage is prepared in each self-organized unit as shown in the Fig. 1.

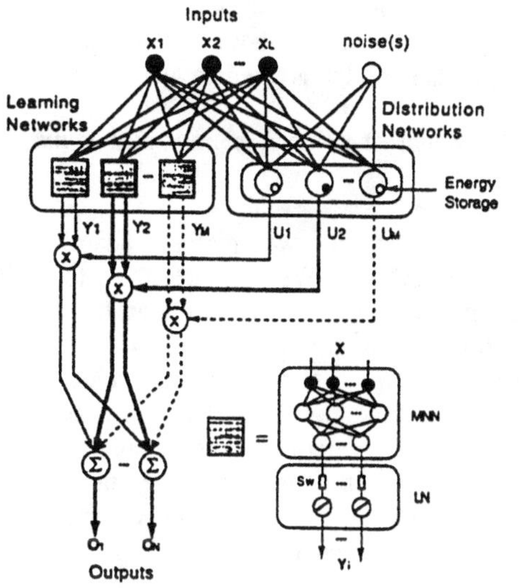

Fig. 1. The Structure of Self-Organized Distributed Networks.

3. IDENTIFICATION AND ADAPTIVE CONTROL USING SODN

3.1 The identification of nonlinear dynamic systems

The problem of identification consists of setting up

a suitably parameterized identification model and adjusting parameters of the model to optimize a performance function based on the error between the plant and the identification model outputs. In this subsection the application of the SODN for identification of nonlinear dynamic systems is discussed. For simplicity, now consider a single input/single output (SISO) plant. Let the input node be fed $u(k)$ at time k, and let the output of the network at the time k be $\hat{y}(k)$. We will use $y_p(k)$ as the target signal at the time k. The identified model is then described by equation

$$\hat{y}(k) = F(u(k), u(k-1), \ldots, y(k-1), y(k-2), \ldots) \quad (13)$$

where F is the nonlinear transformation represented by the SODN. The model represented by (13) is known as series-parallel model for identification. The series-parallel model has several advantages over parallel model. Since the plant is assumed to be bounded-input bounded-output (BIBO) stable, all the signals used in the identification procedure are bounded. Further, since no feedback loop exists in the model, static backpropagation can be used to adjust the parameters reducing the computational overhead substantially. Finally, assuming that the output error tends to small value asymptotically so that $y(k) \approx \hat{y}(k)$, the series-parallel model may be replaced by a parallel model without serious consequences. This has practical implications if the identification model is to be used off-line. In the viewpoint of the above considerations, the series-parallel model is used in this paper.

3.2 The adaptive control of nonlinear dynamic systems

In this subsection, we discuss adaptive control using the SODN. We assume that the plant is stable in the BIBO sense and that there sufficient knowledge about the plant to specify the goal of control in terms of a reference model. Given a plant, a reference model, and a reference input, the problem is to determine the input to the plant which will be the output of a neural network controller so that the output of the plant follows that of the reference model.

Two approaches to the adaptive control of an unknown plant are direct control and indirect control. In direct control, the parameters of the controller are directly adjusted to reduce the output error. In indirect control, the parameters of the controller are adjusted by backpropagating the error between the identified model and the reference model outputs through the identified model. The main difficulty with using neural networks for indirect control is the fact that these are nonparametric identifiers and hence there is no simple relationship between the learned weights of the network and the parameters of the plant (Sastry *et al.*, 1994). This method is discussed in detail by Narendra and Parthararathy(1991) and SODN can be used for indirect control in a similar way.

Fig. 2 shows the block diagram of the method we discussed here, which is called differentiating the model or forward and inverse modeling. Here we keep a forward model of the plant (SODN Identifier in Fig. 2) which is used as a channel to backpropagate the error at the plant output up to the plant input and this backpropagated error is used for training the controller network (SODN Controller in Fig. 2). For the technique to work properly, the forward model of the plant should be accurate at all times. So, at each time step we will update the weights of the plant model based on the error e_I in Fig. 2 between the output of the plant and that of the SODN identifier. Then we calculate the error e_C in Fig. 2 between the output of the plant and that of the reference model, which is then backpropagated through SODN identifier to supply the error signal for the SODN controller.

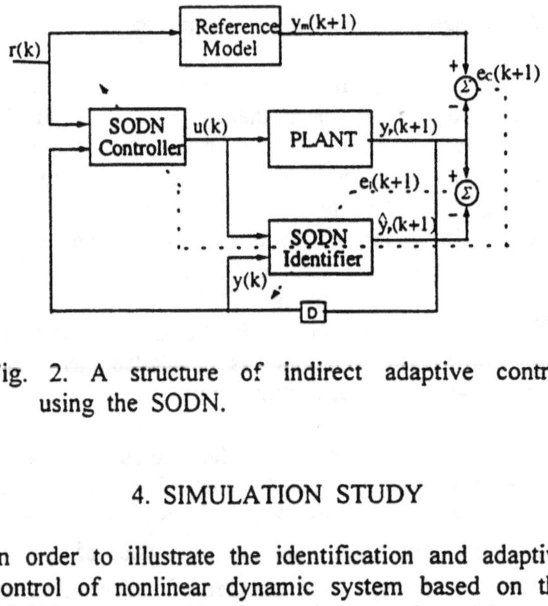

Fig. 2. A structure of indirect adaptive control using the SODN.

4. SIMULATION STUDY

In order to illustrate the identification and adaptive control of nonlinear dynamic system based on the SODN, we consider several examples. Most of these examples are also from (Narendra and Parthasarathy, 1991).

<u>*Example 1*</u>: In this example, the SISO plant is assumed to be of the form

$$y(k) = f(y(k), y(k-1), y(k-2), u(k), u(k-1)) \quad (14)$$

where the unknown function F has the form

$$f[x_1, x_2, x_3, x_4, x_5] = \frac{x_1 x_2 x_3 x_5 (x_3 - 1) + x_4}{1 + x_2^2 + x_3^2}.$$

In the identification model, the SODN that consists of 20 subnets is used to approximate the function f. Identification model has three inputs and one output as following equation

$$\hat{y}(k+1) = F[u(k), y(k), y(k-1)]$$

Fig. 3 shows the output of the plant and the

model when the identification procedure was carried out for 10,000 iterations using 1,000 random input signal uniformly distributed in the interval [-1, 1]. As discussed in above section, during the identification process a series-parallel model is used, but after the identification process is terminated the performance of model is studied using a parallel model. Test input in Fig. 3 is given by

$$u(k) = 0.5(\sin(2\pi k/250) + \sin(2\pi k/100)).$$

Identification result in Fig. 3 shows that error between the output of plat and that of identified model by SODN is very small.

Example 2: This example is MIMO plant with two inputs and two outputs. The plant is described by the equations.

$$y_1(k+1) = \frac{y_1(k)}{1+y_2^2(k)} + u_1(k) \quad (15)$$

$$y_1(k+1) = \frac{y_1(k)\, y_2(k)}{1+y_2^2(k)} + u_2(k)$$

The SODN identification model with 30 subnets has four inputs($y_1(k)$, $y_2(k)$, $u_1(k)$, $u_2(k)$) and two outputs($\hat{y}_1(k+1)$, $\hat{y}_2(k)$). The identification procedure was carried out for 10,000 iterations with 1,000 random inputs $u_1(k)$ and $u_2(k)$ uniformly distributed in the interval [-1, 1]. Fig. 4 show the output of the plant and network for the test signal given by $[\sin(2\pi k/25), \cos(2\pi k/25)]^T$. The identification results are good for the MIMO plant.

Example 3: In this example, We now describe indirect adaptive control. The plant is given by

$$y_p(k+1) = \frac{y_p(k)}{1+y_p^2(k)} + u^3(k) \quad (16)$$

The reference model is a first order linear system given by

$$y_m(k+1) = 0.6y_m(k) + r(k). \quad (17)$$

The SODN used as identifier and controller consists of 20 subnets respectively. Training of the controller proceeds as follows. We start with off-line system identification phase where we train the SODN identifier network, using series-parallel scheme. Then, we have an off-line controller training phase for the SODN controller network where cascade of the SODN controller and the SODN identifier are trained to mimic the reference model. Here we update only the weights of controller network using identifier network as a channel of backpropagation. It should be noted that the plant is not involved in this training. For off-line training, reference input $r(k)$ is taken to be uniformly distributed over [-1, 1] and training is continued for 10,000 iterations. Then we connect the plant and train the controller on-line. For test, the reference input $r(k)$ is given by

$$r(k) = \sin(2\pi k/25) + \sin(2\pi k/10).$$

The output of the controlled plant and that of the reference model are shown in Fig. 5(b). Comparing this with Fig. 5(a), it is easy to see the control is quite effective.

5. CONCLUSION

In this paper an application of Self-Organized Distributed Networks (SODNs) is considered for identification and adaptive control of nonlinear dynamic systems. The SODN is able to identify a variety of complex nonlinear systems. As can be seen from the simulation results, this method is very robust. We have also presented simulations using the SODN for indirect adaptive control. We have shown the effectiveness of SODN controller for tracking the reference model output.

6. REFERENCES

Albus J. (1975). A new approach to manipulator control: The cerebella model articulation controller (CMAC), *Journal of Dynamic System and Measurement Control*, vol. 63, no. 3, pp. 220-227.

Chen, L.-C., W.-C. Chen, and F.-Y. Chang (1993). Hybrid learning algorithm for Gaussian potential function networks, *IEE Proceedings-D*, vol. 140, no. 6, pp. 442-448.

Jacobs, R. A., M. I. Jordan, and A. G. Barto (1991). Task decomposition through competition in a modular connectionist architecture: the whaw and where vision tasks. *Cognitive Science*, vol. 15, pp. 219-250.

Jacobs, R. A. and M. I. Jordan (1993). Learning piecewise control strategies in a modular neural network architecture. IEEE Transactions on Systems, Man, and Cybernetics, vol. 23, no. 2, pp. 337-345.

Kim, H., J.-S. Choi, and C.-S. Lin (1994). Self-organized distributed networks for highly nonlinear mapping. In: *Intelligent Engineering Systems Through Artificial Neural Networks* (C. H. Dagli, etc. (Ed.)), pp. 109-114. ASME Press, New York.

Kohonen, T. (1982) Self-organized formation of topologically correct feature maps. Biomedical Cybernetics, vol. 43, pp. 59-69.

Miller, W. T., R. S. Sutton, and P. J. Werbos (1990). *Neural networks for control,* chapter 1. MIT Press, Cambridge, Massachusetts.

Moody J. and C. J. Darken (1989). Fast learning in networks of locally-tuned processing units. *Neural Computation*, vol. 1, pp. 281-294, 1989.

Narendra, K. S. and K. Parthasarathy (1991). Identification and control of dynamical systems using neural networks, *IEEE Transactions on Neural Networks*, vol. 1, no. 1, pp. 4-27.

Rumelhart, D. E., G. E. Hinton, and R. J. William

(1986). Learning representations by backpropagating errors, *Nature*, vol. 323, pp. 533-536.

Sanner R. M. and J.J.E. Slotine (1992). Gaussian networks for direct adaptive control, *IEEE Transactions on Neural Networks*, vol. 3, pp. ??

Sastry, P. S., G. Santharam, and K. P. Unnikrishnan (1994). Memory neuron networks for identification and control of dynamical systems, *IEEE Transactions on Neural Networks*, vol. 5, no. 2, pp. 306-319.

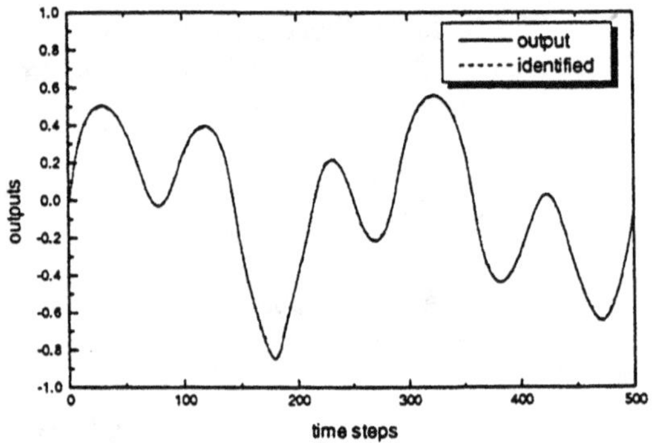

Fig. 3. Example 1: The result of identification for SISO plant.

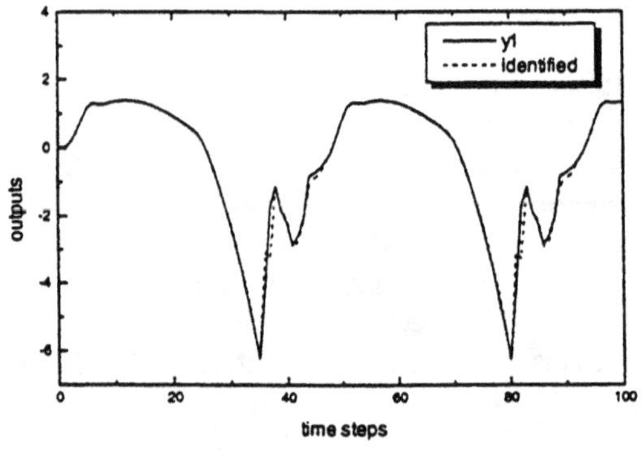

(a)

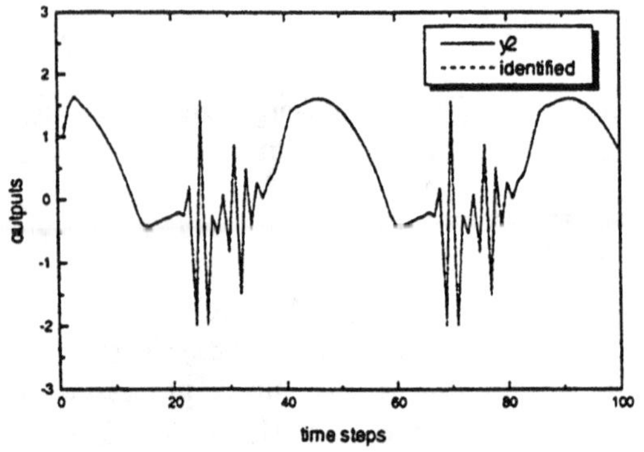

(b)

Fig. 4. Example 2: The results of identification for MIMO plant: (a) y1(k) and (b) y2(k).

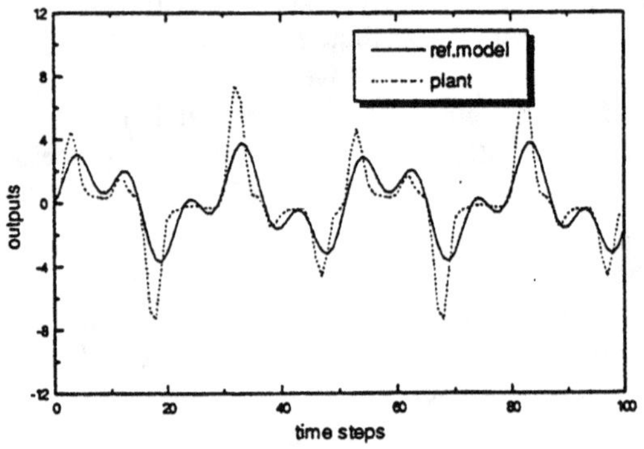

(a)

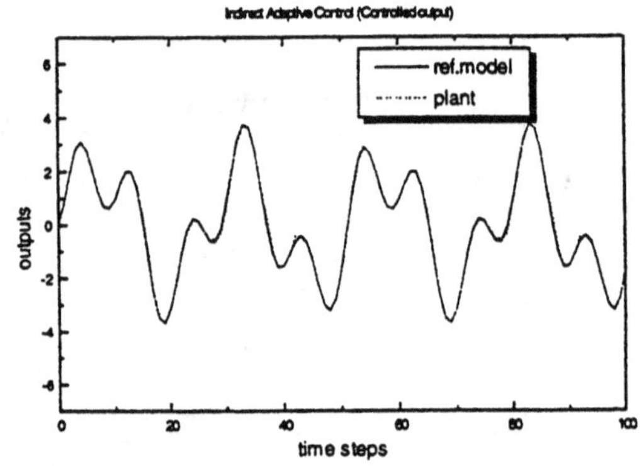

(b)

Fig. 5. Example 3: (a) The outputs of reference model and plant without control and (b) The result of indirect adaptive control for plant.

MODELLING A HOT STEEL MILL COMPLEX USING CONNECTING GRAPH PARADIGM *

Y. Y. Yang , D. A. Linkens, N. Mort

Department of Automatic Control and Systems Engineering
University of Sheffield
Mappin St., Sheffield S1 3JD, UK

Abstract: In this paper a hot steel mill complex model is developed based on the connecting graph paradigm, which includes a connecting graph in the top level and a hybrid automaton in the lower level. Both the discrete events and the continuous dynamics are modelled in a unified framework, allowing a better investigation of the interactions between these two parts. The modelling procedure is then illustrated by a key sub-process of soaking. Corresponding connecting graphs and the associated hybrid automata are given for the critical components involved in the hot steel mill complex. Discussion of the merits and potential application of the resulting model are also given.

Keyword: Steel manufacturing, hybrid systems, connecting graph, modelling.

1. INTRODUCTION

Due to the importance of the hot steel mill process, a considerable amount of research was carried out during the 70's and 80's on their modelling, optimisation, and control [Ashour and Bindingnavle 72, Patel *et al* 76, Lumelskey 83, Rao *et al* 84, Lu and Williams 83]. A hot steel mill complex can be viewed as a typical hybrid system, with the existence of both continuous dynamics and discrete events. However, research on hybrid modelling and control appears sparse until very recently [Yang *et al* 94, Yang and Linkens 95].

The aim of this paper is to establish a hot steel mill complex model with its hybrid dynamics in mind. We believe that by considering the mutual effects of discrete events and continuous dynamics the resulting model can represent the physical process more faithfully. It is also natural to conjecture that control strategies and scheduling policies derived from the enhanced hybrid model will be superior to those based on the more conventional models. The paper is organised as follows. In Section 2 a brief description of the connecting graph modelling paradigm, which consists of a connecting graph level and a hybrid automaton level, is given. Section 3 describes the hot steel mill complex to be modelled. Model development based on the connecting graph paradigm is detailed in Section 4, with the focus on the soaking sub-process. Finally, concluding remarks and future research topics are outlined in Section 5.

2. OUTLINE OF THE CONNECTING GRAPH MODELLING PARADIGM

The original concept of connecting graphs was proposed by Linkens and Yang [96]. It was based on a basic machine model (in a manufacturing system) as shown in Fig 1 below.

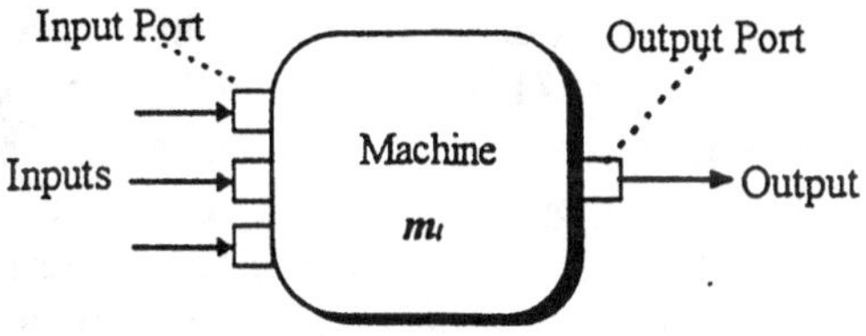

Fig. 1. Basic machine model

* This research is partly supported by an EPSRC Grant GR/H/73585.

The overall system is considered to be a network of a series of machines connected in a certain way according to the production configuration. Each machine can have a certain number of input ports (abbreviated as ports), and the machine dynamics can be represented by either discrete events, or continuous dynamics, or both. When there is at least one unit of input for each of the input ports, a machine can start its production by consuming a unit for each input port, and produce a unit of product through its output port at the end of its production cycle. A buffer can be attached to the output port if needed, and it is assumed that further outputs of a machine are inhibited when the attached buffer is full.

There are two hierarchical levels in a connecting graph model paradigm: a connecting graph which focuses on the inter-machine relationships in a manufacturing system, and a hybrid automaton level which is used to model the machine dynamics with possibly hybrid behaviour.

A connecting graph is composed of nodes and directed arcs, with the nodes representing machines and the directed arcs representing the inter-machine relations among the machines. Machines are connected by directed arcs, which specify the producer-consumer relationships between them. A connecting graph can be formally defined, as given by Linkens and Yang [96]:

Definition 1: A connecting graph M_G is a tuple:

$$M_G = (M, PORT, ARC) \tag{1}$$

where $M = \{m_1, m_2, ..., m_N\}$ is a finite set of machines; $PORT$ is a mapping from machines to their number of ports, i.e., $PORT: M \dashrightarrow I$, with I being a positive integer; the kth port of m_i is denoted by (m_i, k), with $k \leq PORT(m_i)$; and ARC is the mapping from the machines and the ports of machines to the arcs, i.e.,

$$ARC: M \times (M \times PORT) \rightarrow \{0, 1\}$$

Definition 2: The input set of a machine m_j, denoted by $IN(m_j)$, is the set of all machines which have an arc connected to any port of m_j, i.e., $IN(m_j) = \{ m_i \mid \forall k, ARC(m_i, (m_j, k)) = 1\}$. The input set of a port (m_j, k), denoted by $IN(m_j, k)$, is the set of all machines which have an arc connected to the port (m_j, k), i.e., $IN(m_j, k) = \{ m_i \mid ARC(m_i, (m_j, k)) = 1\}$. The output set of m_i, denoted by $OUT(m_i)$, is the set of machines which have an arc connection starting from m_i, i.e., $OUT(m_i) = \{ m_j \mid \forall k, ARC(m_i, (m_j, k)) \neq 0\}$.

Definition 3: A machine that has no input port is a source machine. The set of all source machines is denoted by $SOURCE$, i.e., $SOURCE = \{ m_i \mid max(|(m_i, k)|) = 0 \}$. Similarly, a machine whose product (output) is not used by any other machine is a sink machine, the set of all sink machines is denoted by $SINK$, i.e., $SINK = \{ m_i \mid OUT(m_i) = \phi \}$.

Depending on the configuration of a manufacturing process, the output of a machine m_i can be sent only to a certain number of successor machines for further processing. Fig. 2 shows a simple connecting graph, where the output set of m_4 has two machines (m_7 and m_8), both of the input sets of the last port of m_7 and m_8 have two machines, m_9 is a sink machine, and m_1 - m_6 are the source machines.

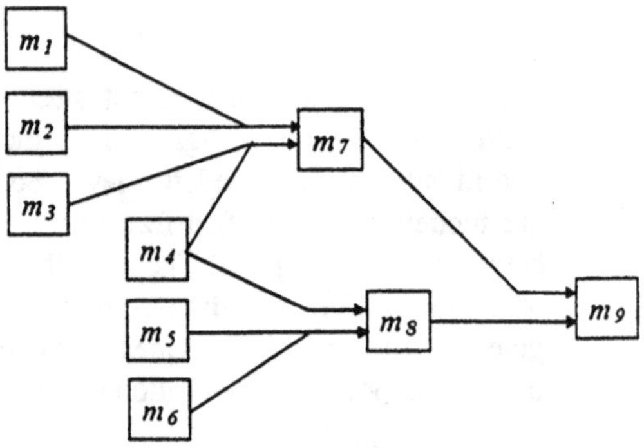

Fig. 2. A simple Connecting graph

The connecting graph proposed is quite general and flexible for modelling the inter-machine relationships in manufacturing systems. It is capable of modelling alternative production routines and group machine operations through the proper use of the ARC configurations. However, there always exists a conflict between the modelling power and the analysis capability. Restrictions can be introduced to increase the analysis ability (Linkens and Yang 96), such as the layered connecting graph and restriction on the number of sink machines, etc.

The real power of the connecting graph model paradigm comes from the employment of hybrid automata in its basic machine model (lower level modelling). With the machine dynamics, a connecting graph paradigm is capable of modelling the overall system behaviour in a hierarchical way.

In this paper we assume that a machine can be treated as a piece-wise continuous process with limited discontinuous points. Each continuous part between the two adjacent discontinuous points is defined as a phase. In other words, the dynamics of a machine can be described by the collection of its phase dynamics, using the following equations:

$$\frac{dx^i}{dt} = f_j^i(x^i, u^i, t), \quad x^i \in \Omega_j^i \subseteq \Re^{n_i}, u^i \in \Re^{m_i}, \tag{2}$$
$$1 \leq i \leq N, \quad j = 1,2,...i_p$$

$$j' = \begin{cases} j, & if\ x^i \notin \partial\Omega^i_j \\ \varphi(j, x^i, u^i), & if\ x^i \in \partial\Omega^i_j \end{cases} \quad (3)$$

where x^i and u^i are the state and control vectors of m_i, respectively; i_p is the number of phases existing in the machine dynamics; f^i_j is a non-linear continuous vector function for m_i at phase j; j' is the next phase to be entered by m_i; and φ is the phase transition function.

Suppose the initial condition of machine m_i is x^i_0, and m_i begins its production at $t = t_0$. The completion of a unit of product at machine m_i occurs after the machine m_i enters its terminal phase and the state x^i belongs to the pre-specified subset Ω^{i*} of the state space. It is also assumed that Ω^{i*} is reachable from x^i_0, i.e., there exists a control u^i such that the solution of equations (2-3) join x^i_0 to some point $x^i(t_f) \in \Omega^{i*}$ within a finite time t_f.

A machine, governed by equations (2-3), can be modelled by a hybrid (phase) automaton. The formal definition of hybrid automaton can be found in Alur *et al* [1993]. At any time instant, the state of a hybrid system can change in two ways: (1) by an instantaneous transition (here phase transition) that changes the entire state according to the successor relation, and (2) by elapse of time that changes only the values of a data variable continuously according to the activities of the current location (phase). The advantage of using hybrid automata is that both the phase dynamics (continuous) and the phase transition dynamics (discrete) can be clearly represented.

3. DESCRIPTION OF THE HOT STEEL MILL PROCESS

Fig. 3 shows the diagram of the hot steel mill complex under consideration, which consists of steelmaking, soaking, and rolling sub-processes. The steelmaking sub-process provides the hot ingots for the soaking and rolling sub-process. Additional cold ingots are ordered from other steel manufacturing units and are stored in a cold ingot bank as a secondary ingot supplier. The soaking sub-process provides ingots with a desired temperature profile (via appropriate heating in the soaking pits) to the downstream rolling mills. The function of the rolling sub-process is to produce semi-final steel products according the customer order specification via a series of rolling activities.

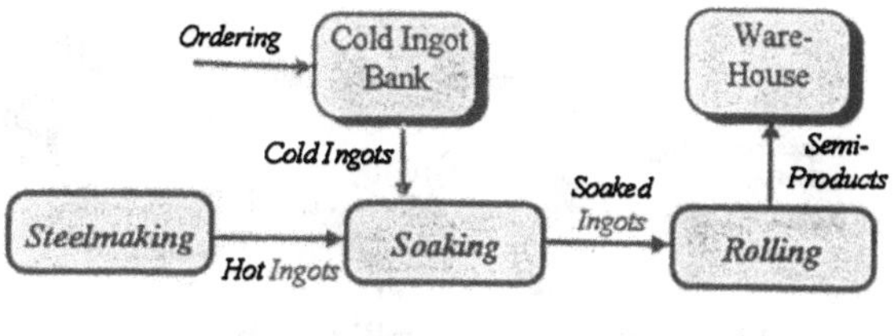

Fig. 3. Diagram of the hot steel mill complex

There are two steel melting furnaces in the steelmaking sub-process. Other facilities include strippers and trains. Hot steel (in liquid phase form) is first generated in the two melting furnaces in a batch production mode. Ingots are then formed by pouring the liquid steel into the moulds. When the liquid steel in the moulds is partially solidified the hot ingots are stripped off their moulds and they are transported to the soaking pit bay by trains through the internal railway network. Fig. 4(a) shows a schematic of the steelmaking sub-process.

The soaking sub-process, as shown in Fig. 4(b), consists of eighteen soaking pits arranged in the soaking pit bay, two cranes to charge and discharge the soaking pits, and two preheat furnaces to supply additional hot ingots from the cold ingot bank when the number of hot ingots from the steelmaking are insufficient.

The rolling sub-process has four mills with different capabilities to produce a variety of shapes and sizes of steel products. #1 and #2 mill are reversible roughing mills, and #3 and #4 are multi-stand continuous finishing mills. Transfer of ingots in the rolling sub-process is carried out by a track conveyor system. Fig. 4(c) shows a schematic of the rolling sub-process.

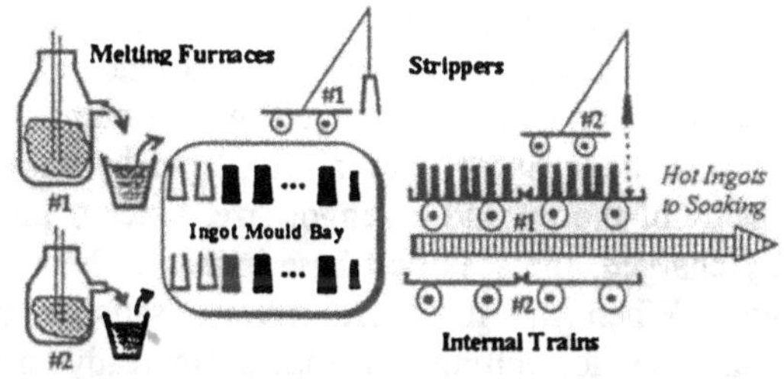

(a) Schematics of the steelmaking sub-process

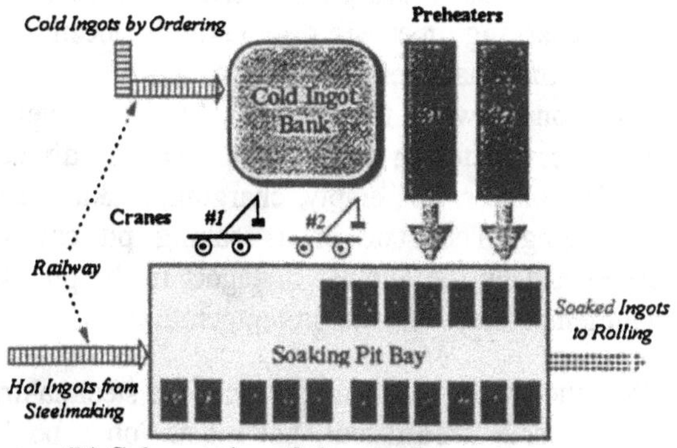

(b) Schematics of the soaking sub-process

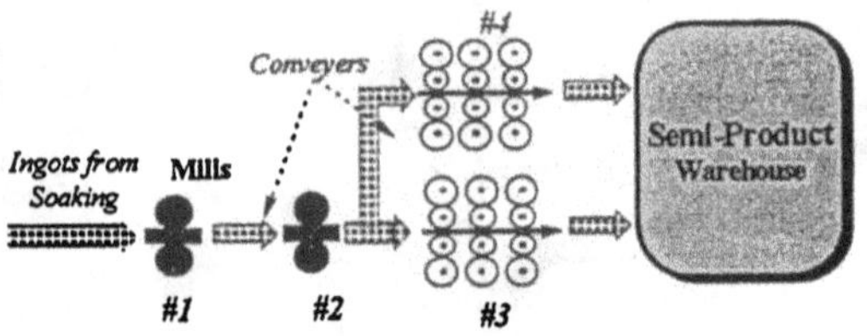

(c) Schematics of the rolling sub-process

Fig. 4 Layout of the hot steel mill complex

4. MODEL DEVELOPMENT OF A HOT STEEL MILL COMPLEX

It is of critical importance to clarify the relationships between different production units (machines) as well as to acquire the internal dynamics of the individual units when the connecting graph model paradigm is used. In the hot steel mill complex considered here, the soaking sub-process is the most complicated since it involves significant hybrid behaviour. The other two sub-processes (i.e., steelmaking and rolling) can be modelled easily by discrete events with only a little information loss as far as optimisation and production scheduling are concerned. Hence we will concentrate our effort on the soaking sub-process.

From Fig. 4(b) it is clear that the soaking sub-process has two sources of ingots, the primary input from the steelmaking sub-process and the secondary input from the cold ingot bank. This latter case is relevant only when the supply of hot ingots from the steelmaking is insufficient. In this case ingots in the cold ingot bank used as a supplementary resource via preheat furnaces for soaking.

It is natural to assume that the initial condition for a soaking pit is empty and waiting for charging. To start its production, it requires the existence of hot ingots and a crane for charging. The total number of ingots to be charged into a pit depends on the capacity of the pit and on ingot size. Once a pit is fully charged, the soaking phase begins to heat the ingots. When the ingot temperature satisfies the requirement for rolling, the ingots are ready to be discharged for rolling. Here again, cranes are required for the discharging activity. The charging and discharging times will depend on the location of the pit, the characteristics of the ingots, and the interaction between the cranes. Thus, the overall activity conducted in a soaking pit can be divided into four phases, i.e., empty, charging, soaking, and discharging. The state of a soaking pit can be represented by the number of ingots in the pit, the ingot temperature and the pit temperature.

When the supply of hot ingots from the steelmaking sub-process is insufficient, hot ingots (on a bogie) are taken out from the preheat furnaces and are charged into the soaking pits. Meanwhile, cold ingots are put into the preheat furnaces by a bogie through the cold end to keep the preheat furnaces fully loaded. During the preheat period, the bogie moves along the furnace, controlled by the demand for the additional hot ingot requirement, and when the bogie has reached the exit point the ingots on it are ready to be removed to a soaking pit by a crane.

Based on the above discussion, the connecting graph for the soaking sub-process has been constructed, and it is shown in Fig. 5.

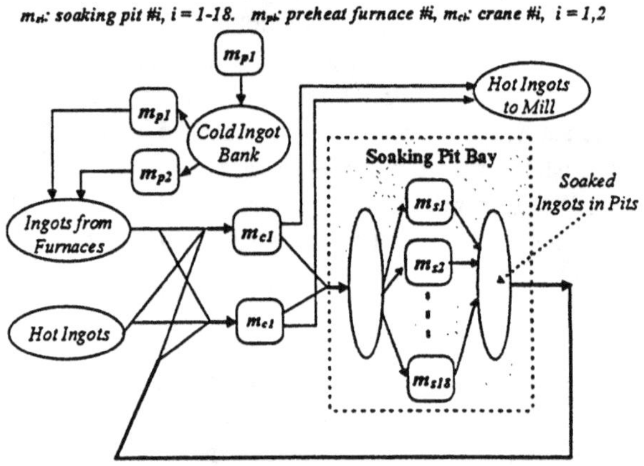

Fig. 5 Connecting graph for the soaking

From Fig. 5 it is clear that all machines involved in the soaking sub-process have only one input port, which means that any process unit requires only one input from its predecessor. However, the practical situation is more complicated. For example, a crane is not permitted to initiate a discharging activity without a request from the roughing mill. It is usually required that a soaking pit should be fully processed (charging or discharging) before another one can start the same procedure. These kinds of constraints can be realised by adding an inhibitor on the relevant machine. A machine can only start its production when its inhibitor is disabled.

The dynamics of a machine depend on the activities carried out by that machine. For example, the preheat furnaces can be treated as pure continuous processes, provided that the computing condition is updated every time when ingots are drawn from the furnaces. The soaking pits, however, should be treated as a general hybrid system which will be described later. The crane dynamics can be described as a discrete event with several states such as busy and wait, and the transition from one state to another will depend on the destination to be reached, the interaction between the cranes, etc.

soaking pit have the same temperature profile. According to Lu and Williams [1983], the temperature distribution within an ingot is assumed to be one-dimensional under a cylindrical co-ordination system. Based on the heat balance principle and basic heat transfer equations, the soaking phase can be modelled by a partial differential equation together with appropriate boundary and initial conditions:

$$
\begin{aligned}
&\frac{\partial T_s(r,t)}{\partial t} = \frac{1}{c\rho r}\frac{\partial}{\partial r}\left(Kr\frac{\partial T_s(r,t)}{\partial r}\right);\\
&\frac{\partial T_s(r,t)}{\partial r}\Big|_{r=R} = q(T_s(R,t),T_p(t))/K\\
&q(T_s,T_p) = f_{gs}[(T_p(t)-T_s(R,t)]+\\
&\qquad e_{wg}\sigma[(T_p(t)+273)^4-(T_s(R,t)+273)^4] \qquad (4)\\
&r<R;\quad t_{0s}\le t\le t_{fs}\\
&T_s(r,t_{0s}) = T_{s0}(r)\\
&t_{fs} = \underset{t}{argmin}(T_s(r,t)\in T_{sf}(r)\pm\sigma T_{sf})
\end{aligned}
$$

where $T_s\ (r,\ t)$ is the ingot temperature; $T_p(t)$ is the pit temperature; K, ρ, and c are the specific conductivity, density and specific heat coefficients of the ingot, respectively; R is the equivalent radius of the ingot; $q(T_s,\ T_p)$ is the heat flow density to the ingot surface; e_{wg} is the equivalent radiative heat transfer coefficient; f_{sg} is the convective heat transfer coefficient; σ is the Stefan-Boltzmann constant; $T_{s0}(r)$ is the initial ingot temperature; $T_{sf}(r)$ is the desired ingot temperature; σT_{sf} is the allowed tolerance of ingot temperature; and t_{0s}, t_{fs} are the time instants when the pit starts and ends its soaking phase, respectively.

The number of ingots in the soaking pit is not changed during the soaking phase, i.e.,

$$N(t)\equiv N(t_{0s}),\quad t_{0s}\le t\le t_{fs} \qquad (5)$$

where $N(t)$ is the number of ingots in the soaking pit at time t.

When a pit is in the discharging phase, the ingot temperature is kept at its desired level although a small fluctuation is inevitable. If there is a request from the roughing mill, an ingot is removed from the soaking pit by a crane and put onto the track conveyer which transfers the ingot to the mill for rolling. This discharging process is repeated until all the ingots in soaking pit have been fully discharged. The phase dynamics can thus be defined by:

$$
\begin{aligned}
&T_s(r,t) = T(r,t_{0d}) = T(r,t_{fs})\\
&N(t) = N(t_{0d}) - \sum_{\tau=t_{0d}}^{t}\delta_d(\tau),\quad t_{0d}\le t\le t_{fd} \qquad (6)\\
&N(t_{fd}) = 0;\quad t_{0d} = t_{fs}^{+};\quad N(t_{0d}) = N(t_{fs})
\end{aligned}
$$

where t_{f0}, t_{fd} are the time instants when the pit starts and ends its discharging phase, and δ_d represents the discharging activity carried out by one of the cranes.

Following the discharge phase a soaking pit enters its empty phase. It ends when there are ingots available for charging and at least one of the cranes is available. During the empty phase, there is no ingot in the soaking pit, i.e. $N(t)\equiv 0$.

The charging activity of a soaking pit is similar to its discharging activity, except that the ingots are transferred in the opposite direction and that it has two separate resources for the incoming ingots. The charging phase is relatively short compared to the soaking phase and hence the ingot temperature can be treated as constant with its initial temperature inherited from its resource. The corresponding phase dynamics can thus be expressed as:

$$
\begin{aligned}
&T_s(r,t) = T(r,t_{0c})\\
&N(t) = \sum_{\tau=t_{0c}}^{t}\delta_c(\tau),\quad t_{0c}\le t\le t_{fc} \qquad (7)\\
&N(t_{fc}) = N_{pit};\quad t_{0s} = t_{fc}
\end{aligned}
$$

where t_{0c}, t_{cf} are the time instants when the pit starts and ends the charging phase, δ_c is the charging activity carried out by one of the cranes, and N_{pit} is the capacity (number of ingots when fully charged) of the soaking pit.

According to the above discussion, the phase dynamics of a soaking pit can be represented by a hybrid automaton as shown in Fig. 6.

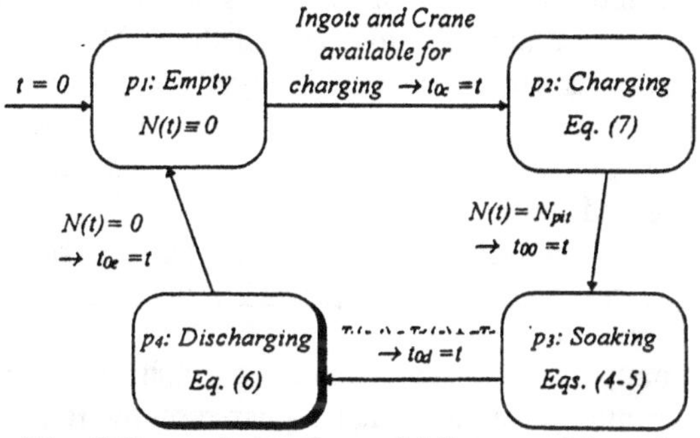

Fig. 6 Phase dynamic model for a soaking pit

Similar procedures can be developed to model other units involved in the hot steel mill complex. For example, a crane can be modelled as a three phase dynamic, i.e., waiting, charging and discharging. Here we treat the crane dynamics as a time-delayed process of transportation by ignoring its continuous dynamics. This leads to an ordinary automaton model as shown in Fig. 7. The connecting graphs for the steelmaking and the rolling sub-processes are shown in Fig. 8 - 9. Due to the imposed space limitation of the paper a detailed description will not be given here.

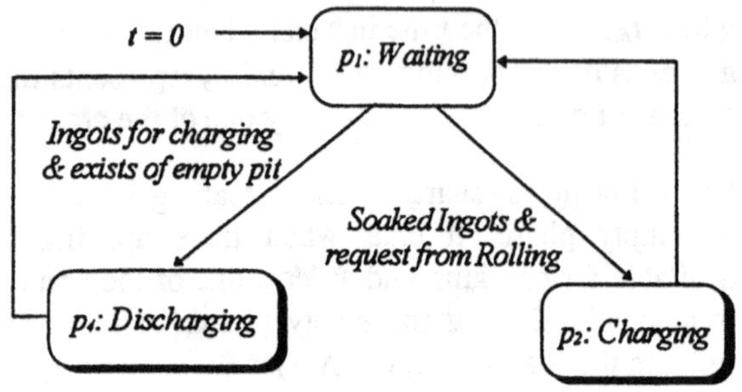

Fig. 7. Automaton model for a crane

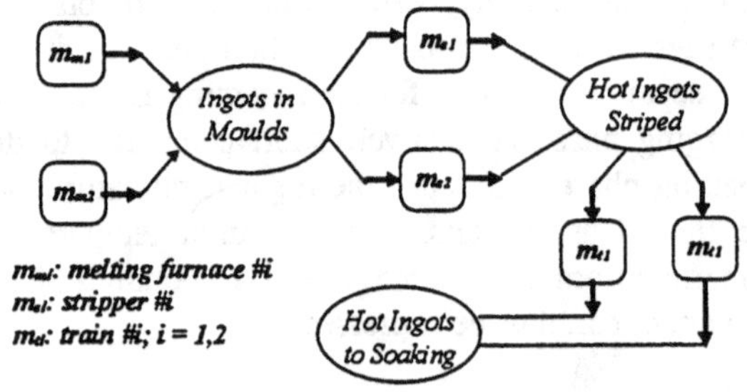

Fig. 8. Connecting graph for steelmaking

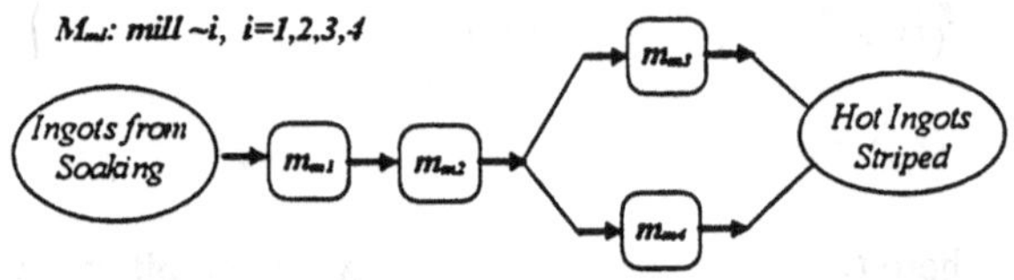

Fig. 9. Connecting graph for rolling

5. DISCUSSION AND CONCLUSIONS

In this paper a hybrid model for a hot steel mill complex has been established based on the connecting graph model paradigm. Principles and procedures for the model development are illustrated via the soaking sub-process modelling. Models of other parts can be obtained straightforwardly using these procedures. One advantage of using a connecting graph model paradigm is that discrete event and continuous variables can be modelled in a unified way that allows full consideration of the interactions between these two different parts.

The application of the connecting graph model can be twofold: theoretical analysis based on the reduced graph and simulation-based analysis. The model can be used to find out critical aspects (bottleneck) of the whole process and to identify the most suitable operating pattern that provides useful design guidance to improve the hot steel mill operation. Other benefits to be gained are model-based optimisation, model-based production scheduling, and on-line simulation. These will be the subject of our future research.

6. ACKNOWLEDGEMENT

The assistance of British Steel Stocksbridge Works and Swinden Technology Centre in generating the case study is gratefully acknowledged.

REFERENCES

Alur, R., Courcoubetis, C., Henzinger, T. A., and Ho, P. H. (1993), Hybrid automata: An algorithmic approach to the specification and verification of hybrid systems, ***Hybrid Systems, Lecture Notes in Computer Science 736***, Springer-Verlag, 209-229.

Ashour, S., and Bindingnavle, S., G. (1972), An optimal design of a soaking-pit rolling-mill system, ***Simulation***, June, 207-214.

Linkens, D. A., and Yang, Y. Y. (1995), Scheduling and optimisation for the heating of steel soaking pits: A case study for hybrid systems, ***IEE Proceedings A: Science, Measurement, and Technology***, **142(5)**, 362-370.

Linkens, D. A., and Yang, Y. Y. (1996), A novel modelling approach for multi-machine manufacturing systems with mixed-mode behaviour, *IEE Colloquium on Hybrid Control for Real-time Systems*, **Digest No 96/256**, 6 Dec. 1996, London, 7/1-7/8.

Lu, Y. Z., and Williams, T. J. (1983), Energy savings and productivity increases with computers -- A case study of the steel ingot handling process, ***Computers in Industry***, **4(1)**, 1-18.

Lumelskey, V. J. (1983), Estimation and prediction of unmeasurable variables in the steel mill soaking pit control systems, ***IEEE Trans. on Automatic Control***, **AC-28(3)**, 388-400.

Patel, C., Ray, W. H., and Szekely, J. (1976), Computer simulation and optimal scheduling of a soaking pit-slabbing mill system, ***Metallurgical Transactions***, **7B**, 119-130.

Rao, T. R. S., Upton, E. A., Rupar, D. L., and Ellis, R. (1984), Optimisation of ingot heating in soaking pits, ***Iron and Steel Engineer***, **61(10)**, 34-38.

Yang, Y. Y., Linkens, D. A., and Banks, S. P. (1994), An unified approach for modelling of hybrid systems, ***Proceedings of the European Simulation Multiconference***, Barcelona, Spain, 1-3 June, 240-244.

Yang, Y. Y., Linkens, D. A., and Mort, N. (1995), Modelling and simulation of a soaking pit/rolling mill process based on extended coloured Petri nets, ***Control Engineering Practice (an IFAC Journal)***, **3(10)**, 1359-1371.

THE USE OF A GRAFCET MODEL FOR SEQUENTIAL SUPERVISORY CONTROL OF A MULTIPASS EXPERIMENTAL ROLLING MILL

D.A. Linkens and Muhammad Ibrahim

Department of Automatic Control and Systems Engineering
University of Sheffield, Mappin Street, Sheffield, S1 3JD, U.K
D.A.Linkens@sheffield.ac.uk cop95mi@sheffield.ac.uk

Abstract: The main purpose of this paper is to present a Grafcet Model for a smale-scale rolling mill. A real-time control strategy is designed and implemented. The rolling mill with process variables and datalogging system are described and the . Grafcet model is overviewed. The implementation of the model on the rolling mill is described in this paper. *Copyright © 1998 IFAC*

Keywords: Sequential control, real-time systems, steel, Grafcet

1. INTRODUCTION

This paper describes a **Grafcet Model,** designed and implemented on a real-time small-scale industrial multipass research/experimental **rolling mill.** Systems which permanently interact with their environment through external information and set specific desired outputs as required are commonly known as **Real-Time Control Systems,** and are widely employed in industry.

The specific laboratory-scale rolling mill can be used for hot and cold rolling. The mill can withstand a maximum rolling load of 50 tonnes. In the **multipass** sequence of rolling, the roll gap and rolling speed can be adjusted as desired by the operator. An induction motor is used for rolling, which gives a variable speed via gearing to the work rolls. The desired roll-gap can be adjusted through the upper roll-bearing by another induction motor.

A typical control system must function at a number of levels. **Supervisory Sequential Control** is concerned with producing the sequence of operations which a system should perform and ensures that the overall objective of the control is achieved. A modern real-time control system might typically involve algorithms for control, simulation, optimisation, filtering and identification and in addition simpler tasks such as event logging and data checking. The more complex the algorithm, the more difficult the problem of performing necessary calculations in the real time. Grafcet is widely applied in different industries for the design and implementation of **automatic systems** in existing and/or future industries e.g. automotive, machining, manufacturing, chemical, electronics/precision equipment. It is also being used for intelligent supervisory applications (Arzenl, 1994). It is being applied to automate several processes eg. refining, welding, rolling, assembly, material transfer, load/unload, production and spray coating. **Programmable Logic controller (PLC)** applications can be seen in cement industry, electrical switch gear,(Mizuno, 1996), transformer tapchanging (Bassett, 1993), construction industry and active gas handling plant process control (Konstantellos, 1992). We present a useful low-cost control method to automate a **steel rolling plant** with increased system reliability and safety. This paper presents a physical description and statement of the control problem i.e. the automation requirement of the plant to be controlled, and explains a Grafcet implementation of a logical controller for the system. This paper gives clear modelling of inputs and outputs with their sequential relations for the rolling process.

2. THE ROLLING PROCESS AND THE MILL

Two types of roll processes, **continuous** and **reversible,** may be employed. The continuous process involves two successive rolls of **sample** or **specimen** metal (slab) in the same direction i.e. after the first pass the sample is returned to the operator, and a new smaller roll gap is set and sample is pushed into the roll for the second pass. In a reversible process, the roll direction is altered between passes i.e. after the first pass the direction of the roll is changed, the gap of the roll is reduced, rolling material is pushed from the second side of the mill and cycle is repeated until the desired number of passes are completed. This paper presents multiple

sequence selection between continuous and reversible rolling.

A physical layout of the rolling mill is shown in the fig. 1.

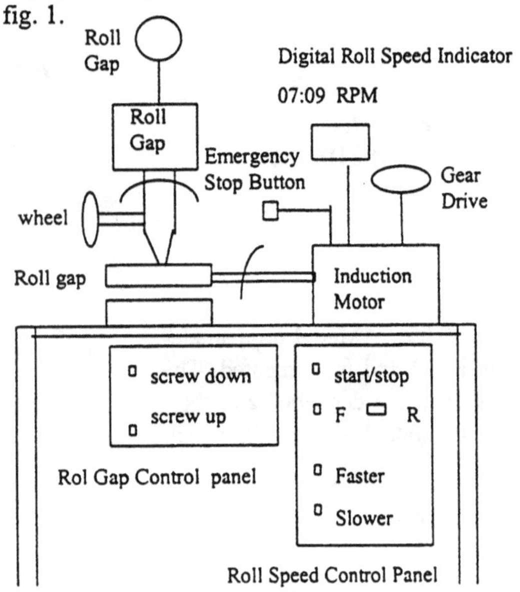

Fig. 1: A physical layout of Rolling Mill

2.1. Roll drive

The drive for the two main rolls is supplied by a three phase, 20 horse power (14.91KW) induction motor through gearing to the work roll whose speed is variable between 2 to 63 rpm, which gives a speed range of 14mm to 0.5 m per second for the rolled specimens.

The output torque from the worm gear reduction is 3467 Nm. The maximum rolling load is 50 tonnes (=500KN). Two roll directions (forward and backward) are controlled by two switches. The desired speed can be achieved by two more buttons "faster" or "slower". The running speed is displayed on a meter.

2.2 Roll gap

The position of the lower roll is fixed, while the upper roll bearing is driven by another induction motor . The required roll gap is adjusted through worm gearing at the rate of 50.8 mm per min. Reading of the roll gap is shown on a circular indicator calibrated in 0.0001 of an in and in mm. The roll gap is increased through " screw down" and screw up " buttons.

2.3. Temperature

A furnace is used to heat the metal. The temperature of the rolled sample is measured directly using a 1.5 mm "Pyrotenax" metal- sheathed mineral insulated thermocouple , which is embedded at different depths in the metal. These thermocouples are able to withstand considerable amounts of deformation. The thermocouple output is fed directly to datalogging system and it is displayed on a monitor.

2.4. The datalogging system

The datalogging system consists of a P.C. using **WORKBENCH** software and an eight channel **Analogue to Digital Converter (ADC)** interface. This measures i. rolling speed, ii. metal temperature, iii. rolling load (both cells), iv. rolling torque (both gauges . Two more channels are available for up to two more thermocouples. The temperature of the metal is logged before the rolling and is continuously measured during and after the rolling. Usually in this rolling mill hot rolling is carried out using one thermocouple but some times more than one are used. Datalogging is a complicated task, because a slow rate is required before and after the rolling but a significant and very high logging rate is implemented during the rolling. This can take as little as 0.1 second. This is achieved using the fast mode of WORKBENCH with a sampling time of of 1 millisecond. This fast mode is only for 5 seconds, after which datalogging is automatically returned to normal logging rate. Rapid temperature changes occurring as the result of deformation heating during the rolling process are recorded with good accuracy.

2.5. Rolling of metal

The rolling mill allows multi-pass rolling of hot or cold metal, for instance aluminium and its alloys or steel, the later of which, this rolling-mill can only roll hot.

The rolling-mill is used for teaching and research purposes. The more prominent utilisation is in research.

The process of hot rolling is of great importance in the metal industry. Many metals are cast into large ingots or slabs from liquid, and must be rolled at high temperature to the correct shape. e.g. plate or strip. A detailed understanding of the behaviour of the metal in this process is desirable for two major reasons:

1. The geometry, temperature and structure at each stage of the process determine the loads and torque required to roll the metal. Prediction of this behaviour is vital for the design of rolling mills.

For instance, an ingot of finished steel from the melting furnace has usually a coarse and relatively fragile internal structure. Consequently, even at high temperature (of the order of 1200 to 1300°C) it is usually well to deform it lightly and slowly at first, so that a new, recrystallized structure is formed, which may then be rolled more vigorously.

2. The microstructural state of the metal at the end of the rolling process largely determines the final product properties. Knowledge of factors that control microstructures at each stage in the rolling process is important for the development of optimum final structures.

At lower finishing temperatures, the process may actually pass out of the **hot working range** and go into the **cold working range**. An arbitrary definition has been set up, according to which all working which is cold enough to exclude spontaneous recrystalisation is defined as cold working. In the case of a grade of steel for which the dividing line would be in the range 600 to 700°C, all rolling would be called 'hot rolling' at temperatures hotter than this range and 'cold rolling' below. However, in common mill conversation, the term 'cold rolling' is generally taken to mean rolling at or near room temperature, and rolling carried out at slightly higher temperatures, say in the range 150 to 500°C, is often referred to as warm rolling.

3. MODES OF OPERATION

Two modes of operation (**Automatic mode** and **Manual mode**) can be selected for the rolling process, as desired by the operator.

The use of the rolling-mill is a rather hectic operation. At high temperatures the metal cools down very quickly, hence the need to perform the operations in little time. In addition, the rolling itself, considering the usual length of the samples and the speed of rotation of the rolls, has a very short duration.

When operating the mill in the manual mode, three persons are needed. One monitors the temperature and time of the processing, so as to indicate to the two others when to proceed with the successive passes. This person also has to switch the datalogger to fast mode when the metal is engaged between the rolls. The two other operators manipulate the metal and the one on the right side of the mill also has to make the roll-gap adjustment between rolling as well as the speed adjustment when there is a change of direction.

The use of the rolling-mill in manual mode requires numerous adjustments or operations to be carried out at the same time, and operations and the sequence of rolling passes occurs at a fast pace. The control sequence and adjustment of the desired parameters described in this paper is automatic , more precise and occurs at faster rate.

4. AN OVERVIEW OF GRAFCET

Grafcet is a modular **graphical tool**, based on sound mathematics and Boolean logic.(David, 1992). It provides control for Real-Time Systems. It is also used to minimise the complexity of large programmes and formalises the **state-machine-structure** to control the sequence of the control signals (Zenger, 1993) i.e. presents specifications for logic controllers. Grafcet provides a compact automation programming structure for real-time discrete systems and real- time hybrid systems. Different control actions can be incorporated in a Grefcet model: **sequential, parallel, concurrent, synchronous** and Boolean with a top-down structural approach, resulting in minimised errors in an overall control system.

An elementary Grafcet is shown in the fig.2.

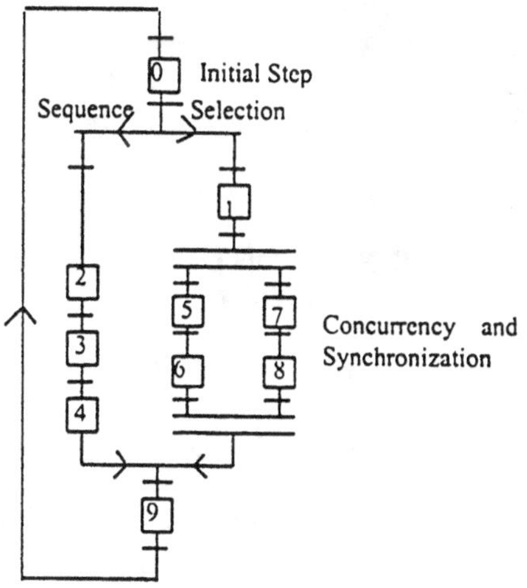

Fig. 2: A Generic Grafcet diagram

The Grafcet model developed in this paper describes the behaviour of the control part of the automated system. It is used for a description of the sequences in the rolling mill.

The Grafcet paradigm is used here to develop a **supervisory sequential strategy**. A Grafcet is a state transition model, based on the notions of the alternate steps and transitions (David and Alla, 1992). Steps model the states of a sequence, while transitions model the logical conditions governing the passage from one state to another. A step is (optionally) associated with "Actions" and transitions with "Conditions". The graphical model consists of alternate rectangles, horizontal bars and links.

Two features of the Grafcet formalism are of fundamental importance:-

i) A **Transition** is associated with a transition **condition.** this is a logic expression defining the condition which governs the evolution of the Grafcet from the transition to the step(s) preceded by the transition. Some transition conditions depend on the continuous variables attaining specified threshold values. Such transition conditions express interactions of a continuous- discrete nature.

ii) **Steps** are associated with **actions,** some of which define operations performed on the continuous part of a process. Such operations might be the priming of a controller set point for the opening of a value, or the switching-off of another. Actions of this kind express interactions of a discrete-continuous nature. Actions and transition conditions thus provide a basis for defining how a discrete model interacts with the complete real-time system.

A **Link** in a Grafcet is a **line** drawn from a step to a transition or a transition to a step, but the sequence of operation is from a upper step to a lower step through the transition between them.

The flow of control tokens from step to step through transitions is called **"EVOLUTION OF A GRAFCET".**

5. IMPLEMENTATION OF THE GRAFCET MODEL

Critical Process Variables are measured by a set of sensors. The temperature of the metal is measured by a thermocouple (T), the gap between the roll is measured by a Variable Linear Displacement Transducer (**VLDT**), roll-separating force is measured from two load cells (**LC**) inserted between the screw down mechanism chocks of the top-roll. A speed sensor (S) measures the roll speed. The four process variables are available during "roll".

The configuration of the control system hardware is shown in fig. 3(a).

We have developed a laboratory full-scale industrial application of Grafcet for a supervisory, discrete-event control system. The control problem is: given a set of requirements and specifications, the plant and environment, and the continuous output from the sensors/transducers, the Grafcet should generate a control strategy such that the behaviour of the rolling mill satisfies the specifications

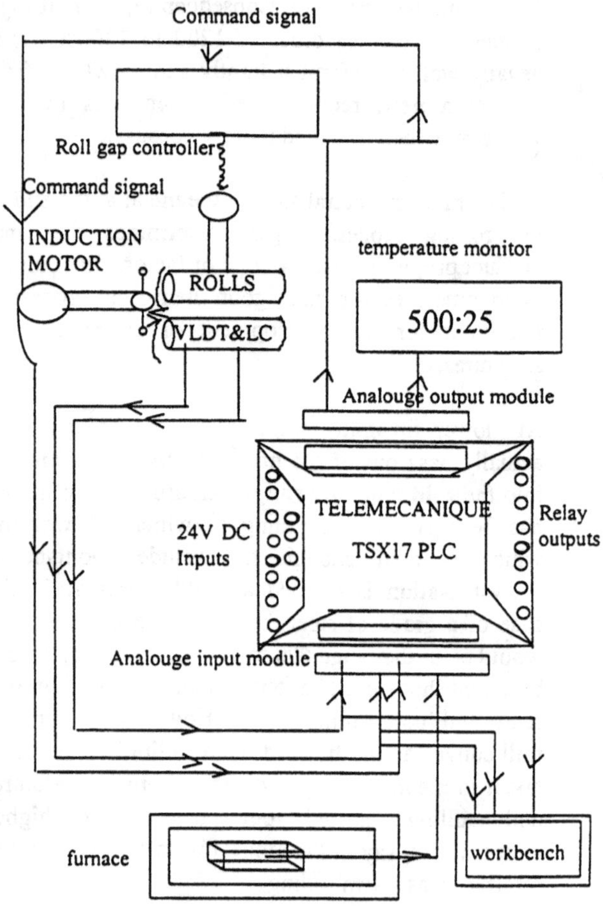

Fig. 3(a) Control system hardware

To start the rolling process, metal is kept in the oven for heating, and the heating may depend on temperature or time. The delay between removal of the metal from the oven and the first rolling is more likely to be temperature-dependent i.e. metal has reached the desired temperature, or it may be time dependent. After execution of the first pass by the Grafcet model, the second pass can be reversible or continuous rolling. The delay between two passes may depend on temperature or time. The Grafcet model permits control option of reversible or continuous rolling for all passes but delay between all passes may only be one i.e. either temperature or time.

The control strategy developed by the Grafcet model in a simplified form is shown in fig. 3(b). The four process parameters mentioned earlier, are fed to a Programmable Logic Controller (PLC), via an **Analogue Input Module (AIM).** The AIM used for the rolling mill consists of 4 channels of input voltage signals of ± 10V. An **Analogue Output Module (AOM)** of 2 channels of output voltage signal of $\pm$ 10V is used for output signals. The temperature of the slab is measured by the is fed to the PLC via one input channel and to

WORKBENCH, which is also a measure of closeness of the

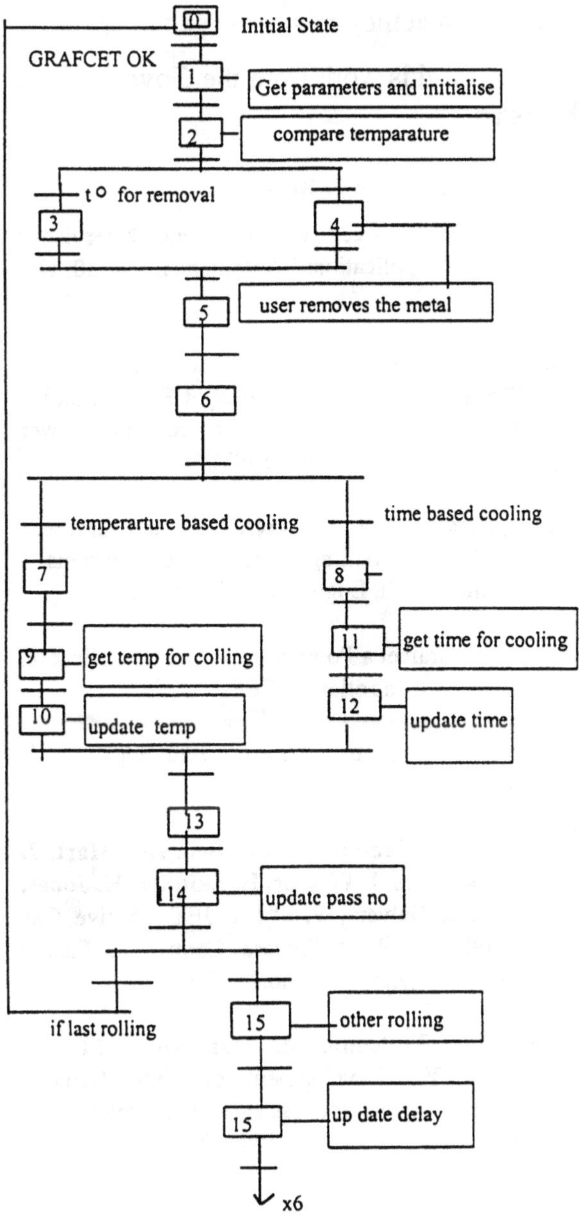

Fig. 3(b)i, ii,iii

thermocouple and is fed to PLC via one input channel and to WORKBENCH, which is also a

temperature to the threshold at which rolling occurs. The Graccet model, when executed, takes the temperature of the sample, and compares the desired temperature with the actual temperature . The desired speed and the roll gap between rolls is obtained by Analogue Input and Analogue Output Modules through the speed (s) and gap (**VLDT**) sensors. When the actual temperature is 10 degrees less then the desired temperature of the rolling dataloggging is switched to the fast mode and a beep is given at the

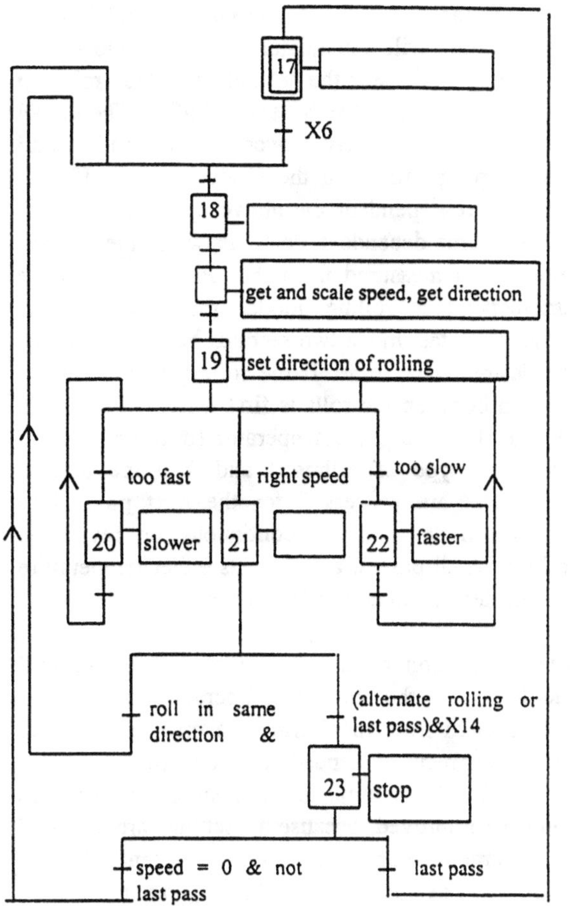

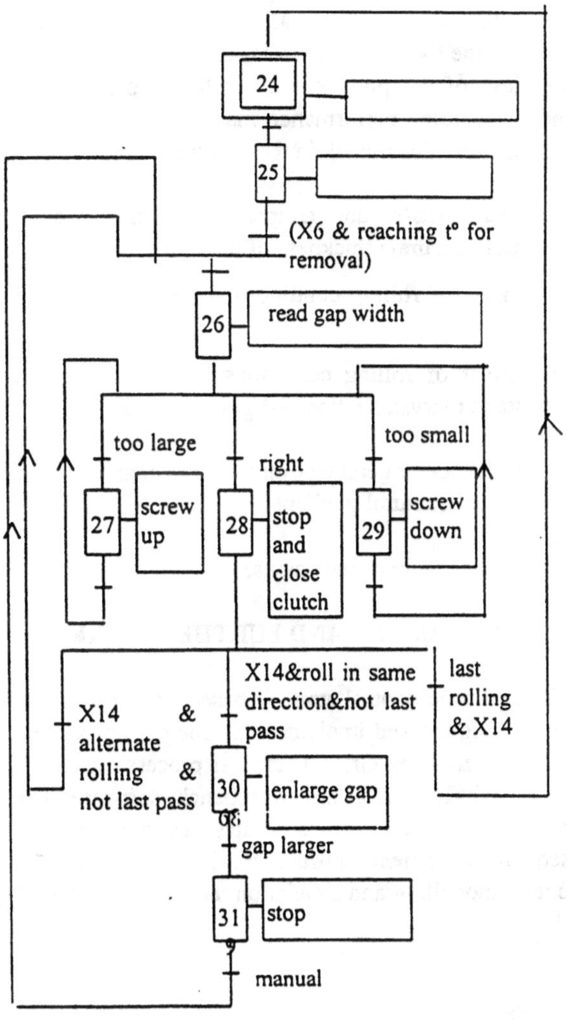

same time Both operations are carried out by the same channel of the Analogue Output Module.Now the operators will immediately push the metal into the rolling mill. For the second pass, the roll gap and speed are immediately set by Grafcet. The model checks the desired delay between the first and second pass as programmed in the Grafcet model. If it is temperature-dependent the above cycle is repeated. For the time-dependent delay between the passes, real-time is measured from the PLC clock. In the time-dependent rolling mode, the datalogger is switched to fast mode two seconds before the end of the desired delay-time. If the rolling is continuous, the gap between the rolls is first increased to return the metal to the process operator (operators should know the type of rolling) and then the desired parameters are processed for the next pass. This Grafcet model controls continuous or reversible rolling for all passes, and permits either temperature or time dependent delays for all passes.

The automation of the experimental rolling mill should reduce the number of operators to two, and relieve them from monitoring the time, temperature and adjustment of speed, gap direction and data logger. The lower limit to the delay between passes could be improved, because all settings are adjusted at the same time and as soon as the current rolling is finished. The only **ON-LINE** operation process operators would have to perform, would be the manipulation of the metal. Therefore we have reduced the likeliness of errors, accident, loss of time and loss of sample piece of metal. Improved and more accurate performance may be obtained from the metal by this model in the following areas:

a) Heat transfer and temperature profile through more accurate thickness of the slab.

b) Effect of rolling conditions on rolling load and torque.

c) Effect of rolling conditions on the process of static recrystalization and grain growth.

d) Influence of geometry on the development of the microstructural gradient.

e) Softening between passes.

CONCLUSION AND FURTHER WORK

The real-time control strategy based on Grafcet has been designed and implemented. The control system strategy involves Grafcet, critical process variables, **PLC ,AMI, AOM** and workbench software The model designed incorporates continuous and sequential elements. Further work is being carried out on modelling and simulation incorporating the **G2 real-time expert system** (Linkens) for the entire process of rolling mill.

ACKNOWLEDGEMENT

The second author acknowledges financial support for this work from the Goverment of Pakistan.

REFERENCES

Arzen, K.E. "Grafcet for Intelligent Supervisory Control Applications" Automatica, vol. 30, No. 10 (1994).

Basset, D.L. "Control of Tape Change Under Load Transformer through the use of Programmable Logic Controller" IEEE Transsection on power Delivery, vol.8 No. 9, (1993).

David, R. and Alla, H. "Petri Nets and Grafcets: Tools for Modelling Discrete Event Systems ", Prentice Hall, Engle wood Cliffs, (1992).

David, R. "Grafcet a Powerful Tool for Specification of Logic Controllers", IEEE

Transaction on Control System Technology, vol.3, no. 3, (1995).

Konstantel, S. Hemmerich, JL. Bell, AC. Mart, J. Yorkshades, L Walker, K. Skinner, K. Jones, G. and Delvert, F. "The JET Active Gas Handling Plant Process Control ", Fusion Technology, vol. 21, no.2, (1992)

Mizuno, K. Tsutsumi, M. Matsumurs, M and Yagi, Y. "Development of High Reliable Control System for Gas Insulated Switchgear", IEEE transaction on Power Delivery , vol. 11, No. 1, (1996).

Linkens. D.A. and Tanyi, E.. "A G2 based hybrid Modelling and Simulation Strategy and its application to a rolling mill", Control Engineering Practice (submitted).

Zenger, " Computer Aided design of Sequential Control in Industrial Process", Third International Workshop on Computer Aided Systems Theory Proceedings, EUROCATS 93, p440-9, (1993).

A 14-CPU DISTRIBUTED AUTOMATIC GAUGE CONTROL SYSTEM FOR A HOT STRIP MILL

Ho-Hoon Lee*

* *Process Automation Research Team*
Research Institute of Industrial Science and Technology
Pohang P.O. Box 135, Seoul 790-600, Korea
email: hhlee@risnet.rist.re.kr

Abstract: This paper describes a 14-CPU distributed real-time automatic gauge control (AGC) system for POSCO's No. 2 Hot Strip Mill at Pohang works. The AGC system uses the well-established gaugemeter AGC and monitor AGC. The AGC system also compensates for process disturbances to prevent performance deterioration. The AGC computer system has been integrated by using VMEbus computer systems along with a commercial real-time operating system. All the application programs and signal input/output are reasonably distributed over the AGC computer system to maximize reliability and effectiveness. The new AGC system is successfully being used in the No. 2 Hot Strip Mill. *Copyright © 1998 IFAC*

Keywords: automatic gauge control, distributed computer control systems, real-time computer systems, real-time operating systems.

1. INTRODUCTION

A hot strip mill is composed of a roughing mill and a multi-stand finishing mill. The thickness control of the hot strip is mainly accomplished in the stands of the finishing mill. Each stand has two work rolls and two large backup rolls inside a mill housing. The work rolls perform the reduction of the strip under the support of the backup rolls that are pushed down by electrical screws or hydraulic cylinders. Loopers are installed between the stands to provide the strip with interstand tension and to measure the variations of the mass flow rate of the strip.

The control of the rolling process includes automatic gauge control, roll speed control, and looper tension control which are closely coupled. The roll gaps and roll speeds are set before rolling based on a nonlinear setup model. The roll gaps are then dynamically controlled during rolling to achieve the desired thickness of the strip. This dynamic roll gap control is referred to as automatic gauge control (AGC). The setup model predicts the desired roll gap, roll speed, roll separating force, and looper tension of each stand for a given temperature of a steel slab. The prediction errors of the setup model are used to update the parameters of the model for the next strips. The AGC is usually performed independently of the control of the roll speed and looper tension in the conventional rolling control.

A control computer system for a hot strip mill consists of a supervisory computer and several real-time control computers. The primary purpose of the supervisory computer is to determine settings of roll gap, roll separating force, roll speed, and looper tension of each stand by using a nonlinear setup model before rolling, and to perform logged-data analysis and storage after rolling. The real-time control computers are provided to perform real-time controls such as AGC, roll speed control, and interstand tension regulation.

The purpose of this study is to develop a new AGC system to replace the existing 10-year-old PDP-11 AGC system of POSCO's No. 2 Hot Strip Mill. The new AGC system is to have at least the same functionality as the existing system in the first

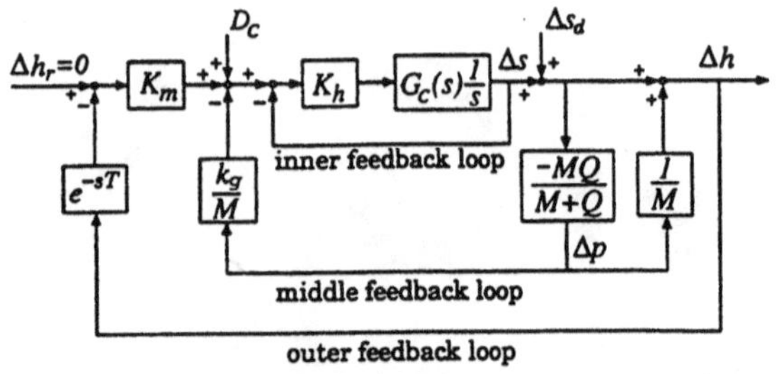

Fig. 1. Block diagram of a single stand AGC system

part of the work and more reliability and functionality in the second part. The new AGC system has adopted the well-established gaugemeter AGC and monitor AGC (Ferguson, *et al.*, 1986). The new AGC system also compensates for process disturbances to prevent performance deterioration.

A new 14-CPU distributed real-time AGC computer system has been developed by using VME-bus computer systems (Peterson, 1992) along with a commercial real-time operating system. The analogue position controllers for the hydraulic cylinders have been also replaced by new digital controllers. All the application programs and signal input/output are reasonably distributed over the AGC computer system to maximize reliability and effectiveness. The new AGC system has completely replaced the existing PDP-11 AGC system.

2. AGC OVERVIEW

Fig. 1 shows the block diagram of a single stand AGC system. In the figure, $G_c(s)$ denotes the transfer function of the servo valve controller; $1/s$ indicates the integrator; e^{-sT} represents the transport delay due to the distance between the stand and the thickness gauge. Δh, Δs, and Δp respectively denote the deviations of the exit thickness, the roll gap, and the roll separating force from the desired values. Δh_r is the reference of Δh. M and Q are the mill modulus and material stiffness, respectively. K_h is the cylinder position controller; K_m is the monitor AGC; k_g is the gain of the gaugemeter AGC. Δs_d and D_c denote the roll gap disturbances and the disturbance compensator, respectively.

2.1 *Cylinder Position Servo Control*

The performance of the AGC is greatly affected by that of the position servo control of the hydraulic cylinder. The position of the hydraulic cylinder is controlled by controlling the flow rate of pressurized fluid into and out of the hydraulic cylinder. The flow rate is controlled by servo valves and their controllers. The plant $G_c(s)/s$ in Fig. 1 has an integrator; hence P, PI, PD, or PID controls can be used for K_h of the inner feedback loop, where P, I, and D are short for "Proportional," "Integration," and "Derivative," respectively

2.2 *Automatic Gauge Control*

The gaugemeter AGC is sometimes called BISRA AGC, named after British Iron and Steel Research Association. The gaugemeter AGC is an indirect method of controlling the strip thickness based on the following gaugemeter equation:

$$\Delta h = \Delta s + \frac{\Delta p}{M} + \Delta s_d. \tag{1}$$

For $\Delta s_d = 0$, the gaugemeter AGC defines a control law $\Delta s_r = -k_g(\Delta p/M)$, where Δs_r is the reference of Δs; $0 < k_g \leq 1$. The gaugemeter AGC constitutes the middle feedback loop in Fig. 1. $\Delta p/M$ is the deviation of mill stretch. In the gaugemeter AGC the roll separating force of each stand is fed back to K_h; thus, the gaugemeter AGC is sometimes called "roll force AGC."

The gaugemeter AGC by nature can not guarantee the desired thickness. This problem can be solved by using the monitor AGC along with the gaugemeter AGC. The monitor AGC is a classical integral control that uses absolute measurements of strip thickness. It is shown as the outer feedback loop in Fig. 1. It guarantees zero steady state thickness error even in the face of low frequency disturbances. Its gain, however, must be limited to avoid the instability due to the transport delay that is proportional to strip speed.

2.3 *Disturbance Compensation*

The gaugemeter AGC assumes that all changes in roll separating force are due to the changes in strip thickness; thus, the gaugemeter AGC magnifies the thickness errors if the changes in the roll separating force are from disturbances. This is because the disturbance Δs_d and the gaugemeter AGC make a positive feedback loop. See the middle feedback loop in Fig. 1. Accordingly, the disturbance should be compensated in order that the gaugemeter AGC may not react to them. The disturbance compensator D_c is designed as

$$D_c = -\Delta s_d. \tag{2}$$

The disturbances are work roll wear, thermal roll expansion, backup roll eccentricity, coolant film thickness, bearing oil film thickness, and oil compression (Ginzburg, 1993).

3. AGC SOFTWARE DESIGN

3.1 *Configuration of No. 2 Hot Strip Mill*

POSCO's No. 2 Hot Strip Mill is a 4-high 7-stand tandem hot-steel-strip rolling mill, which has electrical screws with position sensors in all stands and additional hydraulic cylinders with digital position sensors in stands 4 to 7. All stands have load cells and stand 7 has two thickness gauges. A screwdown master system controls the roll gaps of stands 1 to 3 by using the electrical screws. In stands 4 to 7, the screwdown master system sets the required roll gaps before rolling by using the electrical screws, and then the existing PDP-11 hydraulic AGC system controls the roll gaps during rolling by using the hydraulic cylinders. A separate control system deals with the roll speed control and interstand tension regulation.

3.2 *Control software*

The new AGC system is to control stands 4 to 7 having the hydraulic cylinders. The new AGC system performs cylinder servo position control, gaugemeter AGC, monitor AGC, disturbance compensation, and sequence control, in each stand.

The position servo control is implemented digitally as opposed to the analog circuit of the existing system. The gaugemeter AGC is used in the lock-on mode; that is, the gaugemeter AGC locks onto the head-end thickness of the strip. The disturbances except for roll eccentricity are also compensated. The roll eccentricity is to be taken care of in the second part of this work. The monitor AGC is used to remove the steady state thickness errors of stand 7; the thickness errors of stand 7 are integrated and used as the control inputs to the hydraulic cylinders of stands 4 to 7. The amount of the control input for each stand is determined by a load sharing algorithm.

The sequence control is also implemented. This control monitors various rolling conditions such as control modes, emergency requests, and strip and mill status, and then takes necessary actions.

3.3 *Control gains and sampling period*

The gains of the cylinder position control are selected based on the dynamic specifications of the hydraulic cylinders and servo valve controllers. The gain of the gaugemeter AGC in each stand is limited such that the looper and roll speed controllers can correct any mass flow unbalances due to the thickness variations. The gain of the monitor AGC is determined in consideration of the transport delay in the rolling process.

The sampling periods for the feedback control loops of the AGC system are chosen based on the stability margins, the frequency ranges of disturbances, and the control bandwidths of the feedback loops. The sampling periods are 0.5 msec for the cylinder position control of the inner feedback loop, 10 msec for the gaugemeter AGC and the disturbance compensator of the middle feedback loop, and 50 msec for the monitor AGC of the outer feedback loop.

The sampling period of the sequence control is not critical for system stability and performance if it is reasonably fast, and hence it is set to be the same as that of the monitor AGC (50 msec) for simplicity. The sampling periods are controlled by the interrupt service routines linked to the timer interrupts of CPU boards.

3.4 *Control supporting software*

Data logging processes are designed for collection of mill operating data and subsequent reporting and storage of selected mill operating parameters. Man/machine interface is provided to allow operators to monitor the status of the rolling process and take necessary actions. Data link and communication processes are also provided.

Various sensors including the position sensors of the hydraulic cylinders are calibrated before rolling. The mill modulus having hysteresis is computed for each stand if necessary. Housing modulus and roll bending modulus are the two main components of the mill modulus. The material stiffness Q in each stand is computed on-line using the difference between the entry and exit thickness of the strip in each stand. The thickness of the strip can be indirectly computed based on the gaugemeter equation.

3.5 *Development Environment*

The AGC system uses a commercial real-time operating system for the stand controllers. All the application programs are coded in ANSI C. The real-time control programs including data logging are developed and cross-compiled under the Unix environment, and then they are downloaded to the VMEbus stand controllers. Analysis programs for the logged data are developed and run in a workstation under the Unix environment. The man/machine interface programs are developed and run in VME PC's under the DOS environment. All the application programs and signal input/output are reasonably distributed over the AGC computer system for effectiveness and reliability.

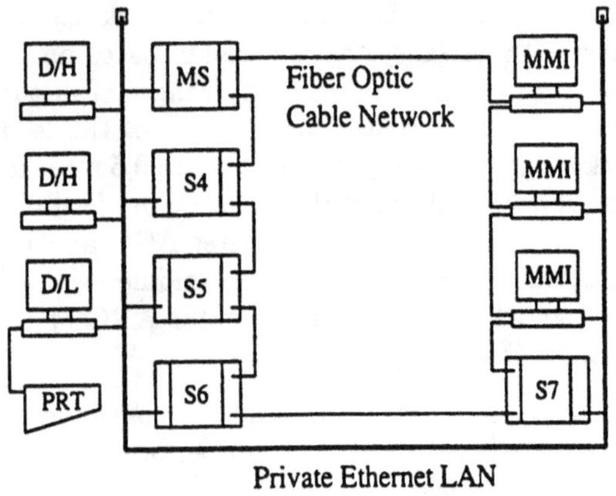

Fig. 2. Configuration of AGC computer system

4. CONTROL SYSTEM INTEGRATION

4.1 *Overall Configuration*

The new AGC computer system has been integrated using industry-standard VMEbus computer systems. The AGC computer system consists of one multi-stand controller and four single stand controllers for stands 4 to 7. The AGC computer system also includes VME PC's and Unix workstations. The Unix workstations are linked to the stand controllers and VME PC's through a private Ethernet LAN.

The configuration of the AGC computer system is shown in Fig. 2. In the figure, MS denotes the VMEbus multi-stand controller. S4, S5, S6, and S7 denote the VMEbus single stand controllers for stands 4, 5, 6, and 7, respectively. D/H and D/L are the Unix workstations used as the development host and the logged-data analysis host, respectively. PRT denotes a printer, and MMI indicates a VME PC for man/machine interface.

Each of the stand controllers and VME PC's has its own intersystem memory board which is linked to the other boards via a fiber optic cable network as shown in Fig. 2. Any changes in the intersystem memory of any board are instantaneously copied to that of the other boards. Thus, the intersystem memory is used as shared memory for data transfer among the stand controllers and VME PC's.

4.2 *Multi-Stand Controller*

The multi-stand controller is a VMEbus system which consists of two MC68040 CPU boards, one intersystem memory board, and several analogue and digital input/output boards. CPU board 1 is used for the control algorithms and input/output signals common to all stands. This board executes the monitor AGC and parts of the sequence control processes such as strip tracking. This board also receives the roll gap data of stands 1 to 3 from the screwdown master system. The push-button signals from the operator control panel are also interpreted in this board.

CPU board 2 takes care of data logging, alarm messages, and data link with the supervisory control computer. A private AGC shell is also provided for command interpretation. Data logging processes run in this board for collection of mill operating data, which are sent to a Unix workstation through a private Ethernet LAN for analysis and storage. The data link with the supervisory control computer is accomplished via RS422 serial links. Both CPU boards can read from and write into the intersystem memory.

4.3 *Single Stand Controllers*

The single stand controllers are VMEbus systems. Each single stand controller has three MC68040 CPU boards, one intersystem memory board, and several analogue and digital input/output boards. The single stand controller takes care of the control algorithms and input/output signals limited to a single stand. CPU board 1 is used as the cylinder position controller, which runs in one of the four modes: control, reset, stop, and gain setting modes. No operating system is used in this board to reduce the control overheads for the memory and task management. The sampling period can be set to be as small as 200 micro seconds in this board.

The main function of CPU board 2 is to execute the conventional lock-on gaugemeter AGC in which the hysteresis and width effects of the mill modulus are compensated. The disturbances such as work roll wear, thermal roll expansion, coolant film thickness, bearing oil film thickness, and oil compression are compensated to prevent performance deterioration. Parts of the sequence control processes run in this board. Various kinds of calibration and initialization processes for a single stand are also implemented in this board. CPU board 3 is reserved for the roll eccentricity control. CPU boards 2 and 3 can read from and write into the intersystem memory and the dual ported memory of CPU board 1.

4.4 *Data Logging and MMI*

Data logging processes run in the CPU board 2 of the multi-stand controller for collection of mill operating data. Five different sampling periods of 0.1, 0.2, 0.3, 1.2, and 5.0 seconds are used for the data logging processes. The logged data are sent to a Unix workstation via the Ethernet LAN under the TCP/IP protocol for subsequent reporting and storage of selected mill operating parameters. The man/machine interface processes run in VME PC's under the DOS environment. The VME PC's are linked to the AGC system through the intersystem memory. The sampling period for the man/machine interface is about 200 msec.

5. GAIN TUNING AND ROLLING RESULTS

Finally, the new AGC system was tuned for No. 2 Hot Strip Mill. As a preliminary step, the signal input/output and the data communication processes were tested. Then the commands from the operator control panel were checked. Memory-mapped displays and logs were developed to monitor and diagnose problems during commissioning.

The cylinder position control was first tuned for each stand. Then the gaugemeter AGC was tuned in each stand. The existing AGC system was partly used as a reference to tune the disturbance compensator. The first rolling test was performed with a thick strip having 8.0 mm exit thickness since thick strips generally cause fewer problems of mass flow unbalance.

Finally, the monitor AGC was tuned. The control gains are the functions of the transport delay in the rolling process. The initial control gains were obtained based on the transport delay and the theoretical bandwidth of the gaugemeter AGC.

The new AGC system is currently being used for No. 2 Hot Strip Mill. Fig. 3 shows the typical rolling results. The first strip has large head-end errors, but the head-end errors of the second strip are substantially reduced because of the parameter update of the setup model based on the rolling results of the first strip. The performance of the new AGC system is quite satisfactory. However, the new AGC system sometimes causes periodic thickness errors due to roll eccentricity and skid marks.

6. CONCLUSION AND FUTURE WORK

In this work, a new 14-CPU distributed real-time AGC system has been developed by using VMEbus computer systems and a commercial real-time operating system. The new AGC system has completely replaced the existing PDP-11 AGC system. The new digital position controllers for the hydraulic cylinders also work better than the existing analogue counterparts. The new system is quite reliable. All these prove the stability and performance of the new AGC system.

As the second part of this work, a roll eccentricity control and a high-speed monitor AGC for head-end errors are now being implemented for better performance. An extra thickness gauge has been installed between stand 4 and stand 5. A feedforward AGC and dynamic resetup of the roll gaps of the downstream stands are also being implemented based on the extra thickness gauge.

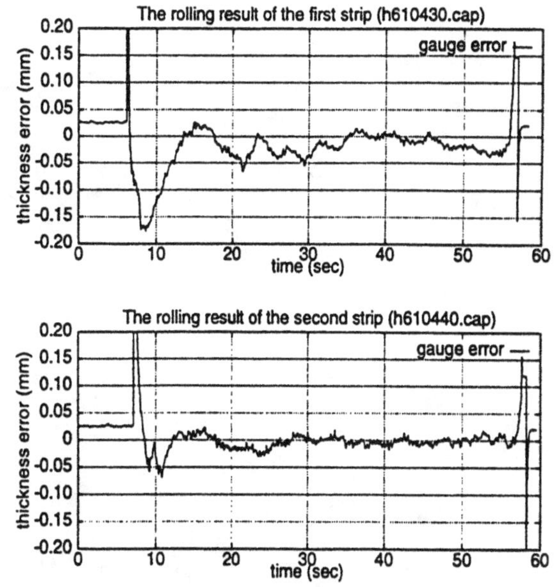

Fig. 3. Rolling results

In future work, the VME PC's for man/machine interface will be replaced by standard industrial PC's having RS422 serial links in order to make the AGC system more reliable and economical. The monitor AGC will be improved by using a smith predictor to overcome the adverse effects of the transport delay within the rolling process. The multivariable control methods such as the H_∞ control are also being considered for integrated control of the strip thickness, roll speed, and interstand tension.

ACKNOWLEDGEMENT

The author is grateful to the members of the measurement and control team of POSLAB and the hot rolling control team of POSCO who participated in this work for their contributions.

7. REFERENCES

Ferguson, I. J., and R. F. Detina (1986). Modern Hot-Strip Mill Thickness Control. *IEEE Trans. on Industry Application*, **Vol. IA-22, No. 5**, pp. 934-940.

Peterson, W. D. (1992). *The VMEbus Handbook Third Edition*, VFEA International Trade Association.

Ginzburg, V. B. (1993). *Gauge Control in Rolling Mills*, Rolling Mill Technology Series, **Vol. 11**, United Engineering, Inc.

A NEW ANTI-SWING CONTROL OF OVERHEAD CRANES

Ho-Hoon Lee *, Sung-Kun Cho,* and Jae-Sung Cho**

* *Process Automation Research Team,*
RIST, Pohang P.O. Box 135, Seoul 790-600, Korea
email: hhlee@risnet.rist.re.kr
** *Measurement and Control Team,*
POSLAB, Pohang P.O. Box 125, Seoul 790-600, Korea
email: gatsby@risnet.rist.re.kr

Abstract: This paper presents a new method for the design of an anti-swing control system for overhead cranes. The velocity servo system of a crane is modeled based on experiments. The model of the velocity servo system is used for the design of the position servo system of the crane via the loop shaping method. The position servo system and the swing dynamics of the crane load are then used to design the anti-swing control system based on the root locus method. In the presence of low frequency disturbances, the new anti-swing control system guarantees both zero steady state errors for position control and fast damping for load swing. Experimental results on a prototype crane show the effectiveness of the new anti-swing control system.

Keywords: anti-swing control, overhead crane, servo systems, root locus method, loop shaping method.

1. INTRODUCTION

Overhead cranes are widely used in factories, steelworks, and dockyards. A major problem with crane operation is that the acceleration of cranes induces unwanted load swing, which frequently causes damages to crane loads and even accidents; thus the crane operation usually requires intensive training. Accordingly, extensive researches have been performed to minimize the load swing in crane operation.

Mita and Kanai (1979) solved an optimal control problem for swing-free velocity profiles of a crane under the constraint of zero swing at the end of acceleration. Their method is an open loop control; thus it can not guarantee the control objective due to model uncertainties and disturbances. Ridout (1987) designed a linear feedback control law based on the root locus method. This method, however, does not guarantee zero steady state error for the position control of a crane. The load swing of a crane was also suppressed by setting the acceleration of the crane proportional to the angular velocity of the load swing (Yoon, *et al.*, 1995; Ohnish, *et al.*, 1981). Yamada et al. (1989) used swing-free velocity profiles and a fuzzy control law to suppress the position error of a crane and the load swing.

This paper proposes a new method for the design of an anti-swing control system, which guarantees not only fast damping for the load swing but also excellent transient responses and zero steady state error for the position control. First, the velocity servo dynamics of a crane is modeled based on experiments. Secondly, the position servo control system of the crane is designed based on the model of the velocity servo dynamics via the loop shaping method. Finally, the anti-swing control system is designed based on the position servo system and the swing dynamics of the crane load via the root locus method. In this study, the control laws and their gains are determined based on the experimental velocity servo dynamics of a crane and hence much effort can be saved in commissioning.

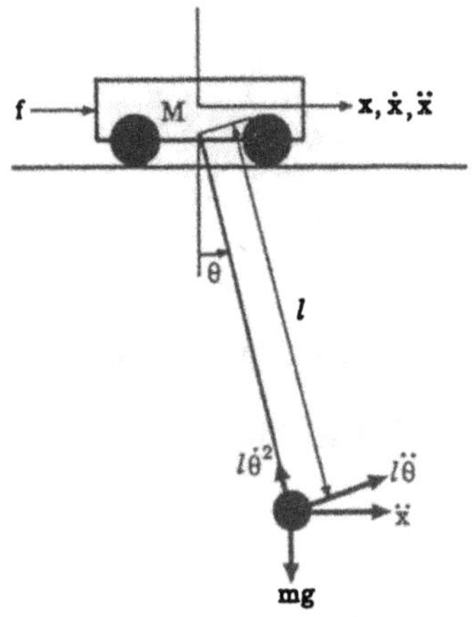

Fig. 1. A plane model of an overhead crane

This paper is organized as follows. In Section 2, the dynamic model of a crane is derived and then the velocity servo dynamics of the crane is obtained via the step response tests. In Section 3, the anti-swing control system is designed and its control performance is discussed. In Section 4, the results of experiments with a prototype overhead crane are presented to show the effectiveness of the anti-swing control system based on the proposed design method. Finally, the conclusion of this study is presented in Section 5.

2. MODELING

2.1 *Dynamic model of an overhead crane*

A plane model of a crane is shown in Fig. 1. In this study, the load is considered as a point mass, and the mass and the stiffness of the rope are neglected. In industry, the rope length is usually kept constant while the cranes are in motion for safety considerations. In this practical case, the equations of motion of the crane and the load can be derived as

$$(M+m)\ddot{x} + D\dot{x} - ml\dot{\theta}^2 \sin\theta + ml\ddot{\theta}\cos\theta = f, \quad (1)$$

$$ml^2\ddot{\theta} + mgl\sin\theta = -ml\ddot{x}\cos\theta, \quad (2)$$

where M and m represent the mass of the crane and the mass of the load, respectively; D is the viscous friction coefficient; θ, x, and l denote the swing angle, crane displacement, and rope length, respectively; f is the force applied to the crane.

In practice, the load swing is kept small for safety while the cranes are in motion. For small motions around the vertical equilibrium (i.e., $\theta \ll 1$), $\sin\theta \approx \theta$, $\cos\theta \approx 1$, and $\dot{\theta}^2 \approx 0$. In addition, most of industrial cranes are driven by high-ratio gear-reduction motors. Then $M \gg m$ holds and hence the $\ddot{\theta}$ term in (1) can be neglected (Ridout, 1987). Accordingly, for the practical cases the equations of motion are simplified as

$$M\ddot{x} + D\dot{x} = f, \quad (3)$$

$$\ddot{\theta} + \frac{g}{l}\theta = -\frac{\ddot{x}}{l}. \quad (4)$$

The Laplace transforms of (3) and (4) yield the following transfer functions:

$$\frac{V(s)}{F(s)} = \frac{1}{Ms+D}, \quad (5)$$

$$G_l(s) \equiv \frac{\Theta(s)}{X(s)} = \frac{-s^2/l}{s^2+g/l}, \quad (6)$$

where s is the Laplace operator; $X(s)$, $V(s)$, $F(s)$, and $\Theta(s)$ are the Laplace transforms of x, v, f, and θ, respectively, with $v = \dot{x}$.

2.2 *Modeling of the velocity servo system*

Most of automatic cranes are powered by AC servo motors with vector controllers. The dynamics of the vector controllers are in general a hundred times faster than those of the cranes. As a result, the dynamics of the vector controllers can be neglected in modeling. That is, the external force $F(s)$ can be written as

$$F(s) = K_m I_m(s) = K_t U_c(s), \quad (7)$$

where U_c is the input to the vector controller and I_m is the motor current; K_m and K_t are the constants that depend on the motor constants and the power train configuration of a crane. Then the equations (5) and (7) result in

$$G_t(s) \equiv \frac{V(s)}{U_c(s)} = \frac{K_t}{Ms+D}. \quad (8)$$

The velocity servo control system $G_v(s)$ is then obtained by setting $U_c(s) = K_v(s)(V_r(s) - V(s))$:

$$G_v(s) \equiv \frac{V(s)}{V_r(s)} = \frac{G_t(s)K_v(s)}{1+G_t(s)K_v(s)}, \quad (9)$$

where $V_r(s)$ represents the velocity command, and $K_v(s)$ is the velocity servo controller for which a PI (Proportional and Derivative) control is usually used.

In this study, the velocity servo dynamics $G_v(s)$ is obtained based on experiments. This will be described with a prototype crane built for this study. The prototype crane is driven by a geared AC servo motor whose angular velocity is controlled by a vector servo controller. Step commands are inputted to the vector servo controller, and then the velocity of the crane is measured as the output. From the input and output relations the velocity servo dynamics $G_v(s)$ can be modeled as

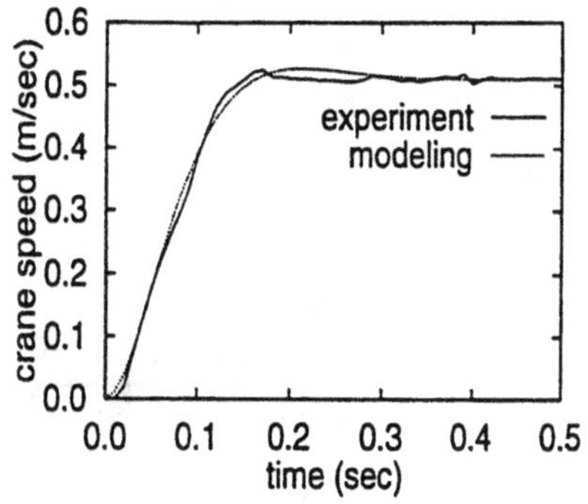

Fig. 2. Modeling of the velocity servo dynamics of the crane

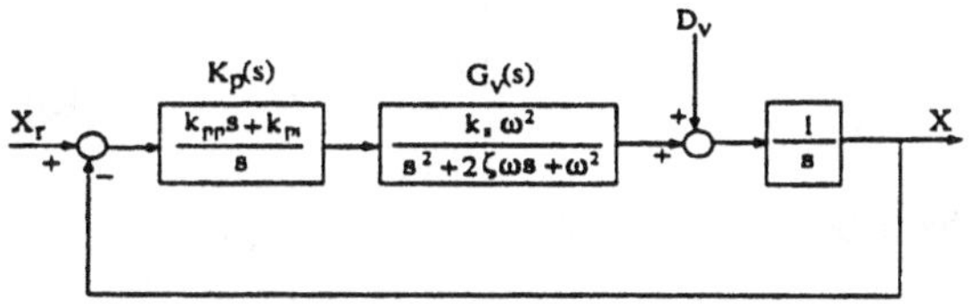

Fig. 3. Block diagram of the position servo system

$$G_v(s) = \frac{0.0567 \cdot 484}{s^2 + 32.6s + 484}. \qquad (10)$$

Fig. 2 shows the step responses: the solid line for the experimental results and the dotted line for the theoretical results from the model (10).

3. CONTROL SYSTEM DESIGN

In this section, a position servo system will be designed, and then an anti-swing control system will be designed based on the position servo system and the swing dynamics $G_l(s)$. The velocity servo dynamics $G_v(s)$ of the prototype crane will be used for the design of the control systems.

3.1 *Design of the position servo system*

The position control loop is shown in Fig. 3. It consists of a position controller $K_p(s)$, the velocity servo dynamics $G_v(s)$, and an integrator $1/s$. An integration control is required to reject the effects of the disturbance D_v on the output X in the low frequency region. The slip of the crane wheels and the offsets of the vector servo controller are the examples of the disturbance.

In this paper, the loop shaping method (Doyle, *et al.*, 1992) is used for the design of the position controller $K_p(s)$. Fig. 4 shows the Bode plot of the open loop transfer function of the position servo system. The magnitude of the open loop transfer function is made sufficiently large in the low frequency region for good command tracking and disturbance rejection. In the high frequency region the open loop transfer function is made sufficiently

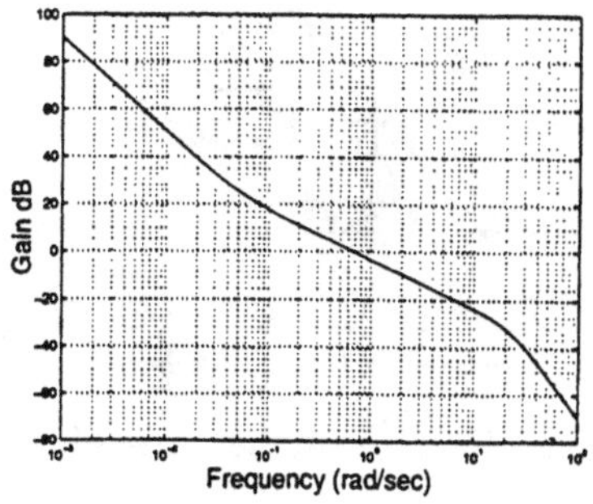

Fig. 4. Bode plot of the open loop transfer function of the position servo system

small for robust stability and sensor noise attenuation. The crossover frequency is selected in consideration of the bandwidth of the velocity servo dynamics $G_v(s)$ and uncertainties in the modeling. Around the crossover frequency the magnitude maintains a gentle slope (-20dB/dec) for a sufficient phase margin. The resulting open loop transfer function $G_{xo}(s) = K_p(s)G_v(s)/s$ is given as

$$G_{xo}(s) = \frac{329.3(s + 0.05)}{s^2(s^2 + 32.6s + 484)}. \qquad (11)$$

From this, the controller $K_p(s)$ can be obtained as

$$K_p(s) = \frac{s \cdot G_{xo}(s)}{G_v(s)} = \frac{k_{pp}(s + k_{pi}/k_{pp})}{s}, \qquad (12)$$

where the control gains k_{pp} and k_{pi} are 12.0 and 0.6, respectively. The position servo system $G_x(s)$ is then given as

$$G_x(s) \equiv \frac{X}{X_r} = \frac{329.3}{(s + 0.054)(s + 0.66)} \times \frac{(s + 0.05)}{(s^2 + 31.8s + 461.3)}. \qquad (13)$$

3.2 *Design of the anti-swing control system*

Fig. 5 shows the block diagram of the anti-swing control system, which consists of the position servo system $G_x(s)$, an angle controller $K_a(s)$, and the load swing dynamics $G_l(s)$. Note that $s/(k_{pp}s + k_{pi})$ of $K_a(s)$ is canceled by $(k_{pp}s+k_{pi})/s$ of $K_p(s)$; thus, the feedback of the swing angle Θ to $G_v(s)$ (i.e., to the vector servo controller) constitutes a simple PD (Proportional and Derivative) control.

The bandwidth of $G_x(s)$ for most industrial cranes is usually smaller than the swing frequency $\sqrt{g/l}$. Therefore the dynamics of $G_x(s)$ should be taken into account when the angle controller $K_a(s)$ is designed.

In this paper, the angle controller $K_a(s)$ is designed via the root locus method. The zeros of

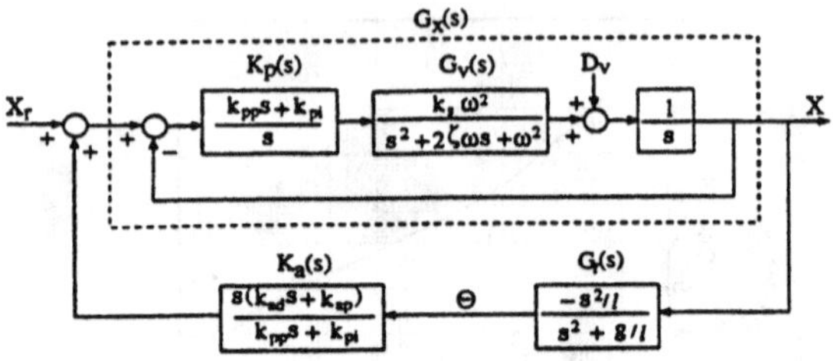

Fig. 5. Block diagram of the anti-swing control system

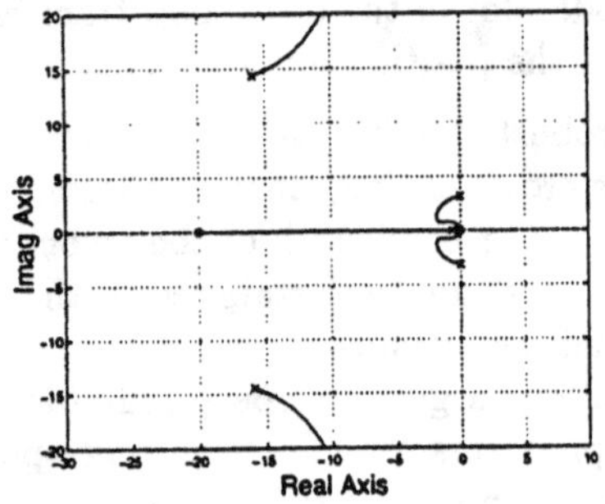

Fig. 6. Root locus of the anti-swing control system

$K_a(s)$ are placed at $s = 0$, $s = -k_{ap}/k_{ad}$, and the pole is placed at $s = -k_{pi}/k_{pp}$ in consideration of the poles and zeros of $G_x(s)$ and $G_l(s)$. The resulting open loop transfer function $G_{\theta o}(s)$ is determined as

$$G_{\theta o}(s) = \frac{-(27.4/l)k_{ad}s^3}{(s^2+9.8/l)(s+0.05)(s+0.66)} \times \frac{(s+k_{ap}/k_{ad})}{(s^2+31.8s+461.3)}. \quad (14)$$

Fig. 6 shows the root locus ($0 \leq k_{ad} < \infty$) for the case of $l = 1.0$ m, where $k_{ap}/k_{ad} = 20.0$. Since two poles are placed on the real axis, the system is always stable regardless of the value of k_{ad}. The system, however, can be unstable for large k_{ad} owing to modeling errors in the high frequency region. The optimal values of k_{ad} having a damping ratio of 0.7 are 3.1 and 3.9 for $l = 1.0$ m and $l = 1.5$ m, respectively.

3.3 *System stability and performance*

The entire control system is stable since it has been stabilized based on the stable position servo system $G_x(s)$. In this section, the control performance of the entire control system is examined by using the transfer functions from each input to each output in Fig. 5. The transfer functions for $l = 1.0$ m can be derived as

$$\frac{X(s)}{X_r(s)} = \frac{27.4(s^2+9.8)(12s+0.6)}{G_c(s)}, \quad (15)$$

$$\frac{\Theta(s)}{X_r(s)} = \frac{-27.4s^2(12s+0.6)}{G_c(s)}, \quad (16)$$

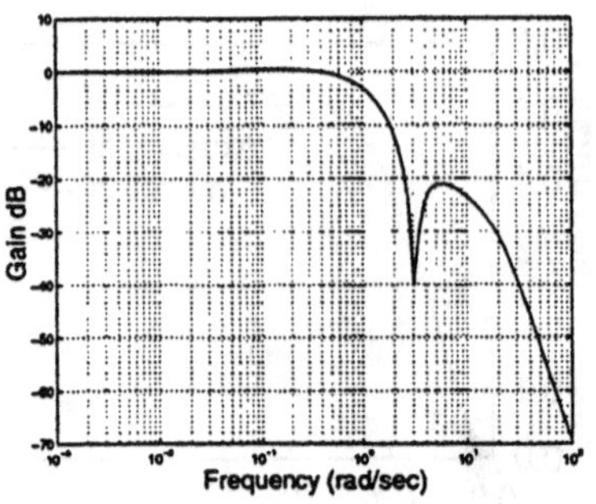

Fig. 7. Bode plot of the closed loop position control system X/X_r

$$\frac{X(s)}{D_v(s)} = \frac{s(s^2+9.8)(s^2+32.6s+484)}{G_c(s)}, \quad (17)$$

$$\frac{\Theta(s)}{D_v(s)} = \frac{-s^3(s^2+32.6s+484)}{G_c(s)}, \quad (18)$$

where $G_c(s)$ is defined as

$$G_c(s) \equiv (s+0.05)(s^2+3.56s+6.33) \times (s+1.07)(s^2+27.9s+437.8), \quad (19)$$

and $G_c(s) = 0$ is the characteristic equation of the closed loop system.

The transfer functions (15) ~ (18) are all stable as expected. Fig. 7 is the Bode plot for X/X_r of (15), which shows excellent command tracking in the low frequency region. The bandwidth of X/X_r is about 1(rad/sec). The notch in X/X_r is due to the swing dynamics of the load. Θ/X_r of (16) shows that the swing angle is zero in the steady state for step or ramp position commands. According to X/D_v of (17) and Θ/D_v of (18), step disturbances have no effect on the crane position and the swing angle in the steady state. Furthermore, the steady state swing angle is not affected by ramp or parabola disturbances.

4. EXPERIMENTAL RESULTS

A prototype crane has been built to evaluate the control performance of the new anti-swing control system. The prototype crane is 5.5 meters long and 2 meters high. The maximum acceleration, speed, and rope length have been chosen such that the dynamics of the prototype crane is similar to that of an industrial overhead crane. The configuration of the prototype crane system is shown in Fig. 8.

The prototype crane is driven by an AC servo motor. If the encoder in the AC servo motor is to be used for measurements of the crane position, the slip of the crane wheels usually causes position errors. In this study, a precision position sensor has been designed. An angle sensor has been also designed and mounted on the bottom of the crane as shown in Fig. 8.

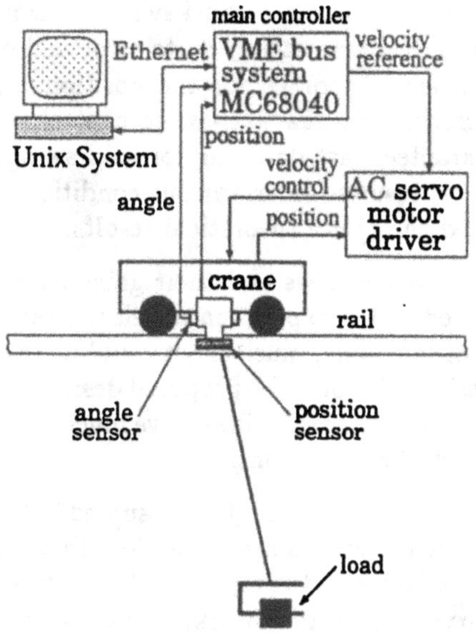

Fig. 8. Configuration of the experimental setup

The controller is composed of VMEbus systems: MC68040 CPU, analogue to digital, digital to analogue, and digital input/output boards. A commercial real-time operating system is used for the VMEbus control system. A Unix workstation is also used as the development host and is connected to VMEbus system through a private Ethernet LAN.

The designed control laws have been implemented through the VMEbus CPU board with 10 msec sampling period. Extensive control experiments have been performed under various conditions. In this paper, the experimental results with $l = 1.0$ m are discussed. The desired trajectories for the crane have been obtained based on the swing-free velocity profiles proposed by Mita and Kanai (1979). Since the control gains have been obtained based on the experimental velocity servo dynamics of the prototype crane, much effort could be saved in commissioning.

Fig. 9 and Fig. 10 show the experimental results for short and long crane displacements, respectively. The control performance is shown to be independent of the length of the crane displacement. The experimental result with $m = 20$ Kg is shown in Fig. 11. The experimental conditions for Fig. 10 and Fig. 11 are the same except for the load mass; the control performance is not affected by the load mass for the cranes powered by high-ratio gear-reduction motors.

The experimental result with about 10° initial swing is shown in Fig. 12. The position control is little affected by the initial swing. In all the experimental results, the steady state position errors are all zero, and the load swing disappears about 2 seconds after the crane reaches the desired positions even with the 10° initial load swing.

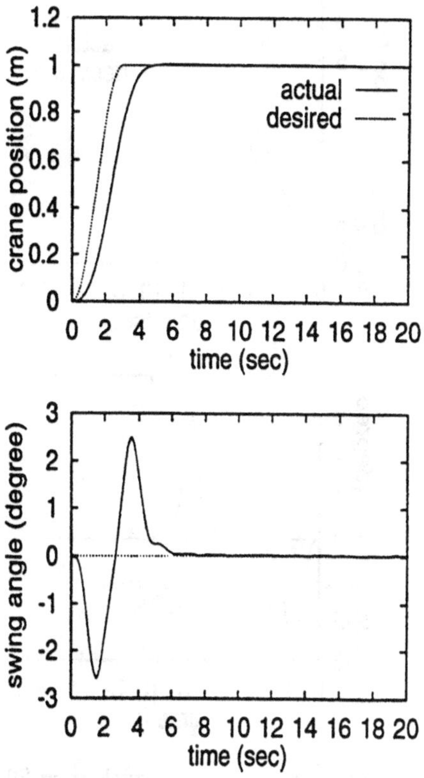

Fig. 9. Experimental results for a short crane displacement

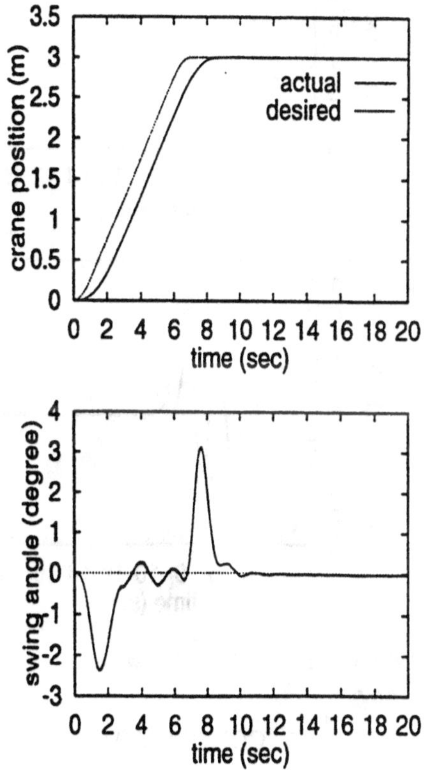

Fig. 10. Experimental results for a long crane displacement

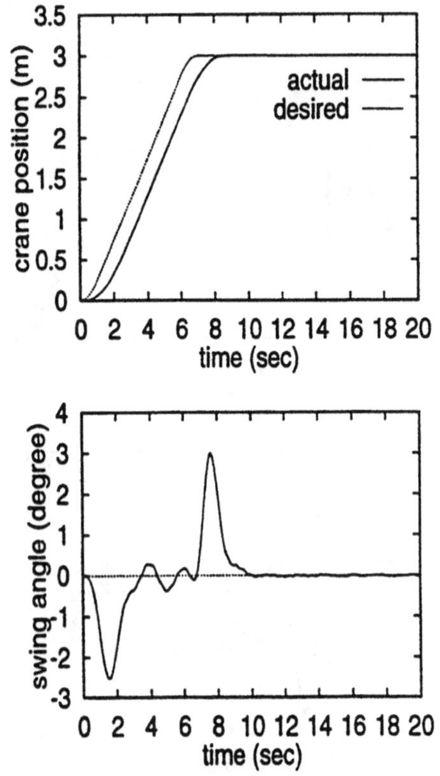

Fig. 11. Experimental results with $m = 20$ Kg

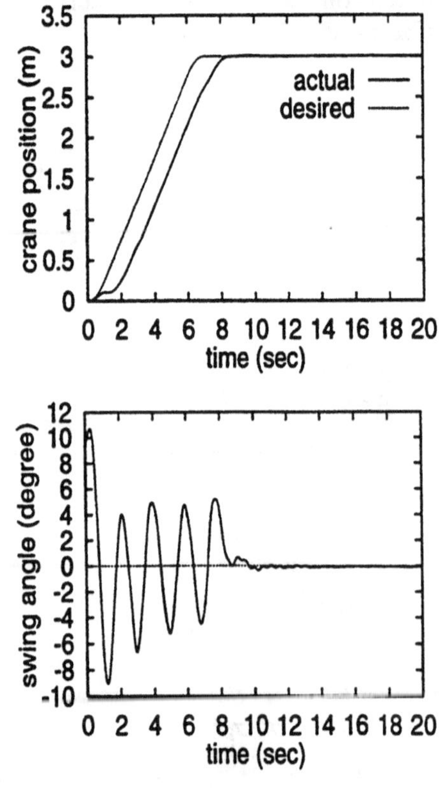

Fig. 12. Experimental results with 10° initial load swing

5. CONCLUSION

In this paper, a new design method has been presented; a new anti-swing control system has been designed based on the experimental velocity servo dynamics of a crane and the swing dynamics of the crane load. The new anti-swing control system guarantees zero steady state error and excellent transient responses for the position control of the crane. The new anti-swing control system also guarantees fast damping for load swing. Experimental results under various conditions have supported the above theoretical results.

Since the control laws and their gains have been determined by the experimental velocity servo dynamics of the crane, much effort could be saved in commissioning. Thus, the proposed design method and the resulting control laws have much potential for industrial applications.

In this study, the rope length is assumed to be constant while the cranes are in motion. In some applications, however, the rope length may be slowly time-varying. In future work, 'a gain scheduling method' will be used for the compensation of the changes in the rope length; an angle gain table for various rope lengths will be derived and will be used afterwards in the real-time control depending on the rope length. The position control is independent of the rope length.

6. REFERENCES

Mita, T. and T. Kanai (1979). Optimal Control of the Crane System Using the Maximum Speed of the Trolley (in Japanese with English abstract). *Trans. Soc. Instrum. Control Engr*, **Vol. 15, No. 6**, pp. 833-838.

Ridout, A. J. (1987). New Feedback Control System for Overhead Cranes. *Electric Energy Conference, Adelaide*, pp. 135-140.

Yamada, S., H. Fujikawa, and Y. Wakasugi (1989). Fuzzy Control of the Roof Control. *IEEE Industrial Electronics Conference Proceedings, Philadelphia*, pp. 709-714.

Ohnish, E., I. Tsuboi, T. Egusa, and M. Uesugi (1981). Automatic Control of an Overhead Crane. *IFAC 8th Triennial World Congress, Kyoto, Japan*, pp. 1885-1890.

Yoon, J. S., B. S. Park, J. S. Lee, and H. S. Park (1995). Various Control Schemes for Implementation of the Anti-Swing Crane. *Proceedings of the ANS 6th Topical Meeting on Robotics and Remote Systems, Monterey, California*, pp. 472-479.

Doyle, J. C., B. A. Francis, and A. R. Tannenbaum (1992). *Feedback Control Theory*, Chapter 7, Macmillan Publishing Company, New York.

CONTROL OF A DOUBLE SIDE SHEAR

Lars Malcolm Pedersen [1] and **Carsten S. Villadsen** [1]

The Danish Steel Works Ltd, 3300 Frederiksværk, Denmark

Abstract: The purpose of this paper is to design a differential position controller for the pinch rolls for a double side shear at The Danish Steel Works Ltd. The purpose of the differential position controller is to ensure that plate edges remain straight despite various disturbances.
First a model is derived for the double side shear. The model is a first order differential equation with two unknown parameters. These parameters are found in the system identification which is based on data from 7 plates. The identification shows that one of the parameters varies considerably from plate to plate.
The differential position controller is designed to handle the parameter variations using a robust stability criterion. Simulations show that the performance of the system is satisfactory, despite the parameter variations.
Comparisons of results from the existing control system with simulations using the new control law indicate a considerable improvement of performance. The control law will be implemented in the near future. The results from the implementation will reveal if the theoretical results are correct. *Copyright © 1998 IFAC*

Keywords:
Steel plates, double side shear, robust control, system identification, modeling.

1. INTRODUCTION

This paper describes the design and implementation of a new controller for the electrical axes for the pinch rolls of the double side shear at The Danish Steel Works Ltd. The double side shear is used for cutting the sides of the steel plates produced by the plate mill.

The pinch rolls are used for moving the plate forward while the plate sides are cut by oscillating shears. By adjusting the distance between the two sides of the shear, and thus the distance between the shears the width of the plate can be adjusted. The width adjustment also includes the pinch rolls. Each pinch roll consists of one drive roll over the plate and one backup roll under the plate. To ensure that the friction between plate and pinch roll is sufficiently large for a proper control of the plate position, the drive and backup rolls are pressed together with a large force. The plate is transported to and from the double side shear by roller tables, see Figure 1.

To ensure that the plate edges are straight it is essential that the plate is transported in a direction parallel to the centerline of the shear. This is obtained by minimizing the differential position of rolls on the two sides of the shear. Beside the conventional velocity control of the pinch rolls, the pinch rolls are therefore equipped with at differential position controller.

To be able to handle the situations when a plate is entering or leaving the shear the differential position of 1st and 2nd set of pinch rolls are controlled separately, see Figure 1. In practice the differential position is controlled by a so called *electrical axes*. The principle is that the difference between the position of the pinch rolls is measured

[1] This work has been supported by The Danish Steel Works Ltd.

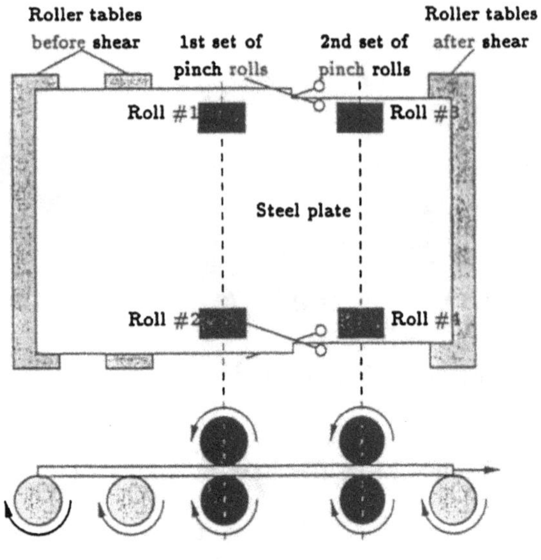

Fig. 1. Diagram showing the plate transport actuators of the double side shear. The plate sides are cut between the first and second set of pinch rolls.

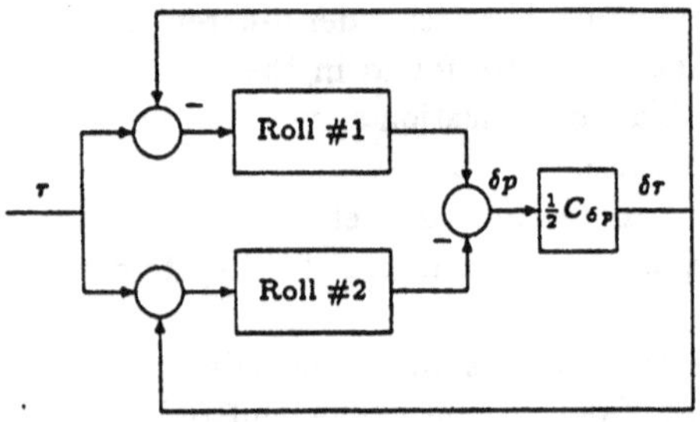

Fig. 2. Diagram showing the principle of differential position controller for the 1st set of pinch rolls. The position difference of the pinch rolls δp is controlled by adding/subtracting a correction δr to the speed reference of the motors r. $C_{\delta p}$ is the speed difference controller. The task of this paper is to design this controller

and a corrective signal is added to the references of the motors driving the pinch rolls, see Figure 2.

Today the differential position of the sets of pinch rolls are controlled by digital P-controllers. To enhance the performance of the double side shear they are replaced by digital PI-controllers. The design of these controllers is the main subject of this paper.

2. MODELING

Depending on which pinch rolls that have contact with the plate three different situations occur

(1) Plate entering shear: Rolls #1 and #2 have contact with the plate
(2) Plate in shear: Rolls #1, #2, #3, and #4 have contact with the plate
(3) Plate leaving shear: Rolls #3 and #4 have contact with the plate

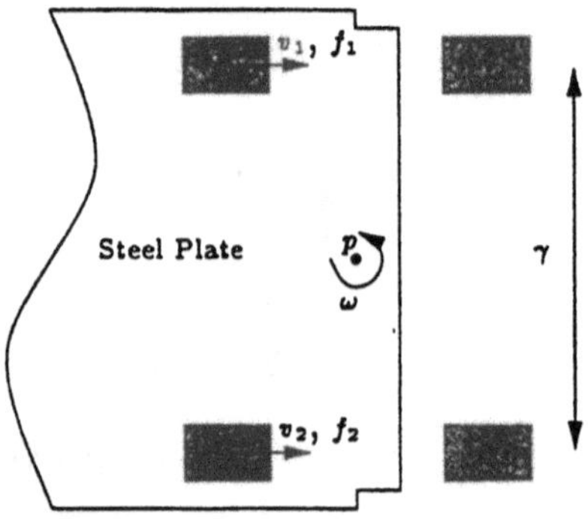

Fig. 3. Principal diagram for the case where only Rolls #1 #2 are used for transporting the plate. ω is the rotational speed around the center point p. v_1 and v_2 are the peripheral speeds of Rolls #1 and #2 respectively and f_1 and f_2 are the forces on the plate in the contact points.

The models structures for case 1 and 3 are identical, we will derive this structure first and return to case 2 later.

Considering Roll #1 and Roll #2 and assuming point contact between rolls and plate the angular acceleration $\dot{\omega}$ of the plate around the center p is given by

$$\dot{\omega}(t) = \frac{\gamma}{2J}(f_2(t) - f_1(t)) \tag{1}$$

see Crandall et al. (1978) and Figure 3. J is the moment of inertia around the plate center, this includes the inertia of the plate, pinch rolls and rotors of the motors. γ is the distance between the centers of the two pinch rolls and f_1 and f_2 are the forces applied to the plate by the pinch rolls. v_1 and v_2 are the peripheral speeds of the pinch rolls. In the modeling the peripheral speeds of pinch rolls and the plate in the points of contact are assumed to be equal.

The pinch rolls are driven by DC-motors. A model for these can be found in Smith (1976). Since the four motors are operating in closed loop it is reasonable to assume that they are identical. A model for the motors are therefore

$$\begin{aligned} f_1(t) &= dk\phi i_1(t) - Fv_1(t) \\ f_2(t) &= dk\phi i_2(t) - Fv_2(t). \end{aligned} \tag{2}$$

In (2), d is the radius of the pinch rolls, $k\phi$ is a motor constant, and F is the dynamical friction of the system. v_1 and v_2 are the peripheral speeds of the pinch rolls, see Figure 3.

Introducing the variables

$$\begin{aligned} \delta i(t) &= i_2(t) - i_1(t) \\ \delta v(t) &= v_2(t) - v_1(t) \end{aligned}$$

and exploiting that $v_1 - v_2 = \gamma\omega$, (2) and (1) can be reduced to

$$\delta v(t) = \frac{dk\phi\gamma^2}{Jp + \gamma^2 F}\delta i(t) \tag{3}$$

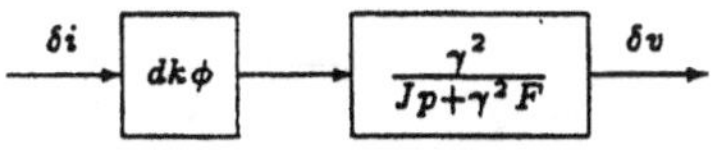

Fig. 4. Block diagram of the model for the differential speed of Rolls #1 and #2. δi is the difference between the currents of the motors of Roll #1 and #2 and δv is the difference between the peripheral speeds of the pinch rolls.

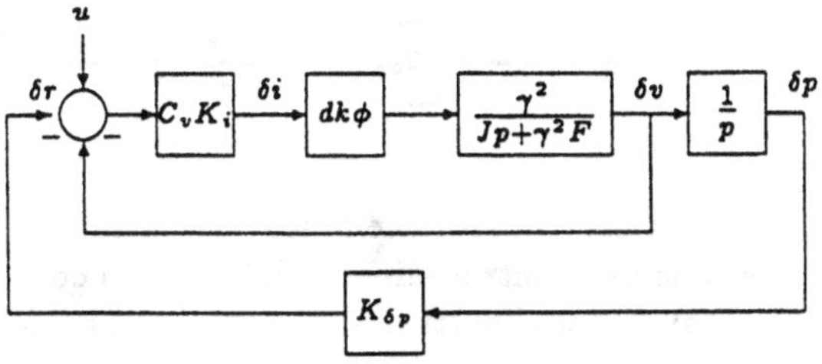

Fig. 5. Block diagram for the existing controller configuration for the differential position of the double side shear. In the figure C_v is the velocity controller, K_i is the gain of the current loop, and $K_{\delta p}$ is the gain of the electrical axes. u is a signal added to ensure identifiability.

where p is the differential operator. A block diagram of the model is shown in Figure 4.

Case 3, where Rolls #3 and #4 are in contact with the plate yields a model structure identical to (3). We therefore return to case 2, where all four pinch rolls are in contact with the plate. In a way similar to the above it can be shown that this model will have the structure

$$\delta\bar{v}(t) = \frac{dk\phi\gamma^2}{Jp + \gamma^2 F}\delta\bar{i}(t)$$

where $\delta\bar{v} = v_2 - v_1 = v_4 - v_3$ and $\delta\bar{i} = i_2 - i_1 + i_4 - i_3$, where i_3 and i_4 are the currents of Rolls #3 and #4 respectively, and v_3 and v_4 are the velocities of Rolls #3 and #4. This shows that the three cases listed in the beginning of this section have the same transfer function, but different inputs and outputs.

3. IDENTIFICATION

Today the differential position of the pinch rolls is controlled by P-controllers and the rotational speed of the rolls are controlled by PI-controllers, see Figure 5. This implies that we will have to identify a system operating in closed loop.

For the identification to be successful we require the system to be *Strongly System Identifiable*, see Gustavsson et al. (1977). This implies that parameters found in the system identification will be correct, provided that the model structure is right. If the input to the process is

$$\delta i(t) = F(q^{-1})\delta v(t) + K(q^{-1})u(t)$$

where u is an external signal which is *persistently exciting* of any order and q is the forward shift operator. The system then is *Strongly System Identifiable* if and only if

$$\text{rank}\begin{bmatrix} K(z) & F(z) \\ 0 & 1 \end{bmatrix} = n_y + n_u \text{ for almost } \forall z \quad (4)$$

where n_u and n_y is the number of inputs and outputs respectively. For (4) to be fulfilled in our case it is sufficient that $K \neq 0$.

In the identification of the parameters of (3) we choose u to be a square wave, which is *persistently exciting* of any order, see Åström and Wittenmark (1995). In our case

$$K(z) = \mathcal{Z}\left(C_v(s)K_i\right) \neq 0$$

where $\mathcal{Z}$ is the appropriate Z-transform of the transfer functions. The criterion is obviously true and the system is thus *Strongly System Identifiable*.

We then proceed with the identification of the parameters of the transfer function

$$G_\delta(p) = \frac{dk\phi\gamma^2}{Jp + \gamma^2 F} = \frac{b}{p+a} \quad (5)$$

where $a = \gamma^2 F/J$ and $b = dk\phi\gamma^2/J$. Except for the inertia J, which may vary with the plate position we expect the other parameters d, $k\phi$, γ and F to be constant for each plate. Using data for one plate at the time we therefore try to regard the system as time invariant in the identification.

The four currents i_1, i_2, i_3, and i_4 and velocities v_1, v_2, v_3, and v_4 have been measured for 7 plates of different width, length and thickness. The values of a and b of (5) are found using the *Optimization Toolbox* in MATLAB which is used for minimizing the criterion

$$V(a,b) = \sum_k (\delta v(k) - G_\delta(a,b)\delta i(k))^2$$

It turns out that the best results are obtained for the case where only Rolls #1 and #2 have contact with the plate, therefore the identification is done using data from this case. This is still expected to give a model valid for all the cases mentioned in the beginning of Section 2, since the transfer function does not depend on which rolls are in contact with the plate. To ensure that we have a good estimate of the parameters a simulation is compared with data not used for the identification, see Figure 6. It is seen that there is good correspondence between real and simulated output.

In the identification it is found that the parameters can be regarded constant for each plate. It is, however, found that a varies a factor 3 from plate to plate, while b is constant for all plates. Since the value of $a = \gamma^2 F/J$ increases with plate weight, the variations might be caused by variations of the

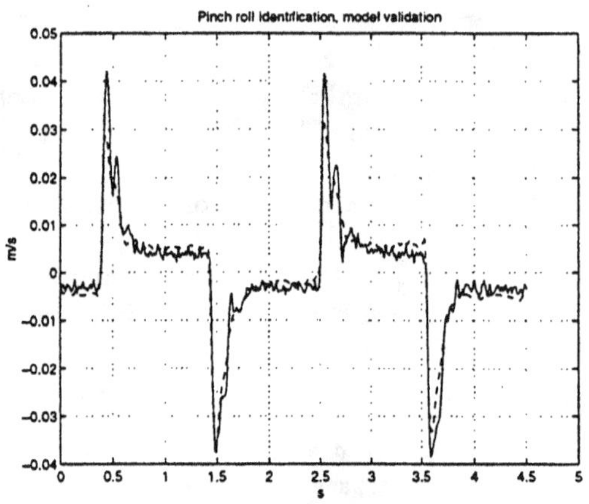

Fig. 6. Real (full) and simulated (dashed) response of the differential speed δv of the pinch rolls. The simulation is done using other data than used for the identification.

static friction of the system. The variations of a will be considered in the controller design.

4. CONTROLLER DESIGN

The control of the double side shear can be divided into the three cases mentioned in the beginning of Section 2. By controlling the 1st set and 2nd set of pinch rolls of Figure 1 by two separate PI-controllers with identical parameters we obtain identical control for case 1 and 3. In case 2 the this control structure will have 2 times the gain of case 1 and 3. This will be considered in the design.

The system identification showed that value of a varied with the plate dimensions. The transfer function for the shear can be represented by

$$G_\delta(s) = \frac{b}{s + a_o + \Delta a} \tag{6}$$

where a_o is the nominal value of a and Δa represents the variations of a. Due to the variations of a it is natural to use a robust design method for finding the parameters of the differential position controller

$$C_{\delta p}(s) = K_{p_\delta} + \frac{1}{T_{i_\delta} s}. \tag{7}$$

Using the polynomial approach described in Kwaakernaak (1993) Δa of (6) is represented as

$$\Delta a(s) = W_1(s)\delta_D(s)$$

where δ_D is an arbitrary frequency dependent perturbation with $|\delta_D(j\omega)| \leq 1 \; \forall\omega \in R$ and W_1 is a frequency dependent function representing the model variations. It is also used to shape the frequency response of the closed loop system. This is done to ensure a proper disturbance attenuation of the differential position control system.

When designing the differential position controller the entire closed loop structure of the double side shear has to be considered, see Figure 7. The transfer function of the velocity controller is

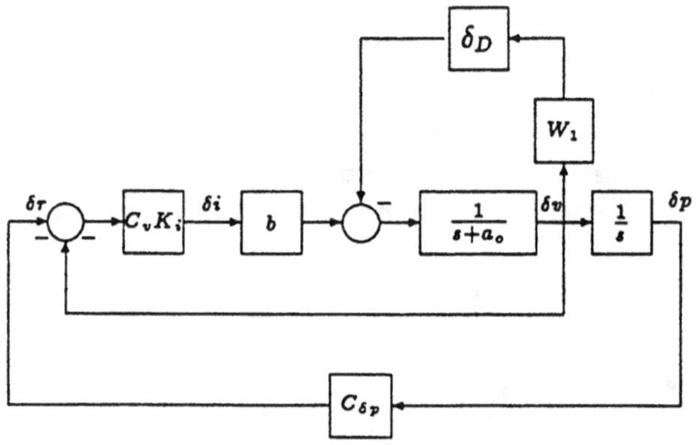

Fig. 7. Block diagram for the pinch rolls including the velocity control, differential position control, and model perturbations. δ_D is an arbitrary perturbation with $|\delta_D(j\omega)| \leq 1 \; \forall\omega \in R$ and W_1 is a combination of the parameter variations and a design tool used to shape the closed loop frequency response of the system. $C_{\delta p}$ is the differential position controller.

$$C_v(s) = K_p + \frac{1}{T_i s}.$$

The closed loop transfer function of the system shown in Figure 7 is

$$\begin{aligned} P(s) &= \frac{bK_i(K_p s + 1/T_i)}{s^3 + (a_o + \Delta a(s) + bK_iK_p)s^2 + bK_i/T_i s} \\ &= \frac{N_o(s)}{D_o(s) + \Delta a(s)s^2} \end{aligned}$$

where N_o/D_o is the nominal closed loop transfer function. An appropriate choice of W_1 turns out to be

$$W_1(s) = \delta a \frac{(s + 10\omega_d)s^2}{s + \omega_d} \tag{8}$$

where ω_d is the desired closed loop band width and $\delta a = \sup \Delta a$. This choice still ensures robust stability since since $W_1 > \delta a \; \forall\omega \in R$.

Robust stability of the closed loop system in Figure 7 is obtained if

$$\left| W_1(j\omega)\frac{1}{D_o(j\omega)}\frac{1}{1 + P_o(j\omega)C_{\delta p}(j\omega)} \right| < 1 \; \forall\omega \in R.$$

where $P_o = N_o/D_o$ is the nominal transfer function. The coefficients of the differential position controller (7) K_{p_δ} and T_{i_δ} are found by minimizing the above performance criterion. This is done using the *Optimization Toolbox* of MATLAB. In the design ω_d of (8) is chosen to maximize the bandwidth of the closed loop system. The results of the design show that the closed loop system has a good stability marginal, despite the variations of the parameter a.

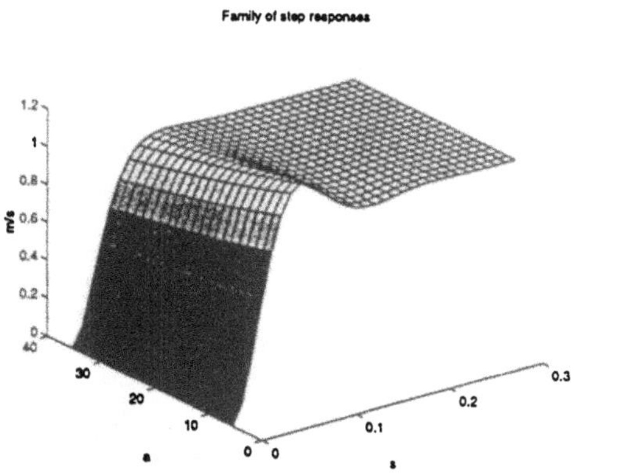

Fig. 8. Step responses for the new differential position control system for the different values of a.

5. SIMULATIONS

The performance of the new position difference controller is verified by computer simulations. The simulations are done using MATLAB. Step responses for the different values of the parameter a is computed, see Figure 8. To ensure maximal gain, the case where all four rolls have contact with the plate is simulated. The performance is considered satisfactory for all values of a.

6. COMPARING PERFORMANCE

Before implementation the differential position controller $C_{\delta p}$ is transformed to discrete time using Tustins approximation, see Åström and Wittenmark (1990). To avoid an excessive deterioration of the phase marginal, the sampling interval h should be chosen such that

$$h\omega_c \approx 0.15$$

where ω_c is the cross over frequency of the open loop system, including the controller. The maximal cross over frequency for the system is $\omega_c = 35$ rad/s, this yields a sampling frequency of 250 Hz.

The measured response of the differential position of the existing P-controller is shown in Figure 9. Note that the system has a time constant of around 1 s and has a stationary error. Simulations of a similar situation for the new position difference controller is shown in Figure 10. It is seen that the performance is improved considerably, with a faster response and zero stationary error.

The new control law will be implemented in the near future. This will show if the simulations shown in Figure 10 are correct. When the controller is implemented it will furthermore be verified that the closed loop system is robustly stable and that the pinch roll motors does not saturate during operation with the new control law.

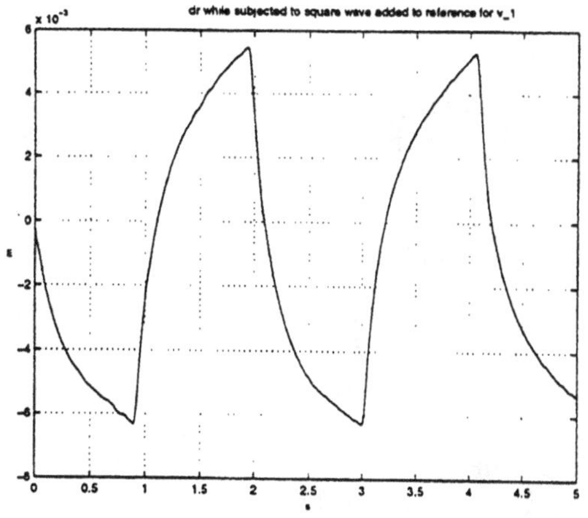

Fig. 9. Plot showing the performance of the existing speed difference controller. The plot shows the position difference δp of Rolls #1 and #2, when a square wave with a period of 2 s is added to the speed reference of Roll #1, as shown in Figure 5. The data in the plot have been measured when Rolls #1 and #2 are in contact with the plate.

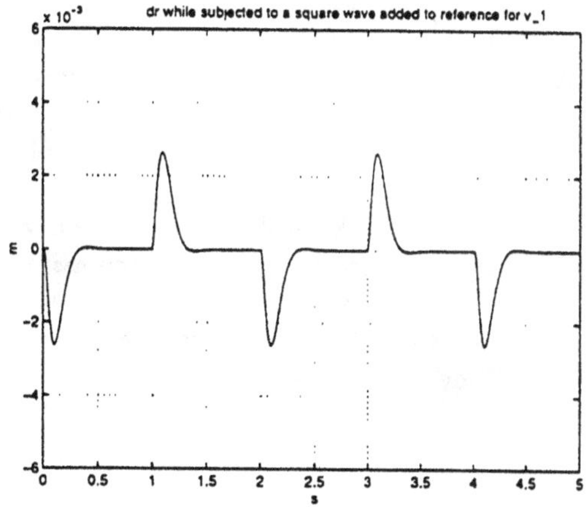

Fig. 10. Plot showing the performance of the new differential position controller. The plot shows the position difference δp of Rolls #1 and #2 when a square wave with a period of 2 s is added to the speed reference of Roll #1, as shown in Figure 5. The plot is based on simulated data, for the situation where Rolls #1 and #2 are in contact with the plate.

7. CONCLUSIONS

The purpose of this paper is to design a differential position controller for the pinch rolls for a double side shear. The purpose of the differential position controller is to ensure that plate edges remain straight despite various disturbances.

First a model is derived for the double side shear, the model is a first order differential equation with two unknown parameters. These parameters are found in the system identification which is based on data from 7 plates. The identification shows that one of the parameters varies considerably from plate to plate.

The differential position controller is designed to handle the parameter variations using a robust stability criterion. Simulations show that the performance of the system is satisfactory, despite the parameter variations.

Comparisons of results from the existing control system with simulations using the new control law indicate a considerable improved performance. The control law will be implemented in the near future. The results from the implementation will reveal if the theoretical results are correct.

References

Karl Johan Åström and Björn Wittenmark. *Computer Controlled Systems.* Prentice-Hall, 1990.

Karl Johan Åström and Björn Wittenmark. *Adaptive control.* Addison-Wesley Publishing Company, 1995.

S. H. Crandall, N. C. Dahl, and T. J. Lardner. *An Introduction to the Mechanics of Solids.* McGraw Hill, 1978.

I. Gustavsson et al. Identification of processes in closed loop – identifiability and accuracy aspects. *Automatica*, pages 59 – 75, Vol. 13 1977.

Huibert Kwaakernaak. Robust control and $\mathcal{H}_\infty$-optimization – tutorial paper. *Automatica*, 29 (2):255 – 273, 1993.

Ralph J. Smith. *Circuits Devices and Systems.* John Wiley & Sons, 1976.

SUBMERGED–ARC FERROSILICON FURNACE SIMULATOR – VALIDATION FOR DIFFERENT FURNACES AND OPERATING RANGES[1]

Anna Soffía Hauksdóttir, Arnar Gestsson and Ari Vésteinsson

Systems Engineering Laboratory, Engineering Research Institute, University of Iceland, Hjardarhagi 2-6, IS-107 Reykjavík, Iceland. E-mail: ash@kerfi.hi.is

Abstract. Models of a 36MW submerged arc ferrosilicon (FeSi) furnace have been developed for the simulation and evaluation of different control schemes in (Hauksdóttir *et al.*, 1995) and (Hauksdóttir and Gestsson, 1996). In this paper, the models developed, describing the furnace controller, the electrode positioning equipment, the dynamics from the electrode positions to the electrode currents and the disturbance environment, effectively forming a FeSi furnace simulator, will be validated for two furnaces, as well as for different operation ranges. *Copyright © 1998 IFAC*

Keywords. Submerged arc ferrosilicon furnace simulator, validation for different furnaces, validation for different operating ranges.

1. INTRODUCTION

Ferrosilicon 75% (FeSi), which is an alloy of silicon (75%) and iron, is one of the basic raw materials in the steel industry. A schematic diagram of a submerged arc FeSi furnace is shown in Fig. 1. The raw materials (mainly quartz, coal and coke) are charged from the top into the cylindrical furnace. Three electrodes are submerged in the charge. An electrical arc creates the high temperature in the crater needed for the chemical reaction forming FeSi. The molten product is tapped through a tapping hole at the bottom of the furnace into a ladle.

The furnaces at Icelandic Alloys Ltd. are typically operated at some 36MW, 150V and 118kA. The present controller is effectively composed of three SISO (single-input-single-output) loops, each controlling one electrode current independent of the others. However, there is a

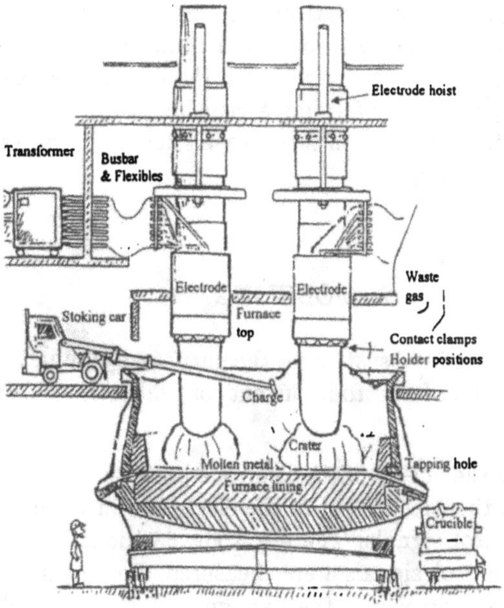

Fig. 1. A schematic diagram of a submerged arc FeSi furnace (only two of the three electrodes are shown).

[1] Supported by Icelandic Alloys Ltd.

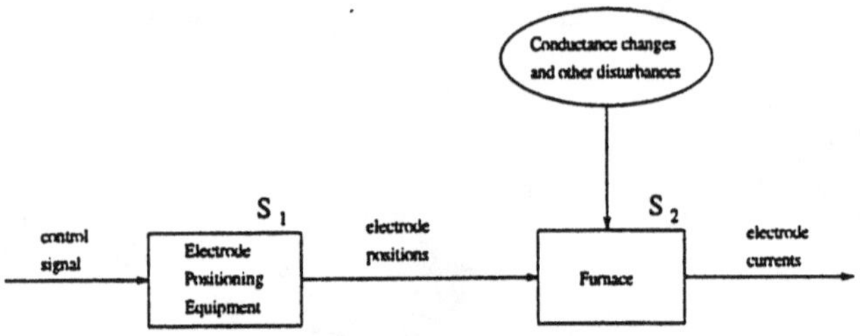

Fig. 2. Important parts of the system for simulation and evaluation of electrode current controllers.

considerable coupling between the electrodes, i.e., moving one electrode strongly affects the currents in the others. Further, the disturbance environment is severe, which is partly due to large conductance changes in the charge as well as the varying electrode to hearth voltage.

Dynamic models based on extensive measurements on the furnace, were developed in (Hauksdóttir *et al.*, 1995). In (Hauksdóttir and Gestsson, 1996), the models developed were used in a closed–loop simulation where the present current controller was included and the results were compared to closed–loop measurements. A new decoupling control algorithm was then developed for the current control loop and its performance was compared to the present current control loop.

In (Hauksdóttir *et al.*, 1995) and (Hauksdóttir and Gestsson, 1996), all modeling and validation data was obtained from the same furnace, furnace 2. Here, data gathered from furnace 1 as well as furnace 2, will be used for additional validation. The models developed, describing the furnace controller, the electrode positioning equipment, the dynamics from the electrode positions to the electrode currents and the disturbance environment, effectively forming a FeSi furnace simulator, will be validated for both furnaces, as well as for different operation ranges.

2. MODELING

Critical parts of the system for the purpose of simulation and evaluation of electrode current controllers are shown in Fig. 2.

S_1 denotes the effects of the electrode positioning equipment, i.e., the positioning of the electrodes in response to the control signals. The electrode positions as well as their control signals were measured directly under closed-loop control and the dynamics of the electrode positioning equipment S_1 were estimated and modelled for furnace 2 in (Hauksdóttir *et al.*, 1995), resulting in

$$G_{\nearrow f_2} = \frac{0.45\exp(-1s)}{s(1.33s+1)} \tag{1}$$

and

$$G_{\searrow f_2} = \frac{1.1\exp(-1s)}{s(2.5s+1)} \tag{2}$$

for the raise and the low movements, respectively. It is known that furnace 1 has a different and faster response, which has been determined in (Gestsson *et al.*, 1997) as

$$G_{\nearrow f_1} = \frac{0.44\exp(-1.7s)}{s(0.32s+1)} \tag{3}$$

and

$$G_{\searrow f_1} = \frac{0.88\exp(-1.2s)}{s(0.85s+1)}. \tag{4}$$

S_2 denotes the current dynamics of the furnace itself, i.e., the effects of the electrode positioning on the electrodes currents, including the coupling effects between electrodes. The electrode positions as well as the electrode currents were measured directly in open loop and a 10 parameter model was determined (Hauksdóttir *et al.*, 1995), given by:

$$\begin{aligned}I_i(t) = {} & [1.069 \pm 0.011]I_i(t-1) - [0.147 \pm 0.011]I_i(t-2)\\ & -[1.055 \pm 0.013]h_i(t) + [1.491 \pm 0.024]h_i(t-1)\\ & -[0.446 \pm 0.016]h_i(t-2)\\ & +[0.203 \pm 0.011]h_{i_{+1}}(t) - [0.211 \pm 0.011]h_{i_{+1}}(t-1)\\ & -[0.818 \pm 0.013]h_{i_{-1}}(t) + [1.169 \pm 0.024]h_{i_{-1}}(t-1)\\ & -[0.361 \pm 0.015]h_{i_{-1}}(t-2).\end{aligned} \tag{5}$$

Here, $I_i(t)$ is the current in electrode i at sampling instant t, $h_{i_{+1}}$ is the next electrode in an increasing cyclic order and $h_{i_{-1}}$ is the next electrode in a decreasing cyclic order.

3. CLOSED–LOOP SIMULATION AND COMPARISON TO MEASUREMENTS

The validation will be done by including the models developed in a closed–loop simulation as shown in Fig. 3 and comparing the results to closed–loop measurements. An estimate of the disturbances affecting the electrode currents, is important in such a simulation setting. The disturbances are mostly due to conductance changes in the charge and the varying electrode to hearth voltage. These disturbances are significant, although mostly slowly varying compared to the process' electrode position/current dynamics. A disturbance sequence generated from closed–loop measurements, referred to as a generated disturbance sequence, is obtained by taking the difference between a measured output of the furnace, I_m, and the output from the developed electrode position/electrode current model, driven by the corresponding measured electrode positions, h_m, see Fig. 3.

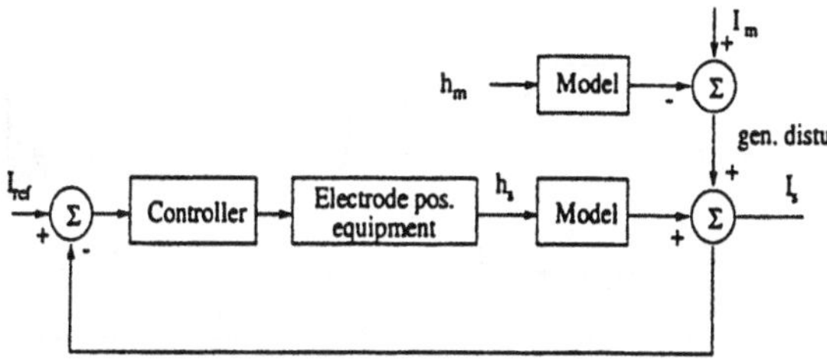

Fig. 3. The simulation setup. I_m and h_m are the measured electrode currents and electrode positions, respectively. I_s and h_s are the simulated electrode currents and electrode positions and I_{ref} denotes the reference current.

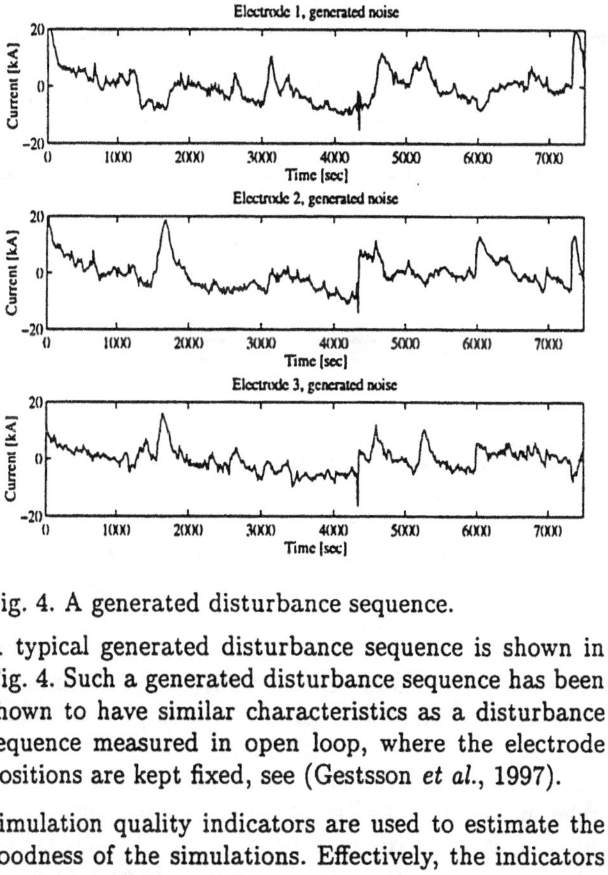

Fig. 4. A generated disturbance sequence.

A typical generated disturbance sequence is shown in Fig. 4. Such a generated disturbance sequence has been shown to have similar characteristics as a disturbance sequence measured in open loop, where the electrode positions are kept fixed, see (Gestsson *et al.*, 1997).

Simulation quality indicators are used to estimate the goodness of the simulations. Effectively, the indicators compare the difference between simulations and the measurements. These indicators are the Current Deviation (CD) and the Electrode Position Deviation (EPD), given by

$$CD = \frac{1}{3}\left(\sum_{i=1}^{3}\sqrt{\frac{1}{N-1}\sum_{1}^{N}(I_{is}-I_{im})^2}\right) \quad (6)$$

and

$$EPD = \frac{1}{3}\left(\sum_{i=1}^{3}\sqrt{\frac{1}{N-1}\sum_{1}^{N}(h_{is}-h_{im})^2}\right). \quad (7)$$

Here I_{is} is the simulated current, I_{im} is the measured current, h_{is} is the simulated position and h_{im} is the measured position of the i-th electrode. Note that initial transients in the simulation (approximately 200 datapoints) are excluded when calculating the simulation quality indicators.

4. PRESENT CURRENT CONTROL SCHEMES

The present electrode current controller controls the current of each electrode by changing the corresponding electrodes position, given a ± 2 kA dead-band around a reference value, typically $I_{ref} = 118kA$. If a current remains under $I_{ref} - 2kA$ for more than 10 sec., a 1 sec. control signal is sent to lower the corresponding electrode, which tends to increase the current. On the other hand, if a current remains above $I_{ref} + 2kA$ for more than 7 sec., a 3 sec. (for furnace 2, 2 sec. for furnace 1) control signal is sent to raise the corresponding electrode, which tends to lower the current. Finally, for extreme over-current situations, i.e., when a current becomes 125 kA or above, a 7 sec. control signal is sent almost immediately to raise the electrode. The controller is set up in each case such that the overall response of the controller and the electrode position equipment is approximately the same for the raise and the low movements for each furnace, see (Þorfinnsson and Hauksdóttir, 1992) and (Gestsson *et al.*, 1997).

There have been a number of guidance rules for operation of FeSi furnaces. Many of them are based on empirical models of the electrical conditions in the furnace. One such method, is Westly's method, which is based on Andreae's formula. The latter was put forth in the 1920's and is given by

$$R\pi D = k. \quad (8)$$

Here, R is the resistance from the electrode to the molten product, D is the electrode diameter and k is a constant. Andreae discovered that the same k value was valid for the same charge composition, even for different sizes of furnaces. Westly adapted Andreae's formula (Westly, 1974), (Westly, 1975), by experimentally noting that k was in fact inversely proportional to the current density i, i.e.,

$$R\pi D \propto \frac{1}{\sqrt{i}} \propto \frac{D}{\sqrt{I}}, \quad (9)$$

where I is the electrode current. Thus,

$$R \propto \frac{1}{\sqrt{I}}. \quad (10)$$

The total active load is given by

$$P = RI^2. \quad (11)$$

Thus,

$$P \propto I^{3/2} \quad (12)$$

or

$$I = CP^{2/3}, \quad (13)$$

where I is in kA, P is the total active load for the 3 electrode furnace in MW and C is Westly's constant.

A control scheme based on choosing Westly's constant, C, measuring the load P and calculating the reference current, I_{ref} from Eq. 13, has gained acceptance. Typically, I_{ref} and C are calculated at each sampling interval of the controller, thus, effectively closing an outer control loop for the current controller. This, however, can lead to undesirable characteristics, even unstability caused by positive feedback, in the closed-loop system. Rather, as Westly intended, Eq. 13 should be used as a long term guidance with slow variations in I_{ref}.

5. VALIDATION OF SIMULATOR MODELS FOR DIFFERENT FURNACES (GESTSSON *ET AL.*, 1997)

The validation of simulator models for different furnaces involves models based on data from furnace 2 in a simulation where closed-loop data from furnace 1 is used in the generation of a disturbance sequence as well as for comparison with the simulation results. Then, the same models are used in a corresponding simulation where the closed-loop data is obtained from furnace 2. This is done for the present current controller in a constant reference current setting as well as in a Westly's based reference current setting.

5.1 *Constant reference current*

The first phase of the validation is a simulation of the present current control loop. Measured electrode positions and currents from furnace 1 are shown in Fig. 5 (dashed/dotted curve), also shown are simulated results (solid curve) where the models based on data from furnace 2 are used in the simulation and closed-loop data from furnace 1 is used in the generation of a disturbance sequence. For comparison, the same simulation was run where models of the current controller and the electrode positioning equipment for furnace 1 are used in

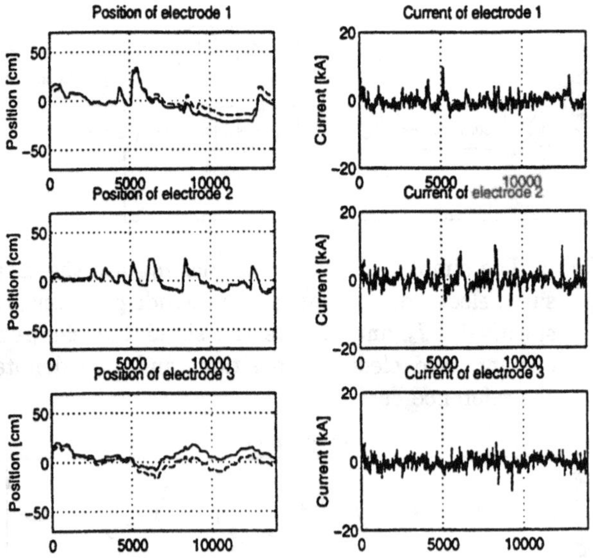

Fig. 5. Simulation using a constant reference current, generated disturbance from furnace 1 closed-loop data and the three models for furnace 2 (solid curve). Corresponding closed–loop measurements (dashed/dotted curve).

Electr. pos. and contr. model	*EPD*	*CD*
furnace 1	4.9cm	0.67kA
furnace 2	4.9cm	0.61kA

Table 1. Electrode position deviation and current deviation for a simulation using a constant reference current, generated disturbance from furnace 1 closed-loop data and the electrode position/current model based on data from furnace 2.

the simulation. There was practially no visible difference between the two situations, and this is further demonstrated in the similarity of the EPD and CD values for the two cases, see Table 1. This suggests that it is irrelevant which controller/electrode positioning model is used in the simulations. This is a natural result in view of the fact, that the controller is set up in each case such that the overall response of the controller and electrode position equipment is similar for both furnaces, i.e., each control signal results in an approximately 1cm raise or low, although each movement is faster in furnace 1 (~2.5 sec. for furnace 1, ~5 sec. for furnace 2). The difference in the speed of the response between the two furnaces is, however, not visible in the much large timescale used in the simulation.

Then, a simulation of the present current control loop is run using models based on data from furnace 2 and closed-loop data from the same furnace, i.e., furnace 2, is used in the generation of a disturbance sequence as well as for comparison with the simulation results. The sim-

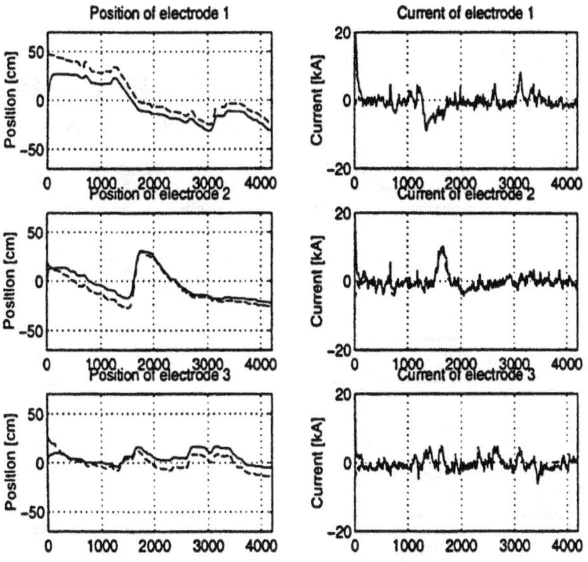

Fig. 6. Simulation using a constant reference current, generated disturbance from furnace 2 closed-loop data and the three models for furnace 2 (solid curve). Corresponding closed–loop measurements (dashed/dotted curve).

Generated disturbance	*EPD*	*CD*
furnace 1	4.9cm	0.61kA
furnace 2	7.5cm	0.69kA

Table 2. Electrode position deviation and current deviation for a simulation using a constant reference current.

ulated and measured positions and currents are shown in Fig. 6.

The EPD and the CD values of the two simulations using models based on data from furnace 2, are shown in Table 2, where closed-loop data from furnace 1 and furnace 2, respectively, are used in the generation of a disturbance sequence as well as for comparison with the simulation results. Here it is evident, that the CD values are practically the same, whereas a slightly higher EPD is obtained when both models and closed-loop data originate from furnace 2.

5.2 *Reference current based on Westly's method*

The second phase of the validation is a simulation of the present current control loop using a reference current based on Westly's method. Measured electrode positions and currents from furnace 1 are shown in Fig. 7 (dashed/dotted curve), also shown are simulated results (solid curve) where the models based on data from furnace 2 are used in the simulation and closed-loop data from furnace 1 is used in the generation of a disturbance sequence. Measured electrode positions and currents from furnace 2 are shown in Fig. 8 (dashed/dotted curve), also shown are simulated results (solid curve) where the models based on data from furnace 2 are used in the simulation and closed-loop data from furnace 2 is used in the generation of a disturbance sequence. The EPD and the CD values for both simulation cases are shown in Table 3. Here, in contrast to the constant current case, slightly better results are obtained when both models and closed-loop data originate from furnace 2, than when the furnace 2 models are used in conjunction with furnace 1 closed-loop data. Further, comparing Tables 2 and 3, the EPD value ranges are similar, whereas slightly higher CD values are obtained in the simulation where the reference current is based on Westly's method. In that case, rapid variations result in the reference current, thus higher CD values can be expected.

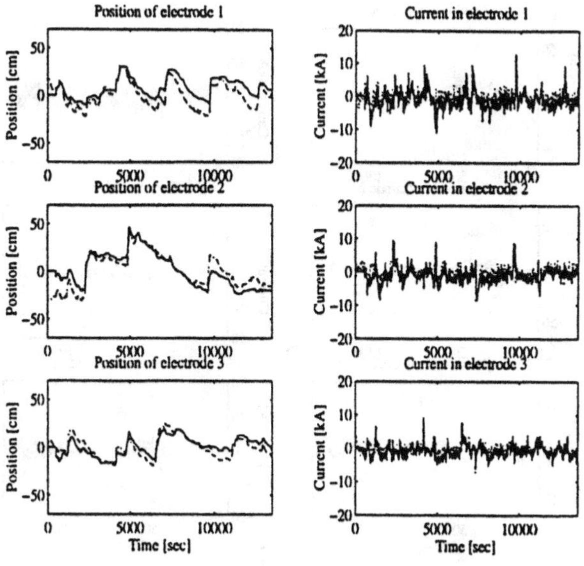

Fig. 7. Simulation using a reference current based on Westly's method and generated disturbance from furnace 1 closed-loop data (solid curve). Corresponding closed–loop measurements (dashed/dotted curve).

Generated disturbance	*EPD*	*CD*
furnace 1	8.2cm	1.72kA
furnace 2	4.3cm	1.41kA

Table 3. Electrode position deviation and current deviation for a simulation using a reference current based on Westly's method.

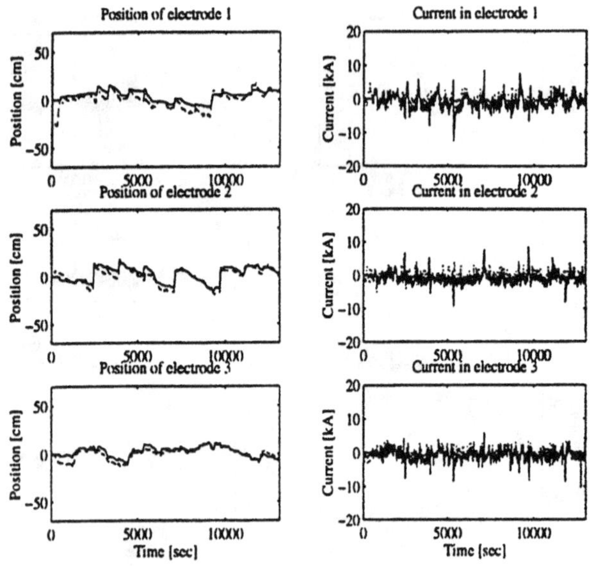

Fig. 8. Simulation using a reference current based on Westly's method and generated disturbance from furnace 2 closed-loop data (solid curve). Corresponding closed–loop measurements (dashed/dotted curve).

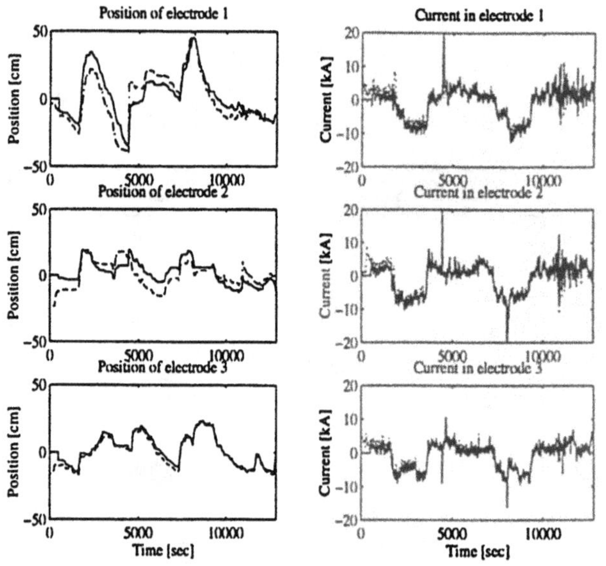

Fig. 9. Simulation using a varying reference current (solid curve). Corresponding closed–loop measurements (dashed/dotted curve).

EPD	*CD*
6.0cm	1.41kA

Table 4. Electrode position deviation and current deviation for a simulation using a varying reference current.

6. VALIDATION OF SIMULATOR MODELS FOR DIFFERENT OPERATION RANGES (GESTSSON *ET AL.*, 1997)

The validation of simulator models for different operation ranges involves models based on data from furnace 2 in a simulation where closed-loop data from furnace 2 is used in the generation of a disturbance sequence, as well as for comparison with the simulation results. Firstly, this is done while varying the reference current between 118.5 and 110 kA. Secondly, this is done under fixed reduced load operation, i.e., during shutdown and startup of the furnaces. During shutdown the load is decreased in steps, first down to approximately 30MW, and then down to 11MW. During startup, the furnace is driven up to 17MW and then to 30MW, until full load is reached.

6.1 *Varying reference current*

The measured electrode positions and currents from furnace 2, where the reference current was varied between 118.5 and 110 kA in steps, are shown in Fig. 9 (dashed/dotted curve), also shown are simulated results (solid curve). Note, that the largest spike visible in the currents is only present in the simulation. At this point, the positioning equipment for electrode 1 had reached it's maximum extension and was manually adjusted. The change is interpreted as a movement of the electrode itself in the simulation and, consequently, a sharp change in the current occurs. Apart from this anomaly, the quality of the simulation is excellent. The EPD and the CD values are shown in Table 4. Here, both values are very similar to the ones obtained in the simulation where the reference current was based on Westly's method (see Table 3), where wide variations in the reference current are also present.

6.2 *Reduced load during furnace start-up and shut-down*

The models forming the FeSi furnace simulator, are based on data taken from the furnace operating almost exclusively at some 36 MW. It is of interest to validate the models under a wider operating range, also in view of the fact that the furnace occasionally operates under reduced load. This can be done by comparing the simulators performance to data gathered under reduced load operation, e.g., during shutdown and startup of the furnaces.

During shutdown the furnace was operating at 30MW load for approximately 24 hrs. Two segments were taken out of this data for simulation. Measured electrode positions and currents for the first segment are shown in

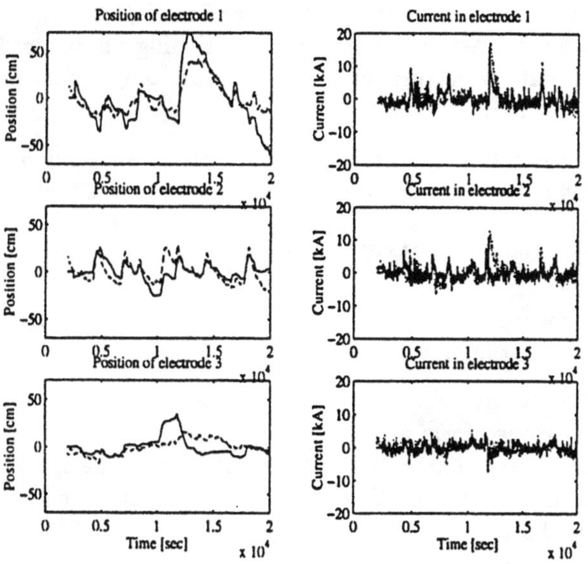

Fig. 10. Simulation segment 1 under 30MW reduced load during shut-down (solid curve). Corresponding closed–loop measurements (dashed/dotted curve).

Fig. 10 (dashed/dotted curve), also shown are simulated results (solid curve). Corresponding results for the second segment, 11MW during shut-down, and 17MW and 30MW during start-up are shown in Figs. 11-14. The EPD and the CD values for all five cases are shown in Table 5. As expected, the furnace operation is calm for very low loads, e.g., the 11 and 17MW, as evident in little current and electrode position changes. Here, the simulation results are excellent, as evident in very low EPD and CD values. Similarly, in the case of the 30MW start-up, the EPD and CD values are quite low and very similar to the results corresponding to the constant reference current case shown in Table 2. On the other hand, both cases corresponding to the 30MW shut-down, result in somewhat poorer simulation results, see Figs. 10 and 11, in particular, there are large deviations in the electrode positions as indicated by the relatively high EPD values. Screening of the results suggests that electrode 3 is moving less than the simulation predicts. In fact, a temporary restriction in the movement of one electrode would cause abnormalities in the position as well as the currents of the other electrodes, and could result in a poor overall simulation. This may have been the case here.

7. CONCLUSIONS

The models forming a submerged-arc ferrosilicon furnace simulator, i.e., describing the furnace controller, the electrode positioning equipment, the dynamics from the electrode positions to the electrode currents and the

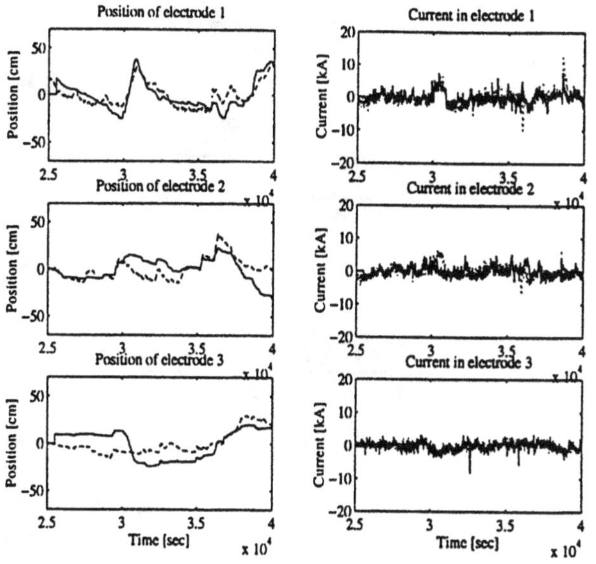

Fig. 11. Simulation segment 2 under 30MW reduced load during shut-down (solid curve). Corresponding closed–loop measurements (dashed/dotted curve).

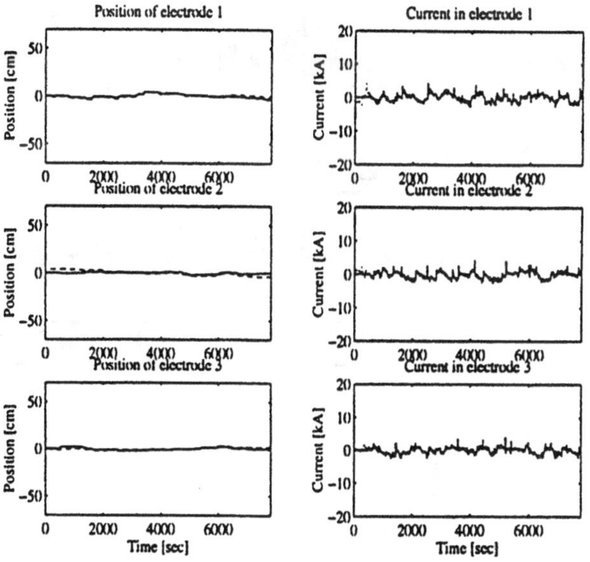

Fig. 12. Simulation under 11MW reduced load during shut-down (solid curve). Corresponding closed–loop measurements (dashed/dotted curve).

Load	*EPD*	*CD*
$30MW_1 \searrow$	12.8cm	1.86kA
$30MW_2 \searrow$	11.2cm	1.26kA
$11MW \searrow$	1.6cm	0.19kA
$17MW \nearrow$	2.0cm	0.31kA
$30MW \nearrow$	3.1cm	0.70kA

Table 5. Electrode position deviation and current deviation for a simulation under reduced load during shut-down and start-up.

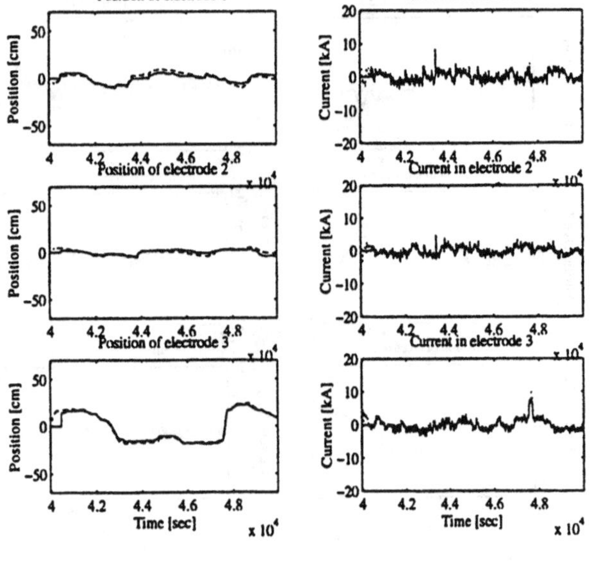

Fig. 13. Simulation under 17MW reduced load during start-up (solid curve). Corresponding closed–loop measurements (dashed/dotted curve).

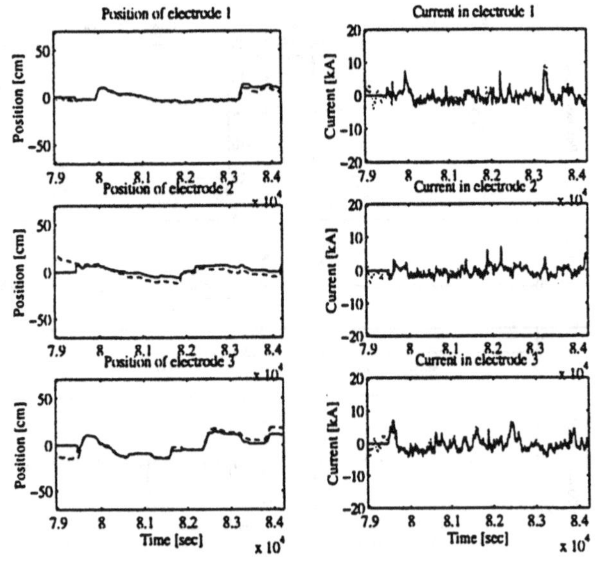

Fig. 14. Simulation under 30MW reduced load during start-up (solid curve). Corresponding closed–loop measurements (dashed/dotted curve).

disturbance environment, have been validated for two furnaces as well as for different operating ranges.

Excellent and comparable results were obtained in simulations, when all models were based on data from furnace 2 and closed-loop data were based on data from furnace 1 and 2, respectively. This was true in the case of a constant reference current as well as in the case when the reference current was based on Westly's method and thus quite widely varying. In the latter case, slightly higher current deviations were obtained. The simulations were insensitive to the use of models for the combined controller and electrode positioning equipment based on data from furnace 1 on one hand and furnace 2 on the other.

Excellent results were again obtained in simulations using a varying reference current as well as in simulations under reduced load during shut-down and start-up. In the case of a 30MW load during shut-down, some discrepancy was evident between the simulations and the measurements. This may be due to a temporary restriction in the movement of one of the electrodes.

The results obtained clearly indicate, that the models developed form a realistic FeSi furnace simulator which may be used for FeSi furnaces of the same type operating over a wide operating range. Such a simulator may be used for training of operators as well as for design and tuning of controllers.

Acknowledgements

The authors would like to thank Icelandic Alloys Ltd., for their continued collaboration and support.

8. REFERENCES

Gestsson, A., A. Vésteinsson and A. S. Hauksdóttir (1997). Icelandic Alloys Ltd.'s FeSi furnace 2 model validation for other FeSi furnaces. Technical Report KVS-970501. Systems Engineering Laboratory, Engineering Research Institute, University of Iceland.

Hauksdóttir, A. S. and A. Gestsson (1996). Current control of a three-phase submerged arc ferrosilicon furnace. In: *Proceedings of the 13th IFAC World Congress*. San Francisco, USA. pp. 421–426.

Hauksdóttir, A. S., T. Söderström, Y. P. Þorfinnsson and A. Gestsson (1995). System identification of a three-phase submerged arc ferrocsilicium furnace. *IEEE Trans. Control System Technology* **3**(4), 377–387.

Þorfinnsson, Y. P. and A. S. Hauksdóttir (1992). System identification of a three phase arc furnace. Technical Report KVS-91003. Systems Engineering Laboratory, Engineering Research Institute, University of Iceland. (in Icelandic).

Westly, J. (1974). Resistance and heat distribution in a submerged-arc furnace. In: *Infacon 74, Proceedings of the First International Ferro-Ally Congress*. Johannesburg. pp. 121–127.

Westly, J. (1975). Critical Parameters in Design and Operation of the Submerged Arc Furnaces. In: *1975 Electric Furnaces Proceeding*. pp. 47–53.

INTELLIGENT COMPOUND CONTROL OF DIRECT CURRENT ELECTRIC ARC FURNACE

Xianwen Gao, Shujiang Li, Tianyou Chai and Xiaogang Wang

Research Center of Automation, Northeastern University, Shenyang,110006,China
E-mail: xwgao@mail.neu.edu.cn

Abstract: Due to steelmaking process of the Ultra High Power Direct Current Electric Arc Furnace (UHP DC EAF) is a complex process with physicochemical reactions, there exist highly nonlinearity、 uncertainty、 time-variant and seriously interconnected (i.e. some of the variables may not be linearly independent) *etc.* It is impossible to establish a precise model for the steelmaking of the UHP DC EAF . According to the concrete practice of the steelmaking process, a intelligent compound control strategy is presented based on the steelmaking process of UHP DC EAF in this paper. This method was successfully used in the steelmaking process of the UHP DC EAF in a factory.

Keywords: intelligent control, bang-bang control, fuzzy control, PID control, predictive control , steel manufacture.

1. INTRODUCTION[1]

In recent years, the steelmaking techniques of electric arc furnace (EAF) have been developed swiftly and violently. In particular, the steelmaking techniques of the ultra high power direct current electric arc furnace (UHP DC EAF) is more outstanding. The UHP DC EAF offers substantial advantages over the AC (Alternative Current EAF) EAF in the following respects. (1).Electrode consumption reduced by about 50%; (2).Electric power consumption reduced by about 5%; (3).For a furnace with capacity for 700000 tons per year, these advantages alone can save up to 2.5 million US$ annually; (4).No hot or cold spots due to central electrode; (5).Uniform heat distribution in the melt due to beneficial stirring effect; (6). In most cases, reduce influence on the feeding network, expensive SVC's are mostly not required; (7). Reduce average noise level due to tighter furnace. Despite UHP DC EAF has, several advantages a lot of electric power is consumed by it. For this reason, many scientists have been taking various measurements to reduce the energy consumption in the steelmaking process of the EAF. Most of the control system for EAF are now based on a mathematical model which is established by means of the analysis in technology or identification of process. Since steelmaking process in the UHP DC EAF is complex process with physicochemical reaction, there exists highly nonlinearity, uncertainty, time-variant and seriously interconnected (i.e. some of the variables may not be linearly independent) *etc.* There fore it is impossible to establish a precise model for the steelmaking process of the UHP EAF. This is why the control of the UHP EAF is difficult. According to the concrete practice of the steelmaking process, researchers have developed some control strategies and have obtained some application results, for example, temperature weighting adaptive controller for electric arc furnace (Billings and Nicholson 1977); the computer control of electric arc furnace steelmaking process based on a mathematical model (Takasi Takawa 1988); intelligent arc furnace had been developed

[1] This work was supported by the National Natural Science Key Foundation of China

(William,1992); intelligent control of electrode and final points adaptive predictive control strategy were presented (1993 and 1994), respectively. All the research work mentioned above are used on the AC EAF, and only some parameters are controlled, such as, the raising and downing control of the electrode, refining process control and final point compositions as well as temperature predictive control etc. These methods may realize local optimum on some controlled parameters, but it can not realize on the whole steelmaking process optimum. In this paper, an intelligent compound control strategy is presented for the steelmaking process of the UHP DC EAF. Our design goal is to allow computers to perform tasks that involve human-like reasoning and intelligence, to obtain the steelmaking process optimizing and supervisory control under operating guide based on the existed equipment of the furnace.

2. CONTROL STRATEGIES AND MATHEMATICAL MODELS

Steelmaking in a UHP DC EAF is a complicated physico-chemical process, which comprises three stages, the melting, the oxidizing and the refining stages. Different procedures and strategies are used in the different stages. It is impossible to establish a unified mathematical model to describe the whole steelmaking process in the UHP DC AEF. The detail description of the models adopted is as follow.

2.1 The Mathematical Model of the Melting Stage

In the melting stage, about 60%--70% of the whole electric energy input is consumed, so the first key is to reduce the electric power consumption and the time required for melting. This can be done by modeling the operators experiences. It is assumed that melting stage is a physical process. Electric energy is a major source of heat. Oxygen is used mainly as an additional source of energy to assist and speed up melting. Thus the effect of 1 m^3 of can be converted into the equivalent to 2.5 kwh of electricity. After that, a melting model may be constructed by the recursive calculation based on the data collected from large amount of normal production records in melting stage (180 of groups of data are used). For example, The model for specified steel GCr15SiMn is as follows

$$E = \sum_{i=1}^{6} a_i x_i \tag{1}$$

where the E (kwh) is the estimated value of the electric energy input required for the melting stage, the x_i (Ton) is weight of each type of the charged material specified for steel RGCr15SiMn, such as scrap iron, steel scrap and others, and a_i is the coefficient for each charged material, respectively (kwh/t) i=1...6. This model is important in a computer control system and it works as a part of the main control program for monitoring the whole steelmaking process.

2.2 The Mathematical Model of the Oxidizing Stage

The main purposes of the oxidizing stage are to remove the impurities, to reduce carbon-content and to raise temperature of the steel melted by oxidative reaction and input of electric energy. The carbon-content and temperature are the output variables. The electric energy、oxygen gas quantity and ore adding quantity are the input variables. The other effect factors are considered as the disturbance.

Generally speaking, the mathematical model in this stage has 3 input and 2 output variables. Thus we have a simplified mathematical model as Fig.1

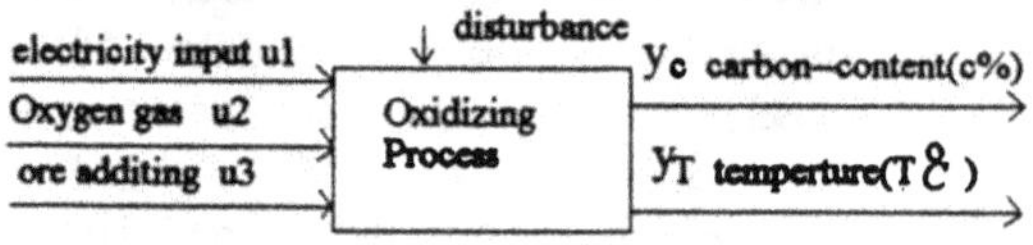

Fig.1. simplified mathematical model of oxidizing stage

A heuristic predictive control strategy is proposed as a dynamic control strategy that allows the model with some uncertainties and updates the outputs of the model by closed loop correction which optimizes the control law in moving horizon *etc.* A predictive control (Kayser *et al*, 1982) is adopted by this paper. It is based on CARMA model.

$$A(z^{-1})y(k) = B(z^{-1})u(k) + C(z^{-1})e(t) \tag{2}$$

where

$$A(z^{-1}) = 1 + a_1 z^{-1} + a_2 z^{-2} + \ldots + a_n z^{-n}$$
$$B(z^{-1}) = b_0 + b_1 z^{-1} + \ldots + b_n z^{-n}$$
$$C(z^{-1}) = 1 + c_1 z^{-1} + \ldots + c_n z^{-n}$$

Where k is the discrete time index, $y(\cdot)$ and $u(\cdot)$ are the process output and input, $e(t)$ is a disturbance signal. $A(\cdot)$, $B(\cdot)$, $C(\cdot)$ are polynomials in the back shift operator z^{-1} of order n. This control system is constructed with 3 inputs and 2 outputs system, it is assumed there is not interconnection between the 2 outputs in practice, then the system control model may be simpled into a 3 inputs and single output. The system models are found in formula (3)

$$A(z^{-1})y(k)=\sum_{j=1}^{3}B_j(z^{-1})u_j(k-d_j)+C(z^{-1})\zeta(t) \quad (3)$$

where u_j is the jth input, y is the ith output, d_j is the time delay of the jth input. The following recursive function can be obtained from (3).

$$\hat{y}(k+p/k)=-\sum_{i=1}^{n_a}A_i\hat{y}(k+p-i/k)$$

$$+\sum_{j=1}^{3}\sum_{i=1}^{n_{bj}} B_{ji}u_j(k+p-d_j-i)+\sum_{i=1}^{n_c}\zeta(k+p-i) \quad (4)$$

When p-i≤0, $\hat{y}(k+p-i/k)=y(k+p-i)$. The parameters of the system are identified in real time with the extended recursive least square estimation, let

$$y(k)=\varphi^T(k-1)\theta(k)+\zeta(k) \quad (5)$$

where

$$\theta(k)=[a_1,a_2,\ldots a_{n_a};b_{10},b_{11},\ldots b_{1n_{b1}};b_{20},b_{21},\ldots b_{2n_{b2}};b_{30},b_{31},\ldots b_{3n_{b3}};c_1,c_2,\ldots c_{n_c}]^T$$

$$\varphi^T(k-1)=[-y(k-1),\ldots,-y(k-n_a);\ u_1(k-d_1),\ldots,u_1(k-d_1-n_{b1});u_2(k-d_2),\ldots,u_2(k-d_2-n_{b2});u_3(k-d_3),\ldots,u_3(k-d_3-n_{b3});\xi(k-1),\ldots,\xi(k-n_c)]$$

where $\theta(k)$ is the vector of parameter, $\varphi^T(k-1)$ is the vector of data. if $\zeta(k)$ in (5) is replaced with $\hat{\zeta}$, then we have

$$\hat{\zeta}(k)=Y(k)-\varphi^T(k-1)\hat{\theta}(k-1) \quad (6)$$

where

$$\varphi^T(k-1)=[-y(k-1),\ldots,-y(k-n_a);u_1(k-d_1),\ldots u_1(k-d_1-n_{b1}),u_2(k-d_2),\ldots,u_2(k-d_2-n_{b2})\ u_3(k-d_3),\ldots,u(k-d_3-n_{b3}),\hat{\zeta}(k-1),\ldots,\hat{\zeta}(k-n_c)]$$

The recursive algorithm of extended least squares is given below.

$$\hat{\theta}(k+1)=\hat{\theta}(k)+K(k+1)\hat{\zeta}(k+1) \quad (7)$$

$$\hat{\zeta}(k+1)=y(k+1)-\varphi^T(k)\hat{\theta}(k) \quad (8)$$

$$K(k+1)=\frac{P(k)\varphi^T(k)}{\lambda+\varphi^T(k)P(k)\varphi^T(k)} \quad (9)$$

$$P(k+1)=\frac{1}{\lambda}\left[P(k)-K(k+1)\varphi^T(k)P(k)\right] \quad (10)$$

When a new pair of input and output values are collected once more, *i.e.* the parameters estimation will be done. Taking the new parameters into formula (5), the estimation values of the output $\hat{y}(k+1/k),\ldots,\hat{y}(k+p/k)$ can be get by recursive calculation. let

$$u_1(k)=U_1(k)-U_1(k-1),u_2(k)=U_2(k)-U_2(k-1)$$
$$u_3(k)=U_3(k)-U_3(k-1),y(k)=Y(k)-Y(k-1)$$

where $U(k),U_2(k),U_3(k),Y(k)$ are total quantity of blowing oxygen gas, electricity supplied, ore adding quantity and the carbon-content (or temperature) respectively. Hence we can get the absolute value predictive function of output.

$$\hat{Y}(k+p/k)=Y(k)+\hat{y}(k+p/k) \quad (11)$$

2.3 The Mathematical Model of the Refining Stage

In the refining stage a new task is to calculate the weights of the added alloy for the refining. The optimal burden model is shown

$$\min\ Z=\sum_{i=1}^{n}c_ix_i \quad (12)$$

where Z is the cost of the burden. c_i is the cost of the ith kind of alloy element, x_i is the weight of the ith kind of alloy element.

The n kinds of the alloy materials are used when the m kinds of compositions of the steel melted are adjusted, the compositions of the steel melted are controlled.

(1) The decision variables

Taking the burden quantity of the ith kind of alloy material as the decision variable x_i (kg) i=1,2,⋯n

Because the burden quantity is not negative i.e,. $x_i\ge0 \quad i=1,2,\cdots\cdots,n.$

(2) The objective functions

Taking the minimum value as cost of the burden materials, the objective function will be described as

$$\min f(x) = c_1x_1 + c_2x_2 + \cdots\cdots c_nx_n = \sum_{i=1}^{n} c_ix_i \quad (13)$$

where c_i is the cost on one kilogram of *i*th alloy material

(3) The boundary conditions

The boundary of burden material compositions

$$\begin{cases} \sum_{i=1}^{n}\left(x_i \cdot a_{ij} \cdot s_{ij}\right) + Q \cdot E_j \geq \left(Q + \sum_{i=1}^{n} x_i \cdot s_i\right) \cdot Emin_j & j = 1, 2, \cdots m \\ \sum_{i=1}^{n}\left(x_i \cdot a_{ij} \cdot s_{ij}\right) + Q \cdot E_j \leq \left(Q + \sum_{i=1}^{n} x_i \cdot s_i\right) \cdot Emax_j & j = 1, 2, \cdots m \end{cases}$$

where Q is the quantity of melted steel in EAF, before the alloy materials added into the EAF. a_{ij} is the *j*th contents of composition in the *i*th alloy material (%). s_{ij} is the gains of the *j*th composition in *i*th alloy material (%). s_j is the total gains of the *i*th alloy material (%). E_j is the present analysis value of the *j*th composition contents in melted steel, before alloy material added into the EAF (%). $E_{\min_j}$ is the low boundary controlled of the *j*th composition in the melted steel (%). $E_{\max_j}$ is the upper boundary controlled of the *j*th composition in the steel melted (%).

$\left(Q + \sum_{i=1}^{n} x_i \cdot s_i\right)$ is the steel melted quantity in the EAF, after the alloy material added into the EAF.

$\sum_{i=1}^{n}\left(x_i \cdot a_{ij} \cdot s_{ij}\right)$ is the actual quantity added into the EAF of the jth composition in the steel melted.

3. INTELLIGENT COMPOUND CONTROL OF THE ELECTRODE

The control performance required for electrode are different according to various steelmaking stage. During the melting stage, the disturbance is strong, the arc is unsatble and the deviation is usually large, it needs to be adjusted rapidly and no overshoot for the control system of the electrode. So the Bang-Bang control strategy is used during this period. The disturbance becomes weak, the arc is stable and the deviation is also small in the oxidizing stage, the speed of the adjustment and the precision of the control are considered at the same time. For this reason, the fuzzy control technique is taken. Finally, for the refining stage, the conditions of the furnace are relatively stable control in this stage becomes very important, therefore, the PID control strategy is used. This is the basic ideas of the intelligent compound control of the electrode, the intelligent compound control system structure diagram is shown in Fig.2. This design philosophy makes the project easily realized and would be easily accepted by the furnace-operators.

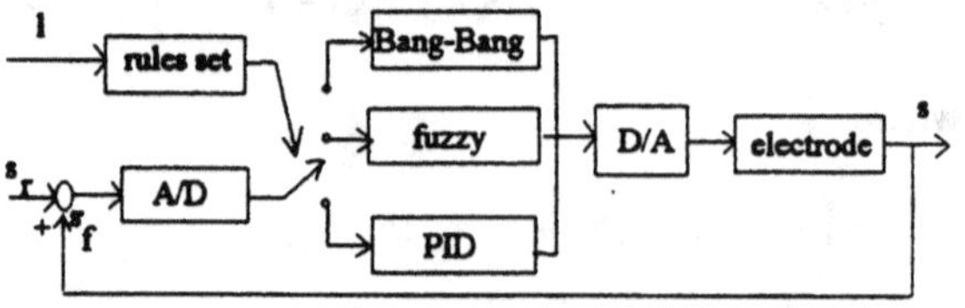

Fig.2. electrode control structure diagram

Because the Bang-Bang control and PID control strategies are easy to understand, it is not described here in detail. The fuzzy control rules based on the design philosophy of fuzzy controller are described in detail. When the deviation of the arc length e is sampled from the actual production process, the rate of the deviation change is computed with the following equations.

$$E = INT\left(e\frac{6}{E_M}\right) + sgn(e) \quad (14)$$

$$\dot{E} = INT\left(\dot{e}\frac{6}{\dot{E}_M}\right) + sgn(\dot{e}) \quad (15)$$

where, the change range of the deviation is limited in $(-E_M, +E_M)$, the rate of the deviation change is limited in $(-\dot{E}_M, +\dot{E}_M)$. Due to the control variable u is broken down into 13 degrees that is -6,-5,-4, -3, -2, -1, 0, 1, 2, 3, 4, 5, 6, but the change range of the control variable *u* is in (-2.5 , +2.5), so that taking the quantified factor *q=2.5/6=0.42,* then we have

$$u_c = qu \quad (16)$$

The program of the fuzzy control can be designed, according to (14), (15), (16)

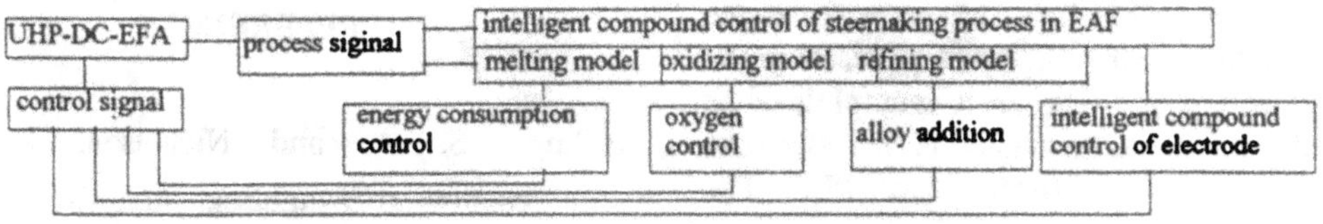

Fig.3. control strategies configuration diagram

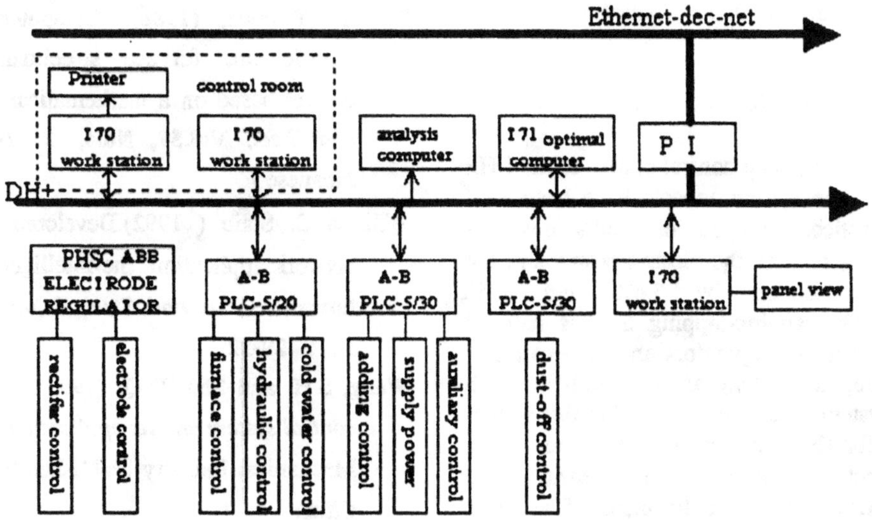

Fig.5 general structure diagram of distribute computer system

The intelligent compound control is realized by means of the simple rules set according to the arc deviation e, the main rules are as follow

$if \quad e_i > 3.1 \quad then \quad u_c = u_m$

$Bang\text{-}Bang \ control, \ the \ electrode \ raised rapidly,$

$if \quad e_i < -3.1 \quad then \quad u_c = u_m$

$Bang\text{-}Bang \ control, \ the \ electrode \ downs \ rapidly,$

$if \quad 0.5 < |e_i| \le 3.1 \quad then \ u_c = fuzzy \ control \ output;$

$if \quad 0.1 < |e_i| \le 0.5 \quad then \ u_c = PID \ control \ output;$

$if \qquad |e_i| \le 0.1 \quad then \ u_c = 0$

A experiment result of the electrode raising and downing intelligent compound control is shown in Fig.4.

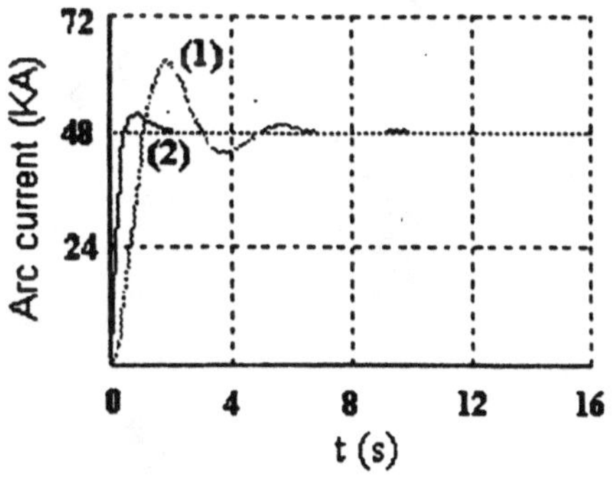

Fig.4 the output curves of the system when input is step signal (in the refining stage)

Curve (1) is the result of the intelligent compound control and curve (2) is the result of the PID control .

It is shown clearly from the experiment result that the intelligent compound control of the electrode takes the advantages each of the Bang-Bang control , fuzzy control and PID control. At the same time, the disadvantages of the three control strategies are avoided, the control precision and tracking index of the system are raised under the same conditions. The response speed of the intelligent compound control is about 5 times fast than the general PID control strategy , the static error of the intelligent compound control is 0.067 mm, but the static error of the PID control is 0.201 mm,there fore, the intelligent compound control is a kind of advance technique for the steelmaking in UHP DC EAF. The control strategies configuration diagram are shown in Fig.3.

4. DISTRIBUTED COMPUTER SYSTEM

The distributed computer control system (DCS) is employed to decompose the multivariable complex process into several single-variable control loops, and the control problems become the tracking and regulating problems in each control loop, (see Fig.5).This is a multi-level computer control system, the scrap steel field, the flux storehouse of the burden and charging of alloy cabin as wells as the operation of EAF are controlled with microcomputer subsystem respectively. The basic control level is constructed with digital regulator of the electrode, auxiliary system PLC, dust-off system PLC. The second level is the process control computer, it is used to supervise and control the whole smelting process. The third level is the center computer, it is used to manage and coordinate production plan.

5. CONCLUSIONS

The intelligent compound control system of the UHP DC EAF and the system models have been well worked in practice. The operation results show that the steelmaking reduces the electric power and the electrode consumption by 5.6% and 12%, respectively. The tapping-tapping time is reduced about 10 minutes. The operators and the experts of the steelmaking technology are very satisfied with the control system. The strategies used in this system are suitable for the industrial process control with slow parameters variation, nonlinearity and uncertainty when the estabblishment of accurate mathematical models is difficult. Therefore, this method is very effective for the complex system with physicochemical reaction industrial process .

REFERENCES

Billings, S. A. and Nicholson, H. (1977). Temperature-Weighting Adaptive Controller for Electric Arc Furnace. *Ironmaking and Steelmaking*, Vol.4, No.4, 216-221

Takeshi Takawa, (1988). Computer control of electric arc furnace steeelmaking process control based on a mathematical model, *Iron and Steel*, Vol.59, No.4, 2122-2129, (in Japanese)

Willima E. Staib (1992).Development in neural network application: the intelligent electric arc furnace. *Iron and Steel Engineer* Vol. 69 No.1, 29-32.

Wang, S.H. and Shu, D.Q. (1995). *The intelligent control system and tis application* ,205-231. Mechanical Industry Published House, Beijing, China

Keyser, D.R.M.C. and Cauwenberghe, V.A.R. (1982). Simple self-tuning multistep predictors, 6th IFAC symp. on Identification and System Parameter Estemation. 1558-1563.

DESIGN OF AN ALGORITHM FOR PLC FAULT DIAGNOSIS SYSTEM

Won-Chul Bang, Zeungnam Bien

Electrical Engineering, KAIST, 373-1, Kusong-dong, Yusong-gu, Taejon 305-701, KOREA

Abstract: In this paper, is developed a diagnosis algorithm, which can be used when a PLC is malfunctioning or has a fault. The algorithm helps operators diagnose a failed system to minimize recovery time when a production line or system is stopped. It is shown that the proposed algorithm can diagnose faults which a typical self-diagnostics in the commercially available PLC cannot. *Copyright © 1998 IFAC*

Keywords: Programmable logic controllers, Fault diagnosis, Sensor failures, Databases, Expert systems

1. INTRODUCTION

A PLC (Programmable Logic Controller) is the most widely used general-purpose controller in many factories. A fault of a PLC may stop a process, which causes the total production line to be paralyzed where a series of processes is performed by stages. To decrease the time to repair of a fault of a PLC is directly connected to the higher productivity and thus important in economic sense. In the field, however, no systematic way to cope with this problem has been prepared except self-diagnostics of PLCs and no previous study has been done. It is required to prevent faults of PLCs from occurring in advance or to cope with the faults when they happen.

The previous study on faults of a system is divided into two fields. One is fault tolerant system in which a reliable system is constructed to tolerate faults and the other is fault diagnosis system where a diagnosis system intervenes when a fault occurs in a system (Siewiorek, 1992).

The typical method to make a control system reliable by using fault tolerant system technique is to apply hardware redundancy, for example, to make a system dual in order to change main system to backup system when faults are detected. Many researchers have been studied in this area. Several related problems and some points to be considered in the redundant control systems were pointed out by Schrodi (1984), a backup control system for fossil power plant was developed by Kim (1987), and fault control tolerant system using switching redundancy was studied by Bien (1993).

Fault diagnosis using fault trees has been studied in detail by Fussell (1973), Bennettes (1975), and Sheen (1995). Quantitatively, analytical-model-based methods have been extensively studied by Willsky (1976), Hosseini (1984) and Frank (1990), while expert systems and other knowledge-based schemes for diagnosis have been proposed by Rich (1987) and Bien (1995).

As for PLCs, the backup system with a dual PLC has been developed during the recent years, however, for all the previous PLC systems with single structure, it is required to add a supplementary system to existing one. Even if it were so, the added system might have effect on the existing system. On the other hand, to make a fault diagnosis expert system for PLCs, it is necessary to systematize the symptoms appearing in the given specific system since expert system confines the problem in a narrow area. In the case of PLCs, the symptoms are to be changed whenever the plants to be controlled by PLCs are changed. Hence it is difficult to develop a fault diagnosis expert system for PLCs and it does not have been developed till now.

It is first proposed, in this paper, a systematic method to diagnose PLCs integrating some techniques. It finds the location and the cause of a fault or helps operators do so. In section 2, PLC systems and the fault examples are analyzed. In section 3, the proposed algorithm is presented. Next, some experimental results with a PLC and a virtual plant are shown in section 4. Finally, in section 5, conclusion of this paper is discussed and further works are considered.

2. ANALYSIS OF PLC SYSTEMS AND FAULT EXAMPLES

A PLC is a digital operated electric apparatus that uses a programmable memory for the internal storage of instructions for implementing specific functions, such as logic, sequencing, timing, counting, and arithmetic, to control machines or processes though digital or analog input or output modules, various types of machines or process (Simpson 1994). There can be various types of faults in PLCs since it does not operate by itself but with a plant to be controlled and I/O devices.

2.1. *Structure of PLC Systems*

Regardless of size, complexity, or cost, all PLCs contain a basic set of components, or parts. Some of these parts are hardware items, and others represent the functional characteristics of the PLC software, or programs. All PLCs have input and output units, memories, a method of programming, a CPU, and a power supply (Simpson 1994). These functions are shown in the functional block diagram of Fig. 1.

The input unit provides a connection to the machine or process being controlled. The principal function of this unit is to receive and convert field signals into a form that can be used by the PLC.

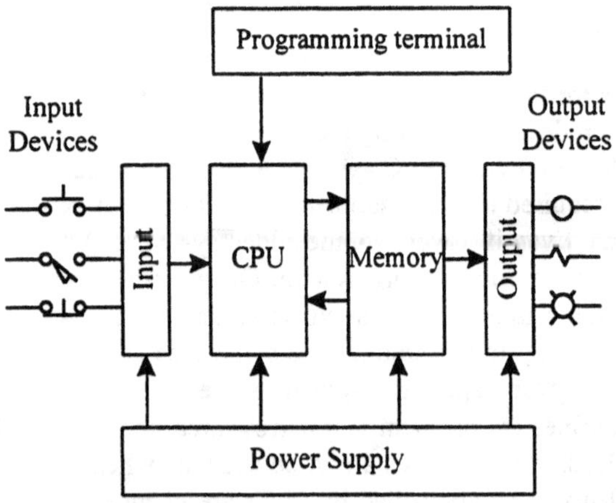

Fig. 1. Functional block diagram of a PLC

The output unit performs the opposite function of the input unit. It takes signals from the CPU and translates then into forms that are appropriate to produce control actions by external devices such as solenoids, motor starters, etc.

The CPU and memory provide the main intelligence of a PLC. Instruction comes from the input section and, based on the stored program, the CPU performs logical decisions and drives output. The CPU continuous refers to the program stored in memory for instruction concerning its next action and for reference data. It also uses memory as a scratch pad to store outside data for future use or for intermediate action when some sort of decision-making operation is involved. The CPU is composed of ALU (Arithmetic Logic Unit) performing all arithmetic and logic operations, timers and counters that allow automatic control systems to make output devices active during specific stages in a process operation.

A program that is written by a user and stored in a PLC's memory is a representation of the actions required to produce the correct output control signals for a given process condition. The programming language of a PLC has a grammar, syntax, and vocabulary that allows the user to write a program starting what the CPU is required to do. Most PLC languages are based on ladder logic, which is an advanced form of relay logic. A programming terminal allows a user to enter instructions into memory in the form of a program.

2.2. *Fault examples*

PLCs are directly connected to plants through I/O units to control processes. Thus, it cannot be distinguished which causes an abnormal operation or a suspension among the PLC, I/O devices or a plant itself. If an input of a PLC received from a plant has a fault, the normal PLC have no choice but to give an abnormal output which makes the PLC seem to have a fault. In this paper, therefore, a fault of a PLC is defined by "an abnormal operation or a suspension due to a PLC, I/O devices or plant itself."

In order to take appropriate actions when a fault occurs, it is important to find the location and the cause of it. The causes of faults of PLCs are summarized as follows. First, aging of parts of PLCs. Poor environment due to temperature, humidity, and dust, etc., decreased performances of parts due to long-term use and use of parts with poor quality make a PLC have a fault. Second, poor contact, which is caused by deformation or deviation due to noxious gas, moisture, and long-term use. Third, inferior design or manufacturing due to improper capacity or poor design for protection against heat, which increases the temperature of parts inside the PLC.

Table 1. Statistics on the faults by causes

	'91	'92	'93	'94	Total
Aging	-	2	-	2	4
Poor Contact	3	5	2	1	11
Inferior design /manufacturing	-	-	1	1	2
Total	3	7	3	4	17

Table 2. Statistics on the faults by locations

Fault type / Source	Fault Distribution
CPU parity error, processor, etc.	5%
Power supply, I/O unit failure	15%
Wring-broken wire, shorted wire	5%
Actuators	30%
Sensor	45%
Total	100%

Fault examples obtained from a very large scaled factory having a thousand of PLCs are analyzed. Table 1 shows the MTTR (Mean Time To Repair) is 3 hours and 22minutes. It can be known that there are most of faults in I/O devices rather than a PLC itself. Another statistics on the faults by locations is shown in Table 2.

Table 2 gives very important information, which the faults in I/O devices (75%) are more frequent than those in a PLC itself (25%) are. This implies that when there occurs a fault in I/O devices, it can be happened that an output unit of a PLC may send wrong output signals to a plant though the PLC is perfectly normal. Thus, it is clear that sensors and actuators of a PLC should be checked first.

2.3. *Categorization of fault examples*

Once the types of faults are categorized, it is possible to collect fault examples systematically. This enables us to accumulate history of faults of PLCs in database and utilize to find the location of faults effectively. Considering the fault examples in the previous subsection, the types are categorized by locations in table 3. First, it can be divided by two, e.g., software failure and hardware failure. There are some sorts of faults in each unit and it includes faults outside the PLC as well as inside it.

Since it is assumed that only one PLC is considered, the faults outside the PLC are excluded in this paper. The faults in the base unit also excluded since it is used to experiment and thus is assumed normal as presented in the next section.

Table 3. Types of faults by locations

<table>
<tr><th colspan="3">Location</th><th>Faults</th></tr>
<tr><td colspan="3">S/W</td><td>Bugs in sequence program</td></tr>
<tr><td rowspan="9">H/W</td><td rowspan="7">Inside PLC</td><td>CPU unit</td><td>Program memory abnormal
Data memory abnormal
Command process abnormal
I/O controller abnormal</td></tr>
<tr><td>Power supply unit</td><td>5V supply abnormal
24V supply abnormal
Fused</td></tr>
<tr><td>Digital input unit</td><td>Inaccessible from CPU
Disagree b/w signal from input unit and signal read by CPU</td></tr>
<tr><td>Digital output unit</td><td>Inaccessible from CPU
Disagree b/w signal written by CPU and signal from output unit</td></tr>
<tr><td>Analog input unit</td><td>Inaccessible from CPU
A/D conversion abnormal
Disagree b/w A/D result and digital data read by CPU</td></tr>
<tr><td>Analog output unit</td><td>Inaccessible from CPU
D/A conversion abnormal
Disagree b/w data written by CPU and digital data to be converted</td></tr>
<tr><td>Base unit</td><td>Connection open/short</td></tr>
<tr><td rowspan="2">Outside PLC</td><td>Input from plant</td><td>Digital input abnormal
Analog input abnormal</td></tr>
<tr><td>Network</td><td>Link to other PLC abnormal</td></tr>
</table>

3. PROPOSED ALGORITHM

In addition to self-diagnostics provided by makers, wrong input backward tracking method, query by keywords of symptoms in database, and test sequence programs by units are used to diagnose. The diagnostic algorithm takes several steps to find the locations and the causes. All steps including all possible methods are properly placed at their orders based on the analysis of fault examples.

3.1. *Proposed algorithm*

Self-diagnostics; The first step is to fetch self-diagnostics results executed by a PLC itself and is a unique method to diagnose until now in the field, and then to show those results to operators with detailed instructions. The main function of self-diagnostics is to write an error code in a pre-specified area of memory when a fault is detected by a PLC itself. The detailed functions of the self-diagnostics, however, are different from one another. To make it generalize, the summarized table including the error codes for each PLC is built in database and only the corresponding part of the table is loaded on memory when a fault occurs to show operator the error codes. Since almost faults which are not serious can be solved by self-diagnostics and the result of self-diagnostics can be obtained immediately, it is referred first of all.

Wrong input backward tracking; The second step is to diagnose input sensoring lines associated with a certain set of output line which shows unexpected performance. Since operator can usually understand which output lines are malfunctioning, it shows an associated set of input sensoring line by analyzing the sequence program loaded in a PLC if an operator selects a set of output lines which are considered suspicious. Suppose an operator step on input sensoring line by mistake to make the input signal wrong which causes wrong output from a PLC. Then abnormal operation can be observed moreover it can be found which output point is wrong at sight. Now, operator keys in the output point number which causes the abnormal operation, then wrong input backward tracking algorithm gives which inputs have an effect on the output point by analyzing the currently running sequence program.

The sequence programs running in factories are usually of a thousand of lines in mnemonic codes, and thus it is required to take an algorithm to find the I/O relation in the shortest time. DLL (Doubly-Linked List) is used to find the corresponding inputs for given output in a sequence program. DLL is a kind of list in which each node contains a pointer to both its successor and its predecessor, so an insertion or deletion requires only the knowledge of one node. By using this concept, all the input, output and internal variables in a sequence program are considered as nodes and the relations between inputs and outputs are stored in the nodes as a pointer to the address. Now the corresponding inputs which directly or indirectly affect the output are to be found by the DLL structure.

For example, a sequence program is shown in Fig. 2. The corresponding DLL structure can be represented as in Fig. 3. If two output points *Y0020* and *Y0021* are assumed suspicious, DLL structure provides the following results. First, input points affecting *Y0020* are *X0000*, *X0001*, *X0002* and *X0003*. Similarly, input points affecting *Y0021* are *X0000*, *X0001*, *X0002*, *X0003* and X0004 though input points directly affecting *Y0021* are *X0000*, *X0004*, *Y0020* and *Y0030*. AND and OR operations for above two give the final results *X0000*, *X0001*, *X0002*, *X0003* and X0004, and *X0000*, *X0001*, *X0002* and *X0003*, respectively. In case of multiple output abnormal like in this example, the result can be presented by AND and OR operation, and they give operator the core and full information, respectively.

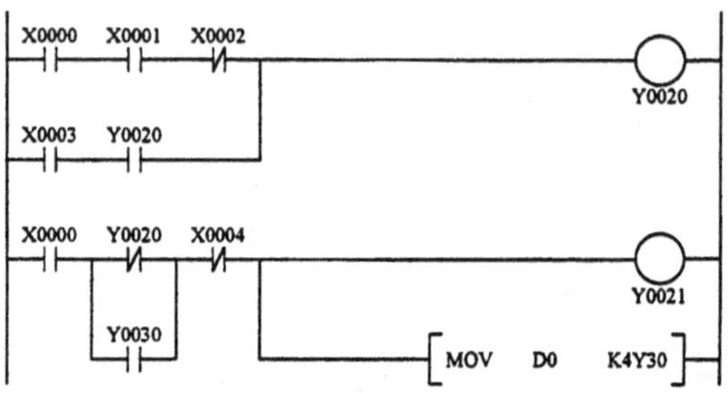

Fig. 2. An example of sequence program

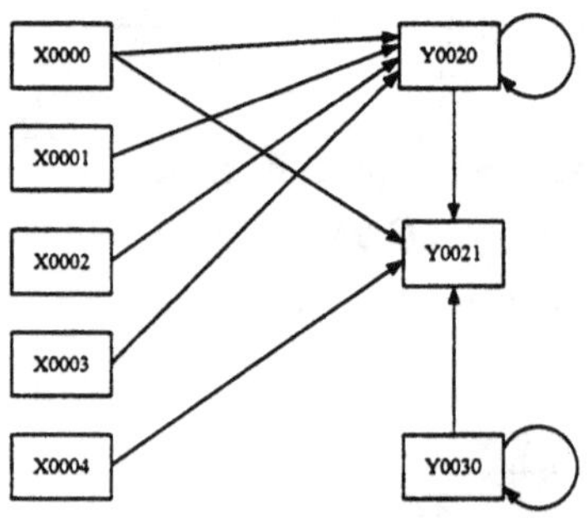

Fig. 3. Structure of DLL for a given example

Query by keywords of symptom; The third step is to search the fault history of a failed PLC based on the query with keywords of the fault symptoms. Operators can search keywords of symptoms in the database which is collected and ordered properly. As the amount of data increases, the system can tell operators more information on deciding possible cause of trouble. It is required to determine keywords of symptoms for each PLC in advance.

Test sequence programs by units; As a final step, it uses a set of specially designed PLC sequence programs by which the system diagnosis a unit by unit. The PLC sequence programs for CPU unit, digital I/O units and analog I/O units are developed for that purpose. It required to backup the original sequence program running currently and download the test sequence programs one by one. It takes more times for this step to be done, and thus this is executed later than anything else.

CPU, digital I/O units and analog I/O units are tested in order. First, as for CPU unit, memory, ALU, timer and counter test sequence programs are sequentially downloaded. Since it is difficult to test ALU, timer and counter without using memory by sequence programs, memory unit should be tested first of all in CPU unit. To test memory unit without ALU, timer and counter, memory test should be conducted directly. Thus it is done by writing to memory via a link unit which enables user to read and write directly and is assumed not to have fault. After testing memory, the others are tested in order by downloaded sequence programs in which all functions are used once to check each part.

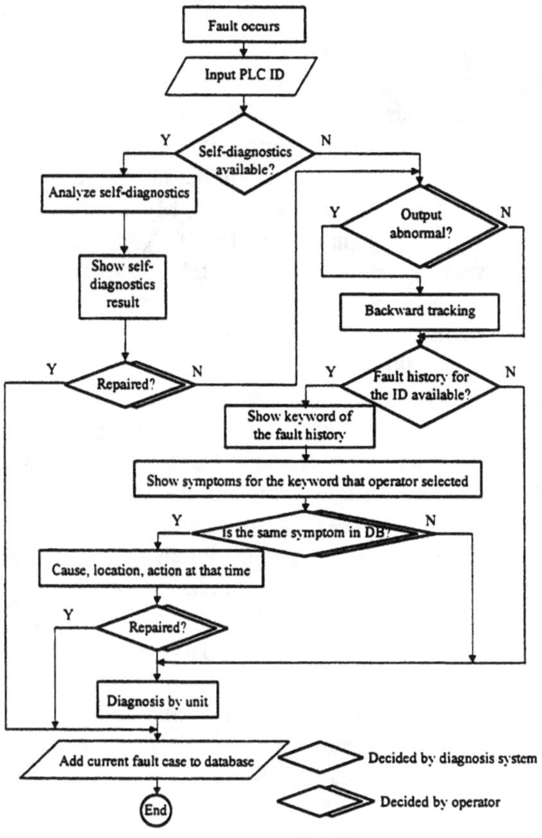

Fig. 4. Algorithm for Fault Diagnosis

3.2. *Fault diagnosis system*

The proposed algorithm presented in the previous subsection can be implemented as in Fig. 4 and is implemented in a Pentium PC. All programs were implemented in C language. The over all configuration for the fault diagnosis system is shown in Fig. 5. The fault diagnosis system has database in which the specifications, fault history and the field environment for all considering PLCs are recorded

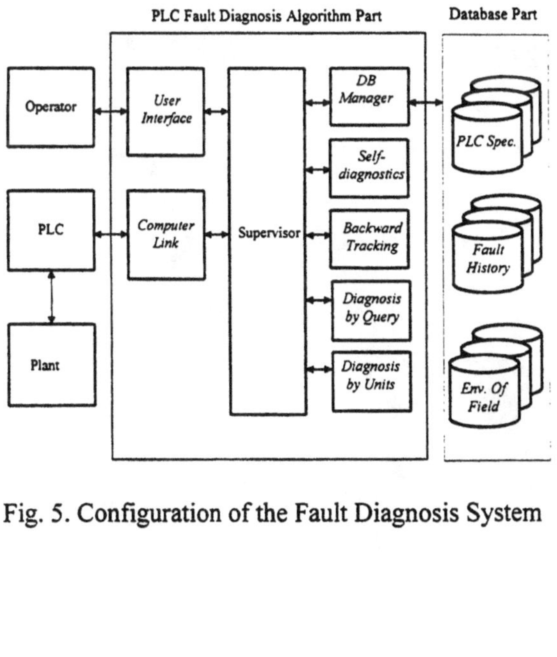

Fig. 5. Configuration of the Fault Diagnosis System

4. EXPERIMENTATION

To validate the proposed algorithm, experimentation is conducted with a real PLC and a virtual plant. A PLC with 512 I/O points and 1.0 μ sec/step processing time is used and a softwarized plant is developed to be controlled by the PLC. In addition to those, the fault generator is needed to diagnose faults.

4.1. *Fault generation*

It is required to occur failures occasionally to evaluate the performance of the proposed diagnosis algorithm. To make intended faults in a real PLC, the virtual faults are used, which occur when the connection lines in a PLC system bus are open or short.

To implement the fault generator, all the connection lines in the system bus of PLC are pre-opened and they are able to be open or short via DIP switch by operator.

4.2. *Virtual plant*

By the definition of the fault of a PLC in this paper, to deal with the external faults of a PLC, a plant is also required. A plant simulator is developed with another Pentium PC using an I/O board. After transformed the voltage from the PLC level to TTL level by a commercially used data acquisition board with A/D, D/A converters and digital/analog I/O ports, the signals through the PLC enter into the PC where the softwarized plant is. The plant has 16 digital input/output points, respectively, and 2 analog input/output channels, respectively.

4.3. *Possible faults*

The various faults in input devices, the PLC and the plant itself are generated by the fault generator and the virtual plant. Table 4 shows the faults list to be considered.

4.4. *Experimentation results*

Fault 1,2,3,4,7 and 8 are diagnosable by both self-diagnostics and the proposed diagnosis algorithm. The others, however, are only diagnosed by the proposed method. Especially, the fault 10 is diagnosed by wrong input backward tracking algorithm and faults 5,6 and 9 are detected by testing sequence programs for the corresponding units. Query by keywords of symptoms is not considered in this experimentation since the plant is not real but virtual.

Table 4. Possible faults list for a considered PLC

	Faults	Generation method
1	*Program bug*	Wrong grammar in sequence program
2	*CPU RAM inaccessible*	Memory protect switch on
3	*DI/DO unit not loaded*	A3 open
4	*DI/DO inaccessible*	D0, D1 open
5	*DI opr. abnormal*	D2, D3~D8 open A6~A13 open
6	*DO opr. Abnormal*	D3, D4~D8 open A6~A13 open
7	*AI/AO unit not loaded*	A3 open
8	*AI/AO inaccessible*	D0~D1, D3~D8 open
9	*AI unit suspension*	A3 open
10	*Sensor failure*	Using plant simulator

5. CONCLUSION AND FURTHER WORKS

The causes of PLC failures are analyzed and the types of faults are categorized from the examples obtained from a thousand of PLCs. The proposed algorithm includes use of self-diagnostics, wrong input backward tracking for sensor failures, searching by query and test by using sequence programs for each unit. The order of them is optimized to minimize the MTTR. Experimentation is conducted with a PLC and a virtual plant implemented by software to show that the proposed algorithm can diagnose faults which a typical self-diagnostics in the commercially available PLC cannot. The more specified knowledge is collected and accumulated in the database by conducting an in-depth study on a target process, the wider range of diagnosis is obtained.

The followings can be thought of as further studies. To extend the range of diagnosis, failures in network as a lower level of CIM is to be considered. In addition, a study by analytic approach is required on this topic.

REFERNECES

Bennetts, R.G. (1975). On the analysis of fault trees. *IEEE Trans. on Reliability* **Vol. R-24**, pp. 175-185

Bien, Zeungnam and Heegyoo Lee (1993). Fault tolerant control with switching strategy. *International Journal of Systems Science* **Vol. 24**, No. 5, pp. 841-855

Bien, Zeungnam, et al. (1995). *Development of predictive diagnostic expert system for thermal power plant.* Final Report, KAIST

Frank, P. (1990). Fault diagnosis in dynamic systems using analytical and knowledge based redundancy - a survey and some new results. *Automatica* **Vol. 26**, pp. 459-474

Fussell, J.B. (1973). A formal methodology for fault tree construction. *Nuclear Science and Engineering* **Vol. 52**, pp. 421-432

Hosseini, S.H., J.G. Kuhl, and S.M. Reddy (1984). A diagnosis algorithm for distributed computing systems with dynamic failure and repair. *IEEE Trans. on Computers* **Vol. C-33**, pp. 223-233

Kim, Jee Hong, Hyun Yong Cho, Myung Jin Chung, and Zeungnam Bien (1987). A case study on fault tolerant control system for power plant boiler controller. *Journal of the Korean Institute of Electronics Engineers* **Vol. 24**, No. 1, pp. 28-34

Rich, S. and V. Venkatasubramanian (1987). Model-based reasoning in diagnostic expert systems for chemical process plants. *Computers & Chemical Engineering* **Vol. 11**, pp. 111-122

Schrodi, E. (1984). Fault-tolerant and fail-safe microcomputer systems for modern automatic process control by redundancy. In: *9th World Congress of IFAC Vol. 2* (J. Gertler and L. Keviczky, (Ed.)), pp.183-188. Budapest, Hungary

Sheen, Seung-Chull, Se-Hwa Park, Jae-Hyeok Lee and Zeungnam Bien (1995). A Study on the Alarm Processing System for Fossil Power Plant. *Journal of the Korean Institute of Telematics and Electronics* **Vol. 32-B**, No. 8, pp. 1-12

Siewiorek, D.P., (1992). *Reliable Computer Systems.* Digital Press

Simpson, C.D. (1994). *Programmable Logic Controllers.* Prentice Hall

Willsky, A.S. (1976). A survey of design methods for failure detection in dynamic Systems. *Automatica* **Vol. 12**, pp. 601-611

DEVELOPMENT OF A HEAVY LOAD HANDLING ROBOT AND ITS ELECTRIC - PNEUMATIC DRIVING MECHANISM

S. D. Park*, Y. Youm and W. K. Chung****

** Researcher, Process Automation Team, RIST at San 32, Hyojadong, Pohang, 790-330, Korea (Tel. : +82-562-279-6735) E-mail:sdpark@risnet.rist.re.kr*
*** Professor, Department of Mechanical Engineering, POSTECH at San 31, Hyojadong, Pohang, 790-784, Korea (Tel. : +82-562-279-2172)*

Abstract : A number of industrial robots have been used in various work task from small parts assembly jobs to heavy load handling ones. To handle heavy load, the manipulator becomes heavy because the payload to wight(p/w) ratio of the conventional robots is very low. In this research, a heavy load handling manipulator whose p/w ratio is over 1/4 was designed and fabricated. An electric-pneumatic driving mechanism was also developed to achieve fine positioning accuracy with heavy payload. The combined driving mechanism was implemented and tested in several aspects. *Copyright © 1998 IFAC*

Keywords : Double parallelogram mechanism, Arm ratio, Balancing

1. INTRODUCTION

Although industrial robots have become widely used in various industrial application fields, its use in heavy load handling jobs has been restricted because of the payload limits of conventional robots. To handle over a 50kg payload, the weight of the robot itself easily exceeds one ton to satisfy the desired motion. For example, payload to weight ratio of PUMA - 560 and 760 robots are 1/20 and 1/48, respectively. Several special mechanisms, therefore, were investigated to achieve high p/w ratio.

Tepper and Lowen (1972) introduced the concept of balancing in a planar mechanism. They verified that if total mass center is stationary by balancing mass than the mechanism becomes force balanced. Time-invariant characteristics of inertia matrix using mass distribution was inroduced by Asada and Toumi (1984). Imam and Levy (1982) showed that the maximum torque at each joint can be reduced by adding dead weight at the rear part of a four-bar linkage manipulator. These methods, however, are not profitable in practice because the total weight of robot itself is increased.

Pneumatic actuators can generate greater output per unit weight than electric ones. The position control systems using pneumatic actuators, however, are less used than the systems using electric ones since the nonlinearity chracteristics such as compressbility, viscosity, low responsibility and output stiffness, friction and the difficulty of linear flow control of compressed air. On the other hand, electric servo motors can guarantee good positioning accuracy. However, industrial robots driven by electric actuators have low p/w ratio because the output per unit weight of the actuators are low.

In this paper, kinematic parameters of double-parallelogram mechanism that maintain moving balance were derived from force equilibrium condition, and a manipulator whose payload to weight ratio is 1:4 was designed. A combined electirc-pneumatic driving mechanism was also developed. The performance of the designed driving mechanism was verified by several experiments.

2. DESIGN OF THE MANIPULATOR

Fig.1 shows the schematic configuration of the designed manipulator using double-parallelogram mechanism. It has 3-d.o.f. to give positioning motion by base rotation and horizontal and vertical movements through the corresponding slots. The kinematics of the robot is derived for the planar 2-d.o.f. joints to obtain the decoupled motion between the horizontal and vertical motions.

If the input translation vector of the slot joints is defined by $\underset{\sim}{s} = [x, y]^T$, the displacement of revolute joints by $\underset{\sim}{q} = [\theta_1, \theta_2]^T$ and the output translation vector of the end effector by $\underset{\sim}{h} = [u, v]^T$, then the time derivatives of these vectors have the following relationship (Youm *et al.*, 1988).

$$\begin{aligned} \dot{\underset{\sim}{h}} &= J_1 \dot{\underset{\sim}{q}} \\ \dot{\underset{\sim}{s}} &= J_2 \dot{\underset{\sim}{q}} \end{aligned} \tag{1}$$

where,

$$J_1 = \begin{bmatrix} -l_4 \sin\theta_1 & -l_5 \sin\theta_2 \\ l_1 \cos\theta_1 & (l_2 + l_5)\cos\theta_2 \end{bmatrix} \text{ and}$$

$$J_2 = \begin{bmatrix} l_3 \sin\theta_1 & l_2 \sin\theta_2 \\ l_3 \cos\theta_1 & l_2 \cos\theta_2 \end{bmatrix}.$$

In the case of $\det(J_2) \neq 0$, there exists a nonsingular manipulator Jacobian J such that $\dot{\underset{\sim}{h}} = J \dot{\underset{\sim}{s}}$, where J can be obtained from $J_1 J_2^{-1}$. Through the careful observation of J, we can diagonalize the Jacobian by selecting k as $k = l_5 / l_2 = l_4 / l_3$ and $l_1 / l_3 = 1 + k$, where k represents the arm ratio. This is the decoupling condition and the Jacobian is now expressed as a function of k only (Youm *et al.*, 1988):

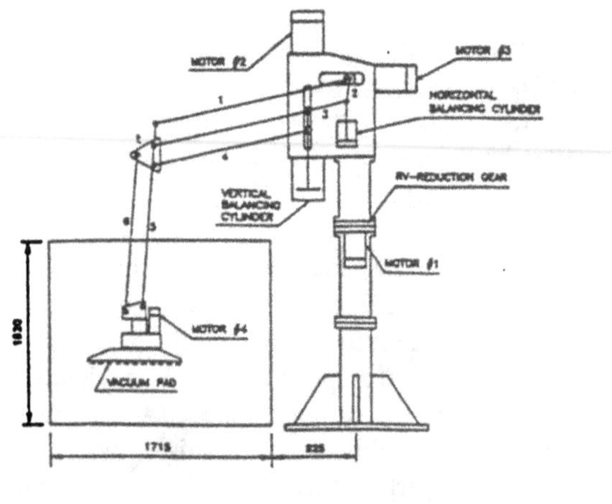

Fig. 1. Schematic of the designed manipulator

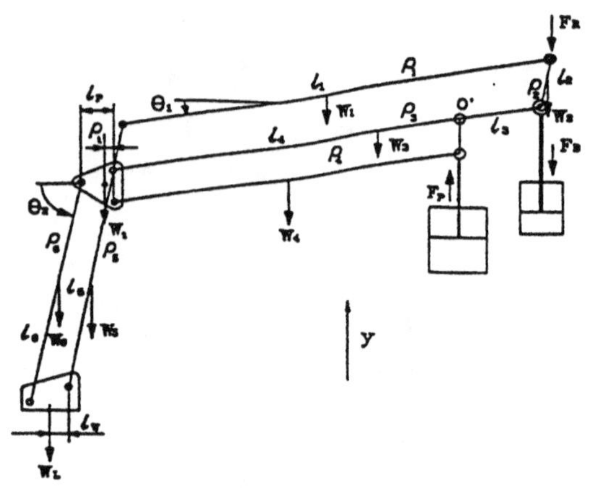

Fig. 2. Free body diagram of the arm of the designed manipulator

$$\begin{bmatrix} \delta u \\ \delta v \end{bmatrix} = \begin{bmatrix} -k & 0 \\ 0 & k+1 \end{bmatrix} \begin{bmatrix} \delta x \\ \delta y \end{bmatrix} \tag{2}$$

Under the consideration of real time control, this kinematic relation is very simple to use in robot control.

In addition to these kinematic considerations, the manipulator must be designed to have a balanced configuration irrespective of robot position and payload variations. In Fig. 2, a free body diagram of the manipulator is shown for the force analysis. By using the force and moment equilibrium conditions, the balancing equation can be obtained as follows (Park, 1993):

$$F_p = \frac{(l_3 + l_4)\cos\theta_1 + (l_2 + l_5)\cos\theta_2}{l_3 \cos\theta_1 + l_2 \cos\theta_2} W_L + \frac{A\cos\theta_1 + B\cos\theta_2 + C}{l_3 \cos\theta_1 + l_2 \cos\theta_2} \tag{3}$$

where

$$\begin{aligned} A &= W_1\rho_1 + W_3(l_3 + \rho_3) + W_4(l_4 + \rho_4) \\ &\quad + (W_5 + W_6 + W_t)(l_3 + l_4), \\ B &= (W_2 + W_3 + W_4 + W_5 + W_6 + W_t)l_4 \\ &\quad - W_2\rho_2 + W_5\rho_5 + F_B l_2 + W_6\rho_6 \text{ and} \\ C &= W_6\rho_6 + W_t\rho_t + W_L l_w. \end{aligned}$$

where l_i and ρ_i represent the length and the length of center of gravity of i-th link, respectively.

To obtain configuration-independent characteristics, F_p, the force exerted by the vertical cylinder, should be almost constant for all robot configuration i.e., F_p is independent of θ_1 and θ_2. These configuration-independent characteristics can be obtained from Eqn. (3) if the following relations are satisfied:

$$\frac{(l_3 + l_4)}{l_3} = \frac{(l_2 + l_5)}{l_2},$$

$$\frac{A}{l_3} = \frac{B}{l_2}, \qquad (4)$$

C is negligible.

Therefore, if all the conditions in Eqn. (4) are satisfied, Eqn. (3) can be rewritten approximately as:

$$F_p = (1 + k)W_L + m \qquad (5)$$

where m is decided by the weight, length and the length of center of gravity of the arms of the designed manipulator. From the second condition of Eqn. (4), the force exerted by the balancing cylinder, F_B, can be represented as:

$$\begin{aligned} F_B = [&W_1\rho_1 l_2 - W_2(l_2 - \rho_2)l_3 + W_3\rho_3 l_2 \\ &+W_4\rho_4 l_2 + W_5(l_2 l_4 - l_3\rho_5) \\ &+W_6(l_2 l_4 - l_3\rho_6) + W_t l_2 l_4)] / l_2 l_3 \end{aligned} \qquad (6)$$

Therefore, if each arm of the manipulator is designed, the balancing forces of the vertical and horizontal axes can be calculated by Eqn. (5) and Eqn. (6), respectively. If one control the pressure in both cylinders to exert the forces calculated by Eqn. (5) and Eqn. (6), the manipulator can get configuration-independent balancing characteristics. The parameters of each arm of the designed manipulator whose arm ratio is 7 are shown in Table 1. Fig. 3 is the photograph of the manipulator.

3. THE ELECTRIC - PNEUMATIC DRIVING MECHANISM AND ITS CONTROL

3. 1. The Electric - Pneumatic Driving Mechanism

Fig. 4 shows the designed electric-pneumatic driving mechanism. The vertical balancing cylinder, in Fig. 4, generates the force proportional to vertical load by

Table 1 Parameters of each arm of the designed manipulator

Arm	l_i (cm)	ρ_i (cm)	W_i (kg)
1	176	88	15.6
2	18	9	2.8
3	22	60	35.6
4	154	77	8.4
5	126	40	23.8
6	126	63	6.8
t		5	4.7

Fig. 3. The developed manipulator

Eqn. (5) and supports the load. The vertical motor controls the position of the load to the direction by a ball-screw. The horizontal balancing cylinder maintains the driving power to the direction constant, regardless of arm configuration, by generating the force calculated in Eqn. (6). The horizontal axis driving motor controls the position of the load to the direction by a ball-screw.

The basic idea of the designed mechanism is to reduce the size of the driving motors and the power transmission parts by dividing the rolls of heavy load supporting and accurate positioning with pneumatic cylinders and electric motors, respectively.

The experimental data of balancing pressure in vertical cylinder with respect to the change of load at end-effector of the manipulator are shown in Fig. 5. From the result of Fig. 5, one can recognize that the balancing pressure, that is balancing force, in the cylinder is proportional to the payload at the end-effector of the manipulator and that the approximate Eqn. (5) works well. One can also know that the payload of the designed manipulator exceeds 250kg under the condition of 6kgf/cm^2 pressure of compressed air. Therefore, the p/w ratio of the designed manipulator is about 1/4 since the weight of the manipulator is 1000kg.

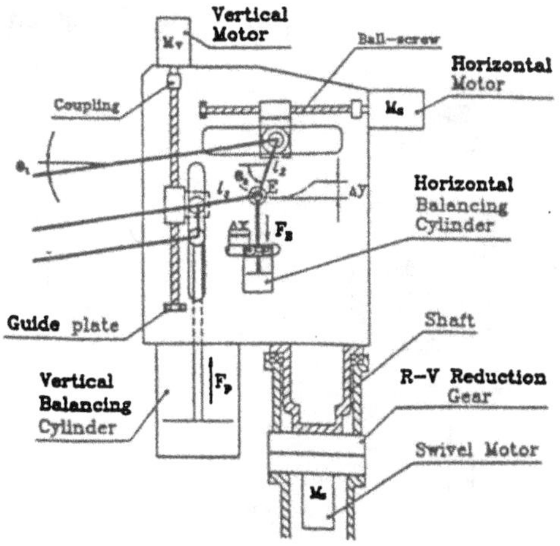

Fig. 4. Schematic of the designed electric-pneumatic driving mechanism

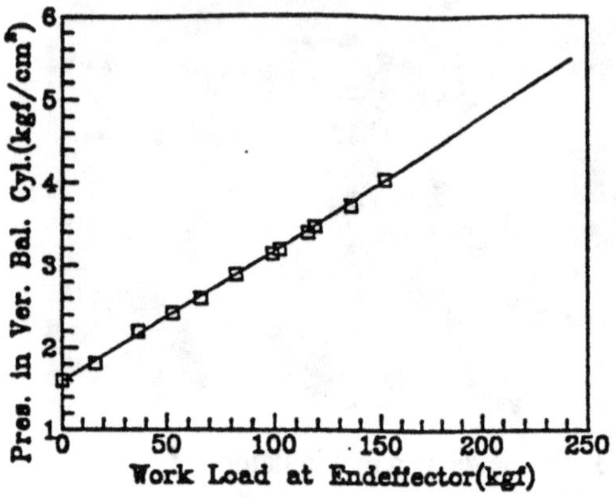

Fig. 5. The change of balancing pressure in vertical cylinder w.r.t. the change of payload

3. 2. Pressure control in each cylinder

Fig. 6 shows the schematic control circuit of pressure in vertical and horizontal cylinders. If the load at the end-effector can be measured by load cell etc., the balancing force, F_p, can be computed by Eqn. (5). The pressure, P, in vertical cylinder can be easily calculated by $P = F_p / A_c$ where A_c is the the effective area of the cylinder. The controller, in Fig. 6, controls the pressure type Electro-Pneumatic proportional valve to generate the calculated pressure P. The pressure P generated by the valve operates the air-operated regulator. The regulator makes the pressure in the cylinder to maintain P, the controlled pressure, even though the piston of the cylinder moves up and down by a ball-screw. In this balanced state, the force applied to the cylinder by a payload at end-effector is offset and a small electric motor can move the load in vertical direction.

In the case, for example, when the designed manipulator lift up 150kgf payload, the force applied to vertical balancing dylinder is 1890kgf by Eqn. (3) or Eqn. (5). If the cylinder is not used, vertical axis electric motor must generate about 300kg-cm of

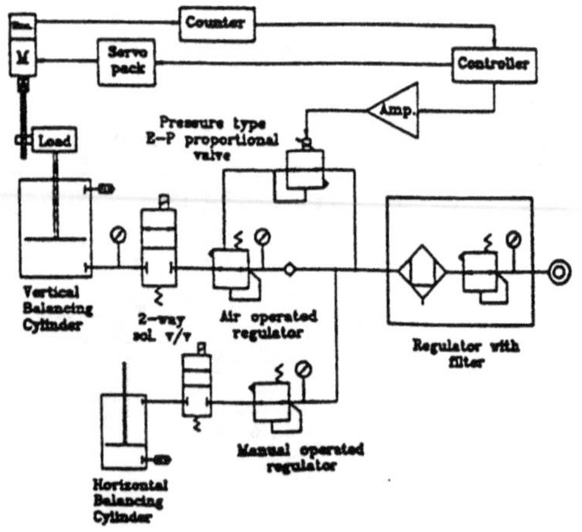

Fig. 6. Schematic of pressure control cuircuit in each cylinder

torque and the necessary diameter of vertical ball-screw is 55mm when 66.72cm/s of velocity at end-effector i.e., 66.72 / (k+1) = 8.3cm/s of velocity at vertical slot, is required under the condition of 5mm lead of the ball-screw and 0.3s of accelerating time. In the same case, however, when the developed electric-pneumatic driving mechanism is used, the required torque of electric motor and the diameter of ball-screw is 73kg-cm and 20mm, respectively. The developed mechanism, therefore, reduces 75% of required motor power and 60% of the size of ball-screw compared with motor-only driving one.

The horizontal balancing force is constant with respect to the parameters of the designed manipulator as represented in Eqn. (6). The pressure in horizontal balancing cylinder, therefore, should be controlled to exert constant required force, F_B, by manual operated regulator.

4. EXPERIMENTS AND RESULTS

Several experiments are carried out to investigate the torque charateristics of vertical axis driving motor with respect to the change of pressure in vertical balancing cylinder. Namely, the torque of vertical axis motor at various balancing pressure is measured with respect to the sine wave position input of which magnitude and frequency are ±1volt and 1Hz, respectively, under the condition of 70kgf payload. One can know, from Fig. 5, that the proper balancing pressure in vertical balancing cylinder at the payload is 2.7kgf/cm^2.

Fig. 7 shows an open loop torque response of vertical axis servo motor with respect to sine wave position input at 70kgf payload and the balancing pressure of 2.7kgf/cm^2.

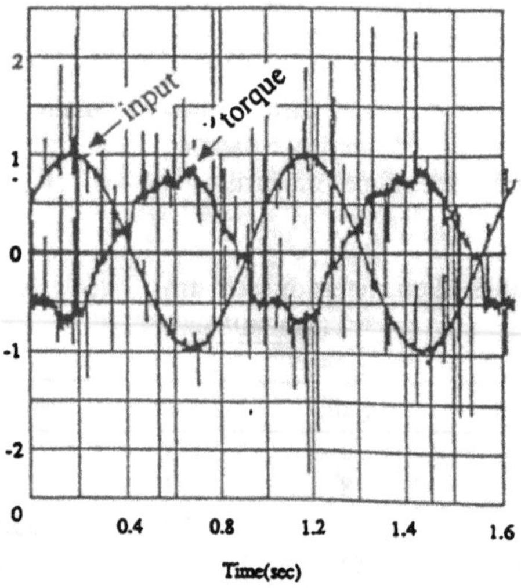

Fig. 7. Open loop torque response w.r.t. sine wave position input at 70kgf payload and the balancing pressure of 2.7kgf/cm^2

The torque signal obtained from torque monitor on the servo drive has 3volt magnitude to 100% torque(88kgf-cm). From the result of Fig. 7, one can know that the torque characteristic is stable in ±0.8volt about lifting and descending motion of the payload.

Fig. 8 is the torque response of vertical axis servo motor with respect to sine wave position input at the balancing pressure of 2.4kgf/cm^2 with the same payload. The measured torque voltage is +0.6volt and -1.5volt at down and up motion, respectively, of the load. It can be known, from the result, that relatively larger torque is required, compared with the previous result, to lift up the load at lower balancing pressure than proper one with repect to payload at the end-effector of the manipulator.

The third experiment is performed under the same conditions except that the balancing pressure is changed to 2.9kgf/cm^2 and Fig. 9 shows the result. The measured torque voltage is +1.2volt and -0.4volt at down and up motion, respectively, of the load. It can be known, from the result, that relatively larger torque is required, compared with the first experimental result, to descend the load at higher balancing pressure than proper one with repect to payload at the end-effector of the manipulator.

Finally, an experiment is executed to investigate the torque charateristics of vertical axis driving motor with respect to the change of payload at end-effector. Fig. 10 is the response when the payload is 100kgf and the balancing pressure in the cylinder is 3.2kgf/cm^2. The pressure is a proper balancing one with respect to the load from the result of Fig. 5. It can be found, from Fig. 10, that the torque characteristic is stable in ±0.9volt about lifting and descending motion of the payload.

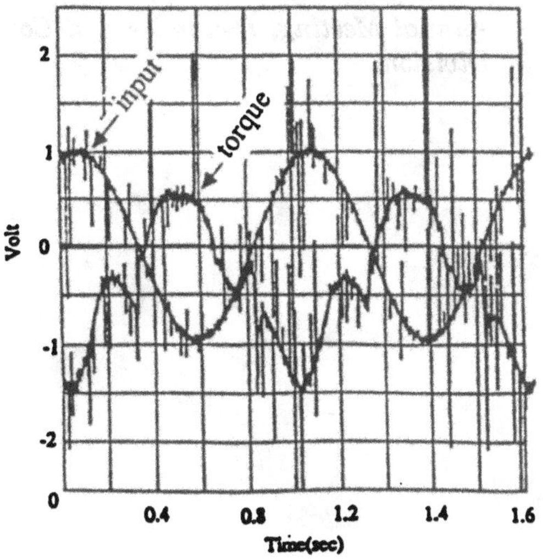

Fig. 8. Open loop torque response w.r.t. sine wave position input at 70kgf payload and the balancing pressure of 2.4kgf/cm^2

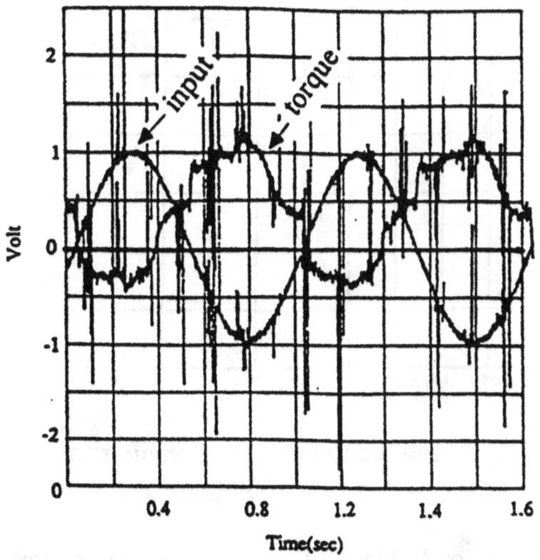

Fig. 9. Open loop torque response w.r.t. sine wave position input at 70kgf payload and the balancing pressure of 2.9kgf/cm^2

It is considered that the increase in torque valtage, compare with the result of Fig. 7, is due to it in payload.

Summarising the results of previous experiments, it can be known that the torque of vertical axis driving motor is minimized and stabilized by appropriate compensation of the pressure in vertical balancing cylinder. In practical implementation, for example, the torque of vertical axis driving motor can be considerably decreased by active control of the pressure in vertical balancing cylinder in the way of raising the pressure higher than the proper one, the value as see in Eqn. (3) and Fig. 4, during lifting the load and dropping it lower than the proper one during descending motion, respectively. The dynamics of the payload at end-effector can be measured utilizing load cell equiped as a feedback device between the end-effector and work load.

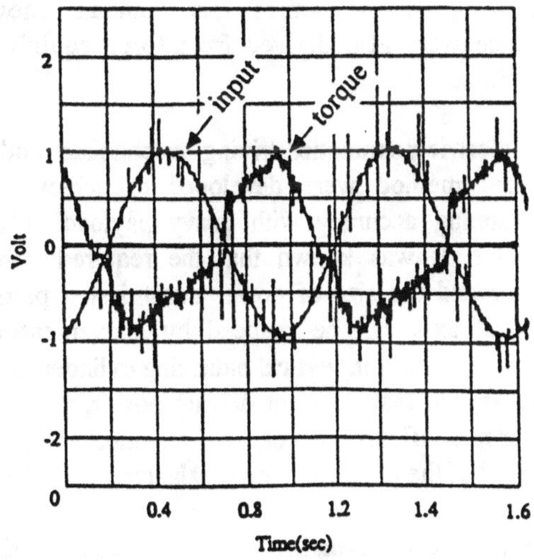

Fig. 10. Open loop torque response w.r.t. sine wave position input at 100kgf payload and the balancing pressure of 3.2kgf/cm^2

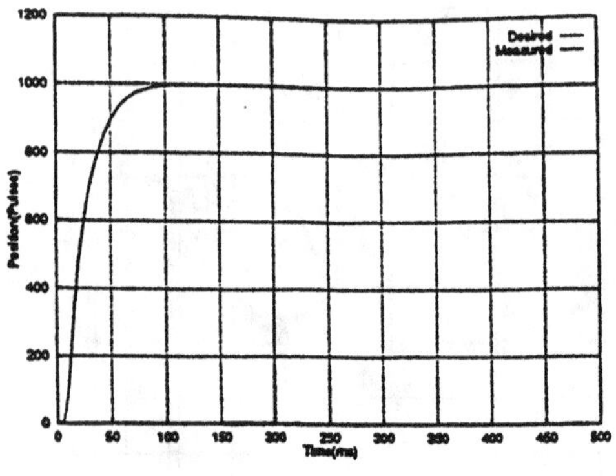

Fig. 11. Servo controlled positioning response of vertical axis w.r.t. a 1000-pulse step input

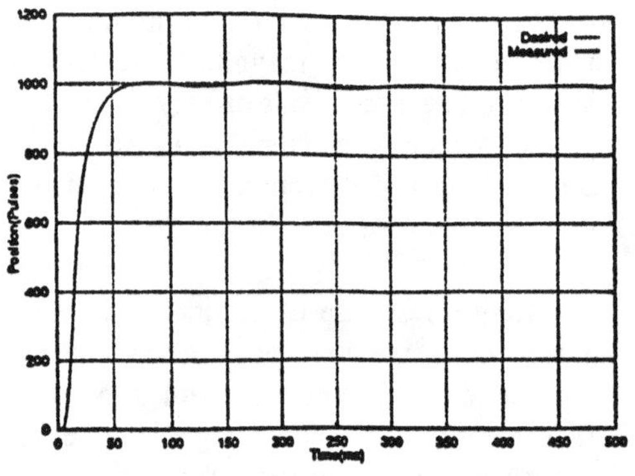

Fig. 12. Servo controlled positioning response of horizontal axis w.r.t. a 1000-pulse step input

Fig. 11 and Fig. 12 are the servo controlled positioning responses of vertical and horinzontal axis, respectively, with repspect to a step input of 1000 pulses. A step input of 1000 pulses is equivalent to a half revolution of the servo motors, 2.5mm dispalcement at both slots and 20mm of vertical and 17.5mm of horizontal displacement, respectively, at end-effector.

In case of the position control of vertical axis, a good response without overshoot was obtained in 150ms, although the experimental results are depend on the system gain. It is considered that the oversoot-free response is due to the damping effect of vertical balancing cylinder. However, a little oscillating response was got in horinzontal directional positioning control due to inappropriate gain selection.

5. CONCLUSION

In this research, a heavy load handling manipulator whose p/w ratio is over 1/4 was designed and fabricated. Kinematic parameters of double-parallelogram mechanism that maintain moving balance were also derived from force equilibrium condition.

An electric-pneumatic driving mechanism and its control method were developed to achieve fine positioning accuracy with heavy payload at end-effector. It was known that the required driving power and the size of power transmission parts of vertical axis can be reduced by abovementioned pressure control in vertical balancing cylinder. It was also known that constant driving power, regardless of arm configuration, can be obtained by proper control of the pressure in both balancing cylinders.

Finally, the performance of position control of the designed manipulator was tested by several experiments, and satisfactory results were acquired.

REFERENCES

Asada, H. and T. K. Toumi, (1984). Analysis and design of a direct drive arm with a five-bar-linkage parallel drive mechanism. *Trans. ASME, J. of DSMC*, Vol. 106, pp.225~230.

Imam, I. and S.Levy (1982). Application of advanced computer aided engineering tools for kinematic and dynamic analysis of robot systems. *AUTOFACT 4*, Vol. 3, pp.28~51.

Park, S. D., K. W. Jeong, W. K. Chung and Y. Youm (1990). Development of Control Method using both Electric and Pneumatic Actuators for a Heavy Load Handling Robot. *J. of the Korean Society of Precision Engineering*, Vol. 10. No. 2, pp. 14 ~ 21.

Tepper, F. R. and G. G. Lowen, (1972). General theorems concerning all force balancing of planar linkages by internal mass distribution. *Trans. ASME, J. of Engg. For Industry*, pp.789~796.

Youm, Y., W. K. Chung and W. J. Chung (1988) Design of a heavy payload manipulator with double-parallelegram. *Conf. on KSME Annual Meeting, Dynamics and Control Division.*

HEAT CONTROL SYSTEM FOR TUBE-MAKING BY THE EXTRUSION PROCESS

Keizo Watsuji, Osamu Sugiyama

*System Planning & Development Section,
Kansai System Planning Department, System Engineering Division.
Sumitomo Metal Industries, Ltd.
1-109, 5-chome, Shimaya, Konohana-ku,
Osaka, 554 JAPAN*

Abstract: In the *Ugine-Sejournet* method of extrusion process tube-making, tube quality is determined mostly by temperature of the materials. An integrated heat control system has been developed which controls material temperature. As well, it controls the pitch of extraction from the furnace and heaters in order to avoid their becoming radiation.

Keywords: Automatic process control, Temperature control, Model-based control, Extrusion, Timing simulator

1.INTRODUCTION

Recently, the demand for higher quality stainless tubes has greatly increased. Tube quality is mostly determined by the temperature of the materials; it is necessary to achieve and maintain adequate temperature in order to obtain processability, resistance of deformation and characteristics of lubrication. However, when customer orders are for small numbers of many different types of tube, temperature control becomes difficult.

In response to this problem, an integrated heat control system has been adopted, which allows the temperature of a variety of materials to be controlled accurately. The system calculates the temperature of materials from ordinary materials before insertion into the furnace. It also calculates the temperature of the materials while they are being conveyed from the furnace to the heaters, from heaters to presses, and from presses to heaters. In addition, the system avoids the radiation of materials and the gathering of materials in the press equipment by controlling the pitch of the extraction. This in turn stabilizes the temperature of the materials at extrusion.

2.OUTLINE OF EXTRUSION PROCESS

Fig.1 shows the outline of extrusion process. The material for the extrusion process, in the form of a billet, has a center hole. Billets are preheated in the rotary furnace, then sent to induction heaters for heating, and then sent to the vertical press for expansion before extrusion at the horizontal press. While the target temperature depends on the type of materials, it is around 1000 degrees Celsius for preheating, and 1200 degrees Celsius for the main heating.

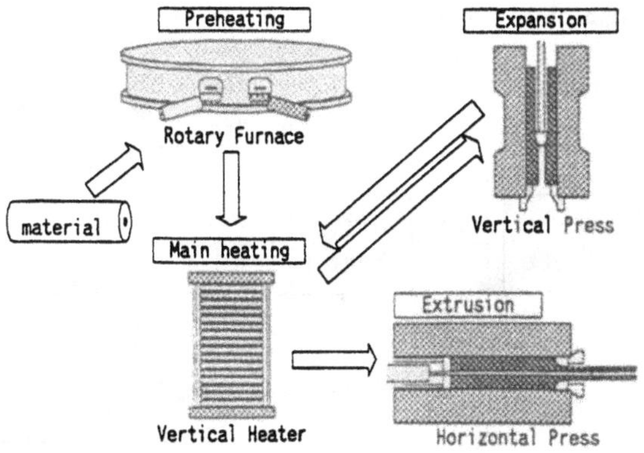

Fig.1 Outline of extrusion process

3.THE TYPES OF EXTRUSION PROCESS

For some types of billets and tubes, an expansion process at the vertical press is required before extrusion. There are three types of process; direct, expansion, and double expansion.

A. Direct process
Billets are preheated in the rotary furnace, heated up to the final temperature at the vertical heaters, and are then extruded at the horizontal press.(See Fig.2)

B. Expansion process
The center holes of billets are expanded before extrusion, after which the billets are reheated at the vertical heaters and extruded at the horizontal press. This process is used to make a large diameter tube.(See Fig.3)

C. Double expansion process
Billets are expanded twice, in order to make larger diameter tubes than is possible with the expansion process.(See Fig.4)

4.THE OUTLINE OF HEATING EQUIPMENT

A rotary furnace and vertical induction heaters are used to heat up billets in this process line;.Table1 shows the specifications. The rotary furnace can heat about 180 billets at once, but in the atmosphere of the furnace the temperature of each billet cannot be accurately controlled. The vertical induction heater, on the other hand, can accurately control the billet temperature but can only heat one at a time. Therefore, the rotary furnace is used for preheating where accuracy is less critical, and the induction heaters are used for the final heating, where achieving a precise temperature is more important.

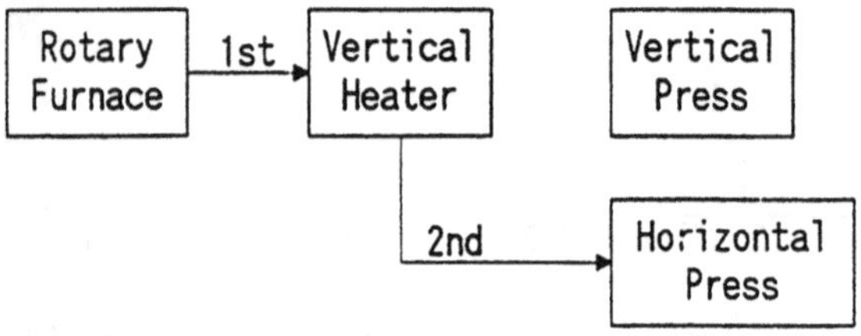

Fig.2 Flow of direct process

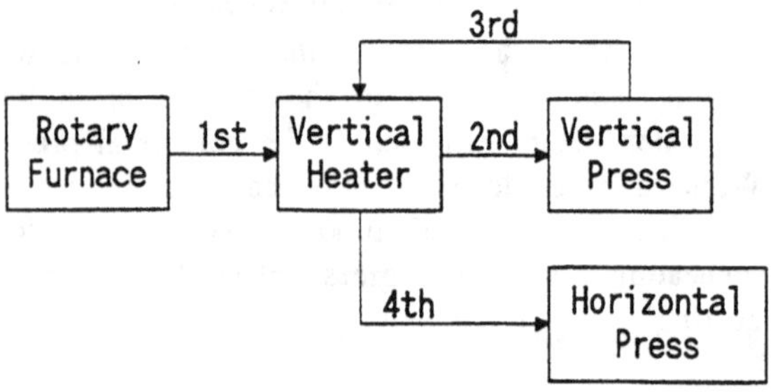

Fig.3 Flow of expansion process

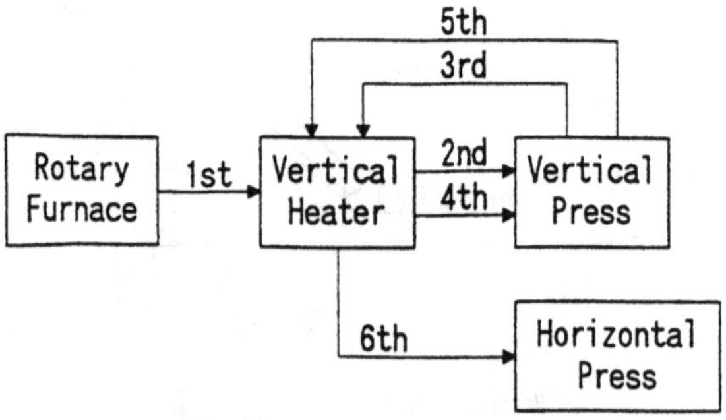

Fig.4 Flow of double expansion process

5.THE IMPORTANCE OF HEAT-TEMPERATURE

Heat-Temperature of the billets is important from the standpoints of hot processability, resistance of deformation, and lubrication characteristics.

A. Hot processability of billet
During the extrusion process, hot processability worsen as the twisting value decrease. As shown in Fig.5, twisting value is closely related to the temperature, and that the range of temperature is very narrow.

B. Resistance of deformation
The working load in the extrusion process is greatly influenced by the resistance to deformation and the billet size(min.175 ϕ～max.350 ϕ). As with twisting value, resistance to deformation is closely related to the temperature (see Fig.6). It is necessary that resistance of deformation be kept small in order to keep the working load from exceeding the maximum capacity of the press.

C. Lubrication characteristics
In extrusion process tube-making, glass powder is generally used as the lubricant. It is melted by the heat of the billet and works as a fluid lubricant, having an adequate viscosity. However, if the billet temperature is not adequate, then the viscosity will in turn be inadequate, resulting in poor lubrication, and surface defects will occur in the tube.

Table1. Specifications of the equipment

Equipment	Specifications
Rotary Furnace	1.inside zones : 4zones
	2.diameter of furnace : 14,000mm
	3.time for heating : 1～3hours
Vertical Heater	1.diameter of heater : 400mm
	2.frequency : 60Hz
	3.time for heating : 2～4minutes
Vertical Press	1.press capacity : 1200tons
Horizontal Press	1.press capacity : 4000tons

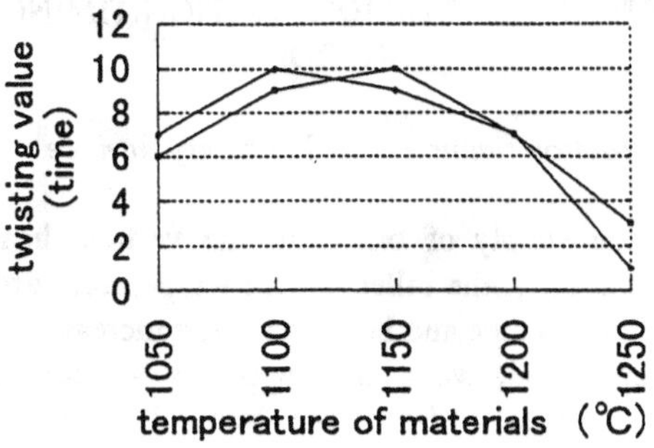

Fig.5 Twisting value

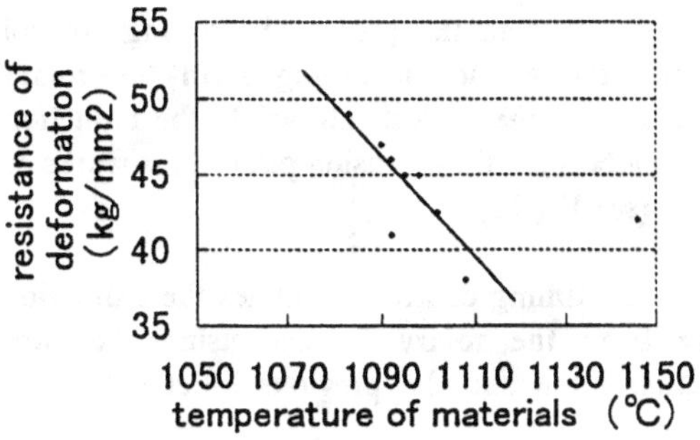

Fig.6 Resistance of deformation
(for austenite stainless steel)

6.CONTENTS OF THE SYSTEM

Fig.7 shows the hardware configuration of this process control system. It has two computers, one for the conveyance control system and one for the heat control system, together achieving a distributed function system. The conveyance control system functions are (a)automatic billet conveyance, (b)automatic setting, and (c)billet tracking on the line. The heat control system functions are (d)extraction timing control of the rotary furnace, (e)extraction timing control of the vertical induction heaters, (f)rotary furnace heat control, and (g)vertical induction heater heat control. Each function is distributed, so that the tracking function is not influenced even if heat control calculation which has high load is periodically running.

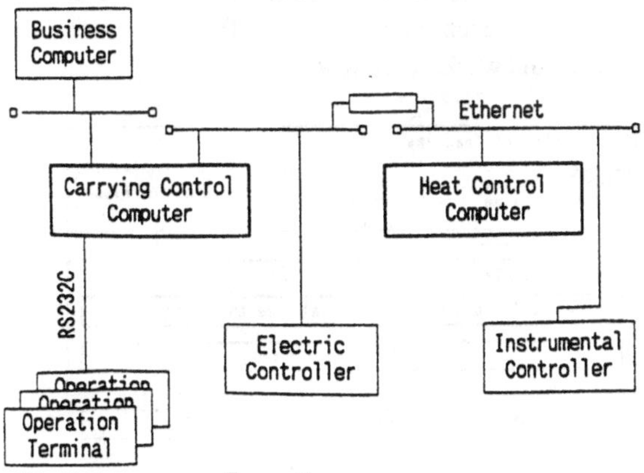

Fig.7 System configuration

7.OUTLINE OF HEAT-CONTROL MODELLING

7.1.rotary furnace heat control

The rotary furnace heat control estimates the billet temperature using the heat transfer calculation method and decides the optimal furnace temperature. It works as follows:
Fig.8 shows the temperature calculation flow chart.

a. The heat control estimates the temperature of all the billets in the furnace by using the actual furnace temperature

b. It estimates the time that each billet needs to stay in the furnace until extraction using the extrusion pitch and the number of billets in the furnace

c. It calculates the control temperature so that the prediction temperature can satisfy the target temperature, and outputs this to the instrumental controller.

7.2. Vertical induction heater heat control

This function estimates and calculates the billet's temperature in the vertical heater using the actual current value. It calculates this temperature frequently and determines the current condition so that the target temperature can be achieved.

A. Temperature calculation during billet conveyance
This function estimates the billet temperature while it is being conveyed (a)between the rotary furnace and the vertical induction heaters, (b)between the vertical induction heaters and the vertical press, and (c)between the vertical press and the heaters. With this function, the temperature accuracy for extrusion from the induction heater is improved.

B. Heat control
This function estimates and calculates the billet's temperature in the vertical heater using the actual current value. It calculates this temperature frequently and determines the current condition so that the target temperature can be achieved. Finally, it outputs an electric current on-off control signal to the instrumental controller, as shown in Fig.9.

C. Modification of the vertical heater characteristics
This process line uses eight induction heaters, each with their own characteristics. Even if the billets are heated at the same heat conditions in each heater, differences in billet temperature distribution appear. This function calculates the correction value so that there are no differences in the temperatures of the billets upon extraction from the heaters. The method works as follows: (see Fig.10)

a. The system measures the surface temperature of the billet with a surface thermometer.

b. It measures the longitudinal temperature distribution by a thermometer below every heater.

c. Finally, it calculates the character value of every heater by using these instrumental data and modifies the instrumental error of the target temperature for the next heating.

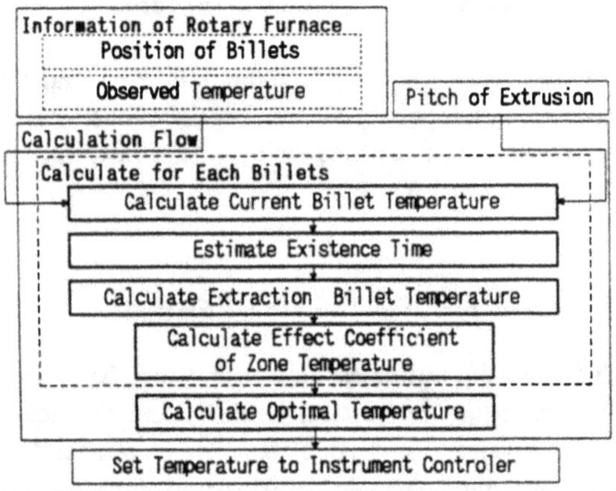

Fig.8 Flow of rotary furnace control

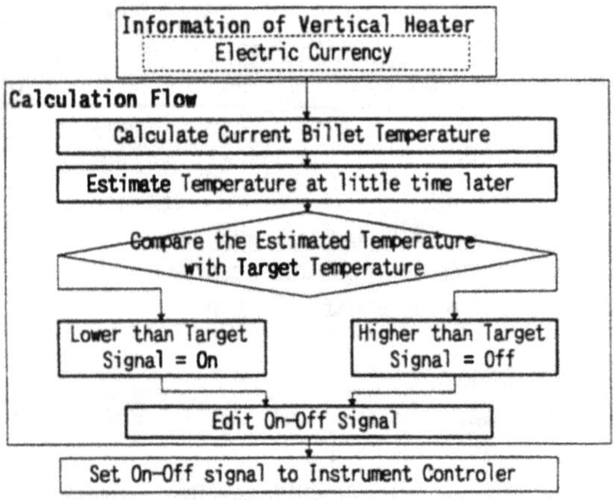

Fig.9 Flow of vertical heater control

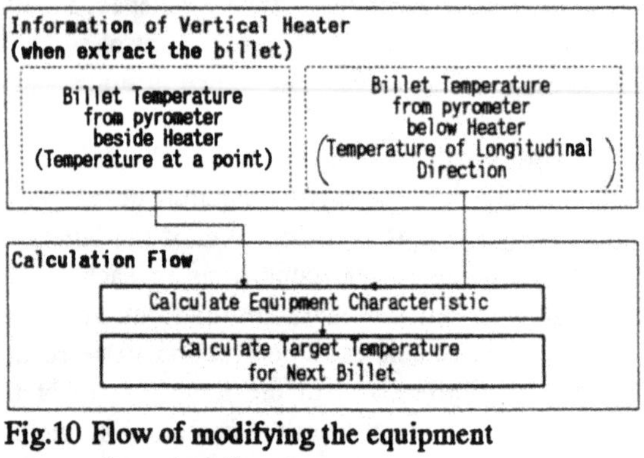

Fig.10 Flow of modifying the equipment characteristics

8.OUTLINE OF THE EXTRACTION TIMING CONTROL

8.1. Extraction timing control of rotary furnace

When the supply of billets to the vertical heater exceed capacity, the billet surface temperature drops during conveyance and heating time is increased. On the other hand, when the supply is too low the extraction timing from the vertical heaters is disrupted because the starting time is delayed. Therefore, it is necessary to optimize the supply of billets to the heater in order to maintain an adequate supply of billets to the press. The timing control calculates the extraction timing form the rotary furnace so that the arrival interval to the induction heater can become the extrusion pitch. This works as follows: (see Fig.11)

a. First, the timing control calculates the extraction timing from the rotary furnace using the tube extraction pitch and the previous billet extraction time.

b. Then, using the position of any other billets present on the conveyer from the rotary furnace to the vertical heaters and the extraction time for any billets in the vertical heaters, the timing control modifies the extraction timing from the rotary furnace so as to not sojourn other billets in the process line.

c. Finally, the timing control outputs the extraction process signal to the electric controller.

8.2.Extraction timing control of vertical heater

When the supply of billets to each press exceeds capacity, the billets sojourn in the line and start to cool. This greatly affects the resistance of deformation in expansion and extrusion and affects the quality as well. On the other hand, when the supply to each press is too low, the tube making pitch becomes late. Therefore, like the rotary furnace timing control, this function calculates the extraction timing from the induction heaters so that the arrival interval to each press becomes the tube-extraction pitch. This works as follows:

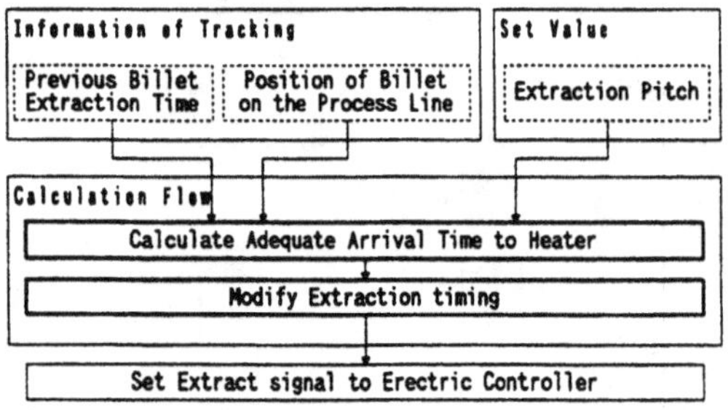

Fig.11 Flow of timing control

a. First, the timing control calculates the extraction timing form the vertical heater, using the tube extraction pitch and the time of the previous tube extraction

b. Then, using the attended status from the vertical heater to each press and the extraction time which is heated, the timing control modifies the extraction time from the vertical heater so that the billet is not delayed in the process line.

9.CONTROL RESULTS

9.1. Results of rotary furnace heat control

Fig.12 shows an example of material temperature controlled by rotary furnace heat control. In this instance, there are various target temperatures, and the control system is able to avoid overheating. These data are for austenite stainless steel.

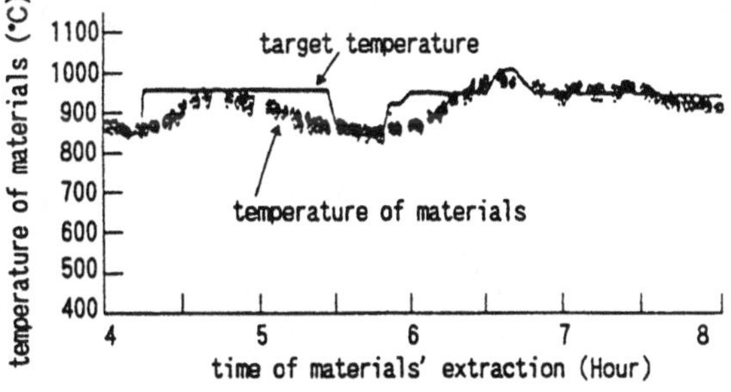

Fig.12 Result of rotary furnace heat control

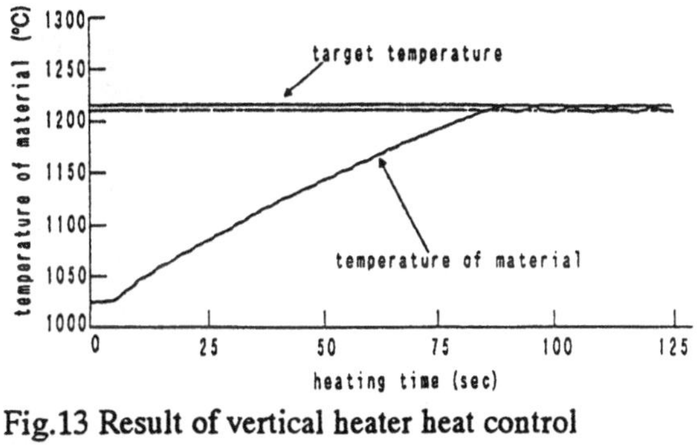

Fig.13 Result of vertical heater heat control

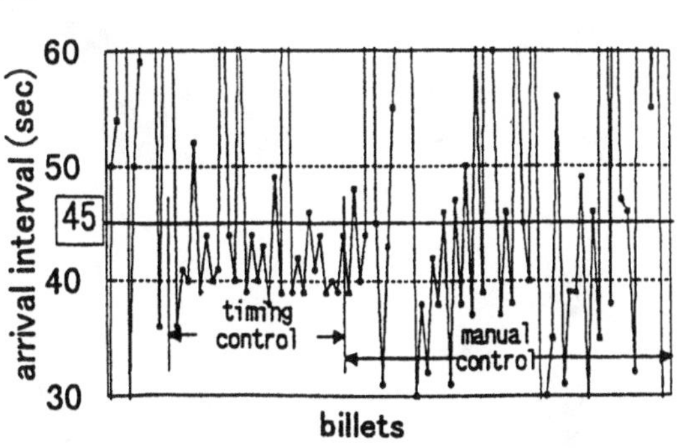

Fig.14 Result of extraction timing control

9.2. Results of vertical heater heat control

Fig.13 shows the billet temperature curve. After achieving the target, the billet temperature is maintained in order to avoid overheating. These data are for austenite stainless steel.

9.3. Results of extraction timing control

Fig.14 shows an example of insertion interval to the vertical press as controlled by the vertical heater extraction timing control. The arrows in Fig. 12 indicate the results of timing control extractions and manual extractions by operators. The timing control extraction inserts are nearer in timing to the set-up pitch (in this case 45 seconds) than are the manual extractions.

10.CONCLUSION

This control system was introduced in September 1993 and is now operating normally.

REFERENCES

Mutsuo Inoue (1979) Tube-making by the extrusion method. Sumitomo Metal Industries, Ltd.

INTERACTION MEASURE OF INTERSTAND TENSION AND THICKNESS CONTROL IN TANDEM COLD ROLLING BASED ON THE STRUCTURED SINGULAR VALUE

Kazuya Asano* and Manfred Morari**

** Technical Research Lab., Kawasaki Steel Corp., 1 Kawasaki-cho, Chuo-ku, 260 Japan*
Tel: +81-43-262-2423 Fax: +81-43-262-2854 E-mail: asano@giken.kawasaki-steel.co.jp
*** Institut für Automatik, Eidgenössische Technische Hochschule, Zürich, Switzerland*

Abstract: Tandem cold rolling should be considered a multi-input multi-output process, but it has been controlled by a series of single-input single-output control loops. To meet tight tolerances for product quality, multivariable control has been applied to the process. However, it inevitably leads to more complicated controllers which are hard to be tuned. Decentralized control can trade off between simplicity and performance deterioration of control systems. In this paper, interactions between interstand tension and thickness control in tandem cold rolling are investigated using the structured singular value to determine the structure of decentralized control for the process. *Copyright © 1998 IFAC*

Keywords: Automatic process control, control schemes, control system analysis, decentralized control, industrial control, interaction, steel industry, structured singular value.

1. INTRODUCTION

Cold rolling is the process where the hot rolled strip is further reduced to the final thickness of products. To achieve the required reduction, final quality and tolerances, several rolling operations are required. This is usually executed by tandem rolling, yielding high productivity.

During tandem cold rolling operations, tension is developed in the strip between stands, which is indispensable for stable rolling operations. In addition to transport delays of the strip between stands, however, the interstand tension gives rise to interactions between rolling operations at each stand, which makes the process a complicated multi-input multi-output (MIMO) system.

The process has been controlled by a series of single-input single-output (SISO) control loops designed based on static analyses of the interactions, but performance is limited because of the interactions. To meet tighter product tolerances, multivariable control has been applied to the process. However, it inevitably brings more complicated controllers which are intuitively incomprehensible and hard to be tuned.

Decentralized controllers, which consist of local controllers corresponding to essential elements of plants, can handle the trade-off between simplicity and performance deterioration of control systems, yielding a good solution to this problem.

Applications of decentralized control give rise to two new issues, namely, how to choose the control structure and how to design the local controllers. This paper focuses on the former issue in the case of tandem cold rolling. The interactions of interstand tension and thickness control in tandem cold rolling are investigated to determine the structure of decentralized control for the process. Off-diagonal elements of the system, which correspond to the interactions, are regarded as uncertainties and evaluated by using the interaction measure based on the structured singular value. The results support the conventional static analyses and empirical knowledge on the interactions.

2. CONTROL PROBLEM IN TANDEM COLD ROLLING

The measures by which the quality of rolled metal is assessed include thickness, width, shape, surface conditions and mechanical properties. The present investigation is concerned with control of interstand

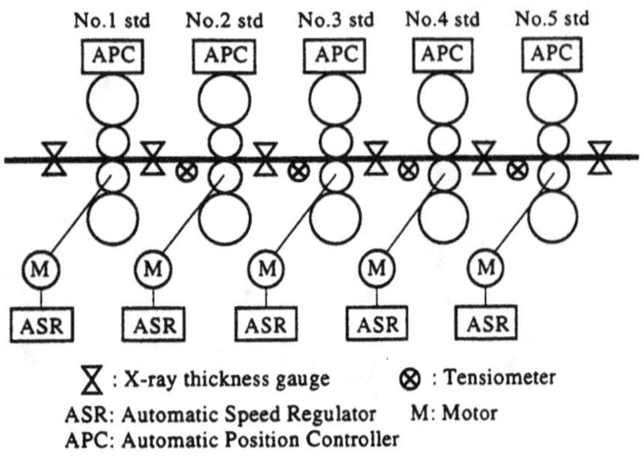

Fig. 1. Schematic diagram of a tandem cold mill

tension and thickness.

A 5-stand tandem cold mill schematically shown in Fig. 1 is considered in this study. The problem considered here is to control thickness of the strip and front tension at each stand by adjusting the gap and velocity of the rolls at each stand according to thickness and interstand tension measured by X-ray thickness gauges and tensiometers, respectively. Excluding front tension at the last stand from the controlled variables and the roll velocity at the last stand from the manipulated variables, respectively, the rolling process can be viewed as a 9-input 9-output system.

Control action at the roll gap or velocity will affect not only the exit gauge of the strip but also the tension in the strip, which means that the rolling process in a single rolling stand is already an interactive system. Furthermore, the rolling processes in the stands are dynamically coupled by material transport delays and interstand tension, so the rolling process as a whole forms a complicated interactive system. In terms of shape control, disturbances to the roll force should be avoided, which imposes additional constraints on tension and thickness control.

3. MATHEMATICAL MODEL OF TANDEM COLD ROLLING

3.1 Nonlinear Model

A lot of studies have been made on mathematical models of tandem cold rolling to describe non-linear characteristics of the process. Misaka (1968) derived fundamental equations of the process and used them for a static analysis of the interactions. Tyler, *et al.* (1996) used a linear model derived from the equations to evaluate a model based fault detection scheme. Their representations of the nonlinear model are used in the further investigation. Variable definitions are summarized in table 1. For each variable, a subscript identifies the stand number.

The strip thickness at the exit is determined by the roll gap and the roll force:

Table 1 Definition of variables

Symbol	Physical meaning
H	Entry thickness
h	Exit thickness
S	Roll gap
P	Roll force
K	Elastic constant
σ_b	Back tension
σ_f	Front tension
m	Coefficient of friction
v_b	Strip velocity at entry
v_f	Strip velocity at exit
f	Forward slip
V_R	Roll velocity
W	Strip width
κ	Defined by (7)
$\bar{k}$	Mean resistance to deformation
D_p	Defined by (9)
r	Reduction in thickness
r_b	Total reduction of ingoing strip
r_f	Total reduction of outgoing strip
R'	Deformed roll radius
R	Roll radius
ϕ_n	Neutral angle
s_b	Yield stress at entry
s_f	Yield stress at exit
H_n	Defined by (17)
H_b	Defined by (20)

$$h_i = S_i + \frac{P_i}{K}. \tag{1}$$

The rolling force is given by the implicit function

$$F_p(P_i, H_1, H_i, h_i, \sigma_{bi}, \sigma_{fi}, m_i) = 0. \tag{2}$$

The strip velocity is determined by the roll velocity and the forward slip coefficient

$$v_{fi} = (1 + f_i)V_{Ri}. \tag{3}$$

The tension is governed by differential equation

$$\dot{t}_{fi} = \frac{E}{L}(v_{b(i+1)} - v_{fi}), \tag{4}$$

where E is Young's modulous of the strip, L is the distance between stands, and v_b is related to v_f through the volume constancy

$$v_{bi} = \frac{h_i}{H_i} v_{fi}. \tag{5}$$

The implicit rolling force function F_p is determined by the following set of equations:

$$F_p = P_i - W\kappa_i \bar{k}_i \sqrt{R'_i(H_i - h_i)} D_{pi} \tag{6}$$

$$\kappa_i = 1 - \frac{(a-1)\sigma_{bi} + \sigma_{fi}}{a\bar{k}_i} \tag{7}$$

$$\bar{k}_i = \alpha(\bar{r}_i + \beta) \tag{8}$$

$$D_{pi} = 1.08 + 1.79r_i\sqrt{1-r_i}\,m_i\sqrt{\frac{R'_i}{h_i}} - 1.02r_i \quad (9)$$

$$r_i = \frac{H_i - h_i}{H_i} \quad (10)$$

$$r_{bi} = 1 - \frac{H_i}{H_1} \quad (11)$$

$$r_{fi} = 1 - \frac{h_i}{H_1} \quad (12)$$

$$R'_i = R_i\left(1 + \frac{cP_i}{W(H_i - h_i)}\right) \quad (13)$$

$$\bar{r}_i = \begin{cases} r_{fi} & \text{for } i = 1 \\ 0.4r_{bi} + 0.6r_{fi} & \text{for } i > 1 \end{cases}, \quad (14)$$

where c is a constant. The forward slip is governed by

$$f_i = \phi_{ni}^2 \frac{R'_i}{h_i}, \quad (15)$$

where

$$\phi_{ni} = \sqrt{\frac{h_i}{R'_i}}\tan\left(\sqrt{\frac{h_i}{R'_i}}\frac{H_{ni}}{2}\right) \quad (16)$$

$$H_{ni} = \frac{H_{bi}}{2} - \frac{1}{2m_i}\log\left(\frac{H_i}{h_i}\frac{1-\dfrac{\sigma_{fi}}{s_{fi}}}{1-\dfrac{\sigma_{bi}}{s_{bi}}}\right) \quad (17)$$

$$s_{fi} = \alpha\,(r_{fi} + \beta)^{\gamma} \quad (18)$$

$$s_{bi} = \alpha\,(r_{bi} + \beta)^{\gamma} \quad (19)$$

$$H_{bi} = 2\sqrt{\frac{R'_i}{h_i}}\tan^{-1}\left(\sqrt{\frac{r_i}{1-r_i}}\right). \quad (20)$$

In the above, α, β, γ and a are material dependent parameters.

3.2 Linearized Model

At steady state, the equation governing tension (4) becomes

$$h_i v_{fi} - h_{i+1} v_{f(i+1)} = 0. \quad (21)$$

In addition, the entry and exit strip thickness are related by

$$H_{i+1} = h_i \quad (22)$$

and the forward and back tension obey

$$\sigma_{b(i+1)} = \sigma_{fi}. \quad (23)$$

Using the tension and thickness schedule and physical parameters such as strip width, roll radius, elastic constant and friction coefficient found in Misaka (1968), a steady state operating point was found as follows. Using (22) and (23), H_i is eliminated for i = 2, ..., 5. Next the exit strip velocity at the last stand v_{f5} is specified. Then a solution S_i, P_i, v_{fi}, to the fundamental equations (1), (2) and (21) is obtained. Finally the roll velocity V_{Ri} is found from (3) and (15). The schedule and obtained parameters in the steady state are shown in table 2.

Table 2 Steady state operating point

Parameters	1 std	2 std	3 std	4 std	5 std
H(mm)	3.20	2.64	2.10	1.67	1.34
h(mm)	2.64	2.10	1.67	1.34	1.20
σ_b(MPa)	0	100.0	125.4	157.8	157.8
σ_f(MPa)	100.0	125.4	157.8	157.8	44.1
v_f(mm/s)	9270	11660	14660	18270	20400
f(–)	0.0526	0.0310	0.0377	0.0349	0.0139
P(10^3kN)	6897	6963	7858	7914	6489
R(mm)	273	273	292	292	292

E: 2.1×10^5(MPa) L: 46,000(mm) W: 930(mm)
μ: 0.07(–) K: 4.606×10^6(kN/mm)
α: 84.6 β: 0.00817 γ: 0.3 (low carbon steel)
a (in (7)): 3.0 c (in (13)): 2.14×10^{-4}(mm/kg^2)

The following dynamic relations exist between the variables. The entry thickness at a given stand is related to the exit thickness at the previous stand through the transport delay

$$H_{i+1}(t) = h_i(t - D_{i,i+1}) \quad (24)$$

where $D_{i,i+1}$ is the time required for the strip to travel the distance between the two stands

$$D_{i,i+1} = \frac{L}{v_{fi}}. \quad (25)$$

In terms of tension, the transport delay is neglected and the steady state relation (23) is assumed to hold.

The roll gap S_i and the roll velocity V_{Ri} can be manipulated according to the following dynamics

$$\dot{S}_i = -\frac{1}{T_S}(S_i - U_{Si}) \quad (26)$$

$$\dot{V}_{Ri} = -\frac{1}{T_V}(V_{Ri} - U_{Vi}) \quad (27)$$

where U_{Si} and U_{Vi} are the reference values for the corresponding manipulated variables. (26) and (27) represent the dynamics of APC and ASR in Fig. 1, respectively.

A local linearization of the nonlinear tandem rolling model around the operating point was carried out to obtain the following state space model:

$$\dot{\boldsymbol{x}} = \boldsymbol{Ax} + \boldsymbol{Bu} \quad (28)$$

$$\boldsymbol{y} = \boldsymbol{Cx} \quad (29)$$

where

$$\boldsymbol{x} = (\Delta\sigma_{f1}, ..., \Delta\sigma_{f4}, \Delta V_{R1}, ..., \Delta V_{R4}, \Delta S_1, ..., \Delta S_5)' \quad (30)$$

$$\boldsymbol{u} = (\Delta U_{V1}, ..., \Delta U_{V4}, \Delta U_{S1}, ..., \Delta U_{S5})' \quad (31)$$

$$\boldsymbol{y} = (\Delta\sigma_{f1}, ..., \Delta\sigma_{f4}, \Delta h_1, ..., \Delta h_5)'. \quad (32)$$

Δx denotes the fractional change of the variable x normalized by its steady state value.

Calculating partial derivatives of the fundamental equations, the following matrices for the state space model were obtained.

$$\boldsymbol{A} = \begin{pmatrix} \boldsymbol{A}_{11} & \boldsymbol{A}_{12} & \boldsymbol{A}_{13} \\ 0 & \boldsymbol{A}_{22} & 0 \\ 0 & 0 & \boldsymbol{A}_{33} \end{pmatrix} \tag{33}$$

$$\boldsymbol{B} = \begin{pmatrix} 0 & 0 \\ \boldsymbol{B}_{21} & 0 \\ 0 & \boldsymbol{B}_{32} \end{pmatrix} \tag{34}$$

$$\boldsymbol{C} = \begin{pmatrix} I & 0 & 0 \\ \boldsymbol{C}_{21} & 0 & \boldsymbol{C}_{23} \end{pmatrix} \tag{35}$$

$$\boldsymbol{A}_{11} = \begin{pmatrix} a_{21} & a_{31} & 0 & 0 \\ a_{12} & a_{22} & a_{32} & 0 \\ 0 & a_{13} & a_{23} & a_{33} \\ 0 & 0 & a_{14} & a_{24} \end{pmatrix} \tag{36}$$

$$\boldsymbol{A}_{12} = \begin{pmatrix} a_{41} & a_{51} & 0 & 0 \\ 0 & a_{42} & a_{52} & 0 \\ 0 & 0 & a_{43} & a_{53} \\ 0 & 0 & 0 & a_{44} \end{pmatrix} \tag{37}$$

$$\boldsymbol{A}_{13} = \begin{pmatrix} a_{61} & a_{71} & 0 & 0 & 0 \\ 0 & a_{62} & a_{72} & 0 & 0 \\ 0 & 0 & a_{63} & a_{73} & 0 \\ 0 & 0 & 0 & a_{64} & a_{74} \end{pmatrix} \tag{38}$$

$$\boldsymbol{A}_{22} = \boldsymbol{B}_{21} = \mathrm{diag}\left(-{}^{1}\!/_{T_{V1}}, -{}^{1}\!/_{T_{V2}}, \ldots, -{}^{1}\!/_{T_{V4}}\right) \tag{39}$$

$$\boldsymbol{A}_{33} = \boldsymbol{B}_{32} = \mathrm{diag}\left(-{}^{1}\!/_{T_{S1}}, -{}^{1}\!/_{T_{S2}}, \ldots, -{}^{1}\!/_{T_{S5}}\right) \tag{40}$$

$$\boldsymbol{C}_{21} = \begin{pmatrix} c_{21} & 0 & 0 & 0 \\ c_{12} & c_{22} & 0 & 0 \\ 0 & c_{13} & c_{23} & 0 \\ 0 & 0 & c_{14} & c_{24} \\ 0 & 0 & 0 & c_{15} \end{pmatrix} \tag{41}$$

$$\boldsymbol{C}_{23} = \mathrm{diag}\,(c_{31}, c_{32}, \ldots, c_{35}) \tag{42}$$

A third order Padé approximation was used for the transport delays denoted by (24) and (25).

Finally, the following transfer function matrix of the process can be obtained from the above state space equations and the Padé approximations.

$$\boldsymbol{y} = \boldsymbol{P}\boldsymbol{u} \tag{43}$$

$$\boldsymbol{u} = (U_{S1}, U_{V1}, U_{S2}, U_{V2}, U_{S3}, U_{S4}, U_{V3}, U_{S5}, U_{V4})' \tag{44}$$

$$\boldsymbol{y} = (h_1, \sigma_{f1}, h_2, \sigma_{f2}, h_3, \sigma_{f3}, h_4, \sigma_{f4}, h_5)' \tag{45}$$

In (44) and (45), 'Δ' before variables is omitted. Note that the manipulated variables $\boldsymbol{u}$ are rearranged so that $\boldsymbol{P}$ becomes diagonal in the case of the conventional controller which will be discussed in the next chapter.

4. STATIC ANALYSIS OF THE INTERACTIONS

4.1 Static Interactions in Tandem Cold Rolling

Table 3 shows the steady state gains of the plant (43), which indicates the static interactions in tandem cold rolling. This table tells us the following:

(1) The effect of the roll gap change on the final thickness is largest at the first stand and negligibly small at Nos. 3~5 stand.

(2) The effect of the roll velocity change on the final thickness is largest at the first stand and comparatively small at other stands.

(3) An increase in the roll gap at the first stand decreases the front tension at every stand. An increase in the roll gap at each of Nos. 2~5 stands increases the back tension at the stand.

(4) An increase in the roll velocity at the first stand decreases the front tension at every stand. An increase in the roll velocity at each of Nos. 2~4 stands increases the back tension and decreases the front tension at the stand.

4.2 Conventional Control Schemes for Tandem Cold Rolling

The tandem cold rolling process has been controlled by a series of SISO controllers designed based on experimental and theoretical investigations of the interactions such as above. At the upstream stands, thickness is controlled by manipulating the roll gap and front tension is controlled by manipulating the roll velocity. At the down stream stands, on the contrary, thickness and front tension are controlled by the roll velocity and roll gap, respectively.

Table 3 Steady state gains of the process

	U_{S1}	U_{V1}	U_{S2}	U_{V2}	U_{S3}
h_1	0.177	0.027	-0.042	-0.030	-0.004
t_{f1}	-0.045	-0.129	0.200	0.144	-0.018
h_2	0.152	0.114	0.079	-0.076	-0.053
t_{f2}	-0.037	-0.044	0.030	-0.083	0.169
h_3	0.150	0.126	0.036	0.012	0.036
t_{f3}	-0.024	-0.021	-0.002	-0.010	0.029
h_4	0.154	0.130	0.032	0.021	-0.0003
t_{f4}	-0.030	-0.025	-0.006	-0.005	0.003
h_5	0.153	0.130	0.032	0.022	-0.003

	U_{V3}	U_{S4}	U_{V4}	U_{S5}
h_1	0.003	-0.0004	-0.0004	0.0001
t_{f1}	-0.015	0.002	0.002	-0.0003
h_2	-0.045	0.006	0.006	-0.001
t_{f2}	0.144	-0.019	-0.019	0.003
h_3	-0.106	-0.045	-0.044	0.007
t_{f3}	-0.075	0.134	0.133	-0.019
h_4	-0.010	0.032	-0.103	-0.034
t_{f4}	-0.007	0.036	-0.118	0.122
h_5	-0.004	0.003	-0.011	-0.004

This conventional decentralized controller inherently has performance limitations due to the interactions. To this end, several MIMO controllers have been proposed for the process. Tanuma (1973) applied the optimum control theory. Edwards (1978) developed an non-interactive controller based on static decoupling of the interactions. These control schemes, however, lead to complicated controllers which are hard to be tuned and maintained.

5. INTERACTION MEASURE BASED ON THE STRUCTURED SINGULAR VALUE

The use of decentralized control invariably leads to performance deterioration when compared to the system with a full controller, but the deterioration depends on the neglected off-diagonal elements of the plant, namely, interactions. Therefore, it is very important to evaluate the effect of the interactions to determine the control structure.

In Fig. 2, $\boldsymbol{P}$ is an $m \times m$ transfer function matrix and $\dot{\boldsymbol{P}}$ is the block diagonal part of $\boldsymbol{P}$, which has the same structure as the decentralized controller $\bar{\boldsymbol{C}}$. The controller $\bar{\boldsymbol{C}}$ is to be designed so that the block diagonal closed-loop system with the transfer matrix

$$\dot{\boldsymbol{T}} = \dot{\boldsymbol{P}}\bar{\boldsymbol{C}}(\boldsymbol{I} + \dot{\boldsymbol{P}}\bar{\boldsymbol{C}})^{-1} \tag{46}$$

is stable($\delta = 0$ in Fig. 2). Consider the following matrix which expresses the relative error arising from the approximation of the full system $\boldsymbol{P}$ by $\dot{\boldsymbol{P}}$:

$$\boldsymbol{L}_T = (\boldsymbol{P} - \dot{\boldsymbol{P}})\dot{\boldsymbol{P}}^{-1} \tag{47}$$

Then the next theorem holds.
Theorem(Grosdidier and Morari 86):
Assume that $\boldsymbol{P}$ and $\dot{\boldsymbol{P}}$ have the same right-half plane poles and that $\dot{\boldsymbol{T}}$ is stable. Then the full closed-loop system $\boldsymbol{T} = \boldsymbol{P}\bar{\boldsymbol{C}}(\boldsymbol{I} + \boldsymbol{P}\bar{\boldsymbol{C}})^{-1}$ ($\delta = 1$ in Fig. 2) is stable if

$$\bar{\sigma}(\dot{\boldsymbol{T}}(i\omega)) < \mu^{-1}(\boldsymbol{L}_T(i\omega)) \quad \forall \omega, \tag{48}$$

where $\bar{\sigma}(\boldsymbol{A})$ and $\mu(\boldsymbol{A})$ express the largest singular value and structured singular value of the matrix $\boldsymbol{A}$, respectively. $\mu(\boldsymbol{L}_T(i\omega))$ is calculated with respect to the controller structure.

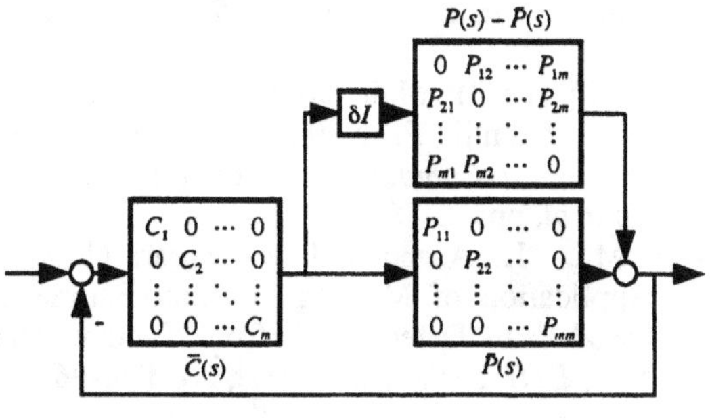

Fig. 2. Structure of decentralized control with additive uncertainty representation of interactions

Requirement (48) expresses the constraints imposed on the choice of $\dot{\boldsymbol{T}}$, which depends on the control structure and gives the tightest sufficient condition for the stability of the decentralized control system. $\mu^{-1}(\boldsymbol{L}_T(i\omega))$ is called the μ-interaction measure(μ-IM) and can be used as a powerful tool for screening control structures.

6. INTERACTION MEASURE OF THICKNESS AND TENSION CONTROL

6.1 μ-IM of Tandem Cold Rolling

The control structures in Fig. 3 are considered for possible decentralized control for the process. The 9×9 blocks in the figure denote the 9-input 9-output tandem cold rolling process $\boldsymbol{P}$, and gray blocks indicate that the decentralized controll structure includes the corresponding subsystems.

In the diagonal case Fig. 3(a), thickness is controlled by manipulating the roll position at the 1st, 2nd and 3rd stands and by the roll velocity at the 4th and 5th stands. Forward tension is controlled by roll velocity at the 1st, 2nd and 3rd stands and by the roll position at the 4th and 5th stands. This case can be regarded as one of the conventional control schemes. In the block diagonal case Fig. 3(b), thickness is controlled by manipulating the roll position at the 1st stand and at the rest of stands, controllers with 2 inputs and 2 outputs are used so that the interactions within these subsystems can be taken into consideration. In both cases, thickness at each stand is assumed to be obtained without delay by using the X-ray gaugemeters and calculated thickness based on massflow.

Fig. 4 shows μ-IM for both cases. The plots of μ-IM lies under the value of 1 in both cases, which indicates that both of the control structures are not applicable due to the strong interactions.

6.2 μ-IM with Feedforward Compensation

As the static analysis on the interactions shows, changes in the roll velocity affect both back and front tension because it affects the massflow balance both before and after the stand. Some feedforward loops are required to compensate this interaction. A feedforward

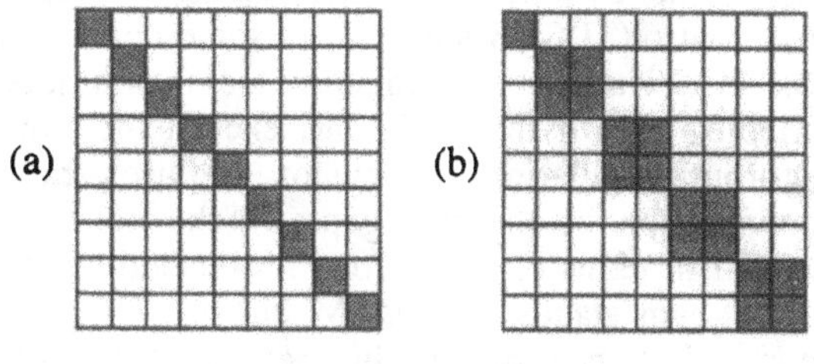

Fig. 3. Control structures
(a) Diagonal (b) Block diagonal

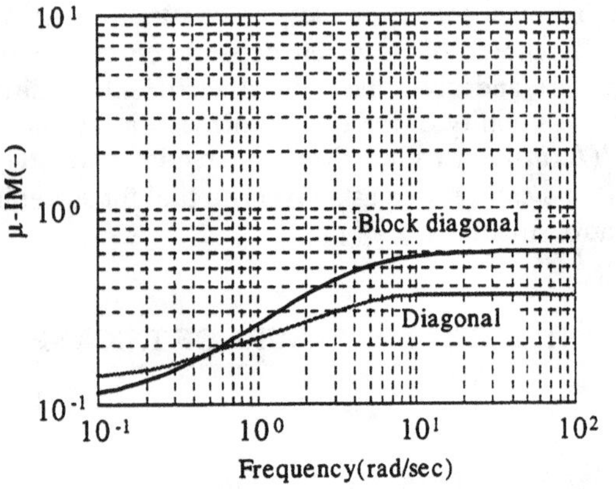

Fig. 4. μ-IM of thickness and tension control in the diagonal and block diagonal cases

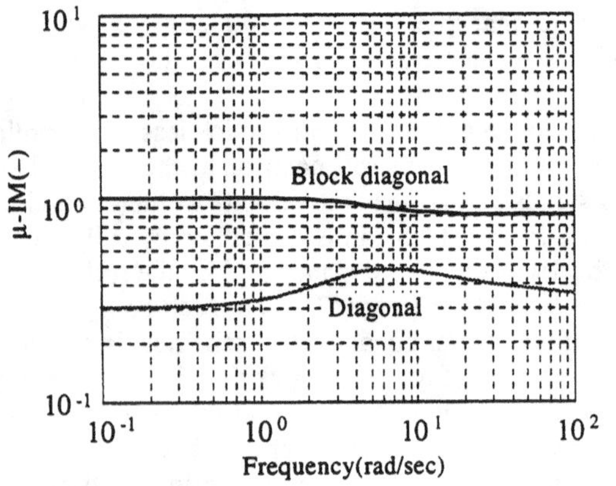

Fig. 5. μ-IM of thickness and tension control with feedforward compensation of the roll velocity to the upstream stands

compensation of the roll velocity change to the roll velocity at the upstream stands is widely used in tandem rolling. Fig. 5 shows μ-IM for both control structures when this feedforward compensation is introduced to the roll velocity at each stand.

In the diagonal case, the plot indicates that the structure is still not applicable. In the block diagonal case, however, the plot lies over the value of 1 in the frequency area under 6(rad/sec), which means that this structure is applicable although the interactions impose constraints on performance of subsystems.

Actually, the diagonal structure has been used to conventional control schemes in actual tandem cold mills. In those cases, however, tension control is executed not by a rigid feedback loop but by tension limit control(TLC), which allows some amount of tension deviations and changes the manipulated variables only if the deviations exceed the limit. Rigorous regulation of tension requires rapid manipulation of the roll gap, which should be avoided in terms of shape control especially at the downstream stands. The μ-IM shows that it is impossible to control both thickness and tension by the diagonal controller and it is considered to be one of the reasons why TLC has been applied.

The block diagonal structure in Fig. 3(b) can be found in Kitamura, *et al.* (1991), though the reason for the choice of the structure is not described. As the μ-IM indicates, it is considered to be the minimum decentralized structure for the process. Block diagonal structures with larger subsystems lead to more complicated and intuitively less incomprehensible controllers, so this structure is considered to be the most practical for the process.

7. CONCLUSIONS

Interactions between interstand tension and thickness control in tandem cold rolling were investigated to find possible decentralized control structures. Off-diagonal elements of the rolling process were excluded from the nominal model of the process and regarded as structured additive uncertainties. The structured singular value was applied to evaluate the uncertainties, yielding a measure of the interactions which indicates restrictions on performance of subsystems.

The interaction measure showed that the diagonal controller was not applicable to the process, but the block diagonal controller with one 1×1 and four 2×2 subsystems was applicable. The block diagonal control structure is considered to be the most suitable for the process.

The remaining problems for decentralized control of interstand tension and thickness control in tandem cold rolling include methodologies for designing controllers for subsystems with robustness against changes of operating conditions.

REFERENCES

Edwards, W.J. (1978). Design of entry strip thickness controls for tandem cold mills, *Automatica*, **14**, pp. 429-441.

Grosdidier, P. and M. Morari (1986). Interaction measures for systems under decentralized control, *Automatica* , **22**, pp. 309-319.

Kitamura, A., M. Konishi and Y. Naito (1991). Multivariable thickness control system for tandem cold mills based on decentralized technique, *Trans. Inst. Syst. Cont. Inf. Eng.*, **4**, pp. 140-154. (in Japanese)

Misaka, Y. (1968). Control equations for cold tandem mills, *Trans. Iron Steel Inst. Japan*, **8**, pp. 86-96.

Tanuma, M. (1973). Multivariable control of a five-stand cold mill, *Proc. 4th IFAC/IFIP Int. Conf. on Digital Computer Application to Process Control*, pp. 75-86.

Tyler, M.L., K. Asano and M. Morari (1996). Application of Moving Horizon Estimation based Fault Detection to Cold Tandem Steel Mill, *ETH Technical Report*, **AUT96-06**.

NEURAL NETWORK TECHNIQUES AND ITS APPLICATIONS IN LADLE FURNACE BURDEN

Shujiang Li, Xianwen Gao, Tianyou Chai and Xiaogang Wang

Research Center of Automation Northeastern University, Shengyang 110006
E-mail xwgao@mail.neu.edu.cn

Abstract: In this paper, a new kind of optimal burden technique of ladle furnace is presented. Based on the single pure form method of linear programming, the optimal calculation of the addition quantity of each kind of alloy in the minimum cost is carried out, the percentage of alloy may be decided by means of neural network technique under the different liquid steed and slag conditions. So the small scale control in composition of liquid steel can be realized . At the same time, the cost of the steel making will be lower than ever before, the method presented in this paper is effective on concrete practice of steelmaking in ladle furnace. *Copyright © 1998 IFAC*

Keywords: Neural Network, Linear programming, Optimal control, Predictive control, Steel Manufacture.

1. INTRODUCTION[1]

The technique of the refining out of the furnace has been known as one of the three key techniques in the metallurgical industry for 40 years, however, most of the refining furnace used in metallurgical industry is ladle furnace, which is widely used in modern steelmaking of electric are furnace.

One of the important functions with ladle furnace is to adjust the chemical compositions of the liquid steel and satisfy the requirement of the index . When the alloy steel is metallurgied, especially for multi-elements alloy steel, many kinds of alloys are added into the liquid steel, in the same time, each kind of the alloy is made of different content. It is very difficult for us to analyze the effect of alloy profit from the oxygen and the slag because of the strong nonlinearity. On the other hand, the competition on the market is becoming more intense than ever before. The requirement on the composition of steel is very strict by users. For this reason, increasing percentage of hits of composition and choosing optimal scheme of alloy addition is becoming one of the urgent problem to be solved in metallurgy industry. Zhang, J.M (1995) used K factor method to deal with this problem, but he only considered the effect on the liquid caused by one of the main elements of alloy and regardless of other ones . The minimum target function strategy was proposed (Zhao, B. T. 1993, Li, S. Q. and Li, W.l. *etal* 1995), in which the linear programming technique is utilized to realize the static control for alloy addition, however, this kind of method takes the percentage of alloy as a constant, and without considering the effect on the liquid steel and the slag on the percentage of obtaining under different conditions. A method proposed in (Fukunaga, M. 1994, Seike, M. 1994), which is combined with expert system and linear computation, carry out alloy adding computation. But it is difficult for us to deal with this problem when the kinds of steel is more than one or some situations which are not included in the knowledge base, thus the correct result will not be reached. in order to realize the optimal computation of alloy adding, the proposed method must has the following two characteristics : one is the adjusting

[1]1 This work was supported by the National Natural Science Key foundation of China

on line of percentage of obtaining of the alloy, the other is the optimal computation of adding the alloy.

This paper adopts the technique of neural network to learn on line the varying the percentage of obtaining the alloy, after the learning is satisfied with the required accuracy, then the linear programming method is utilized to compute the optimal quantity of adding alloy. If the prediction error is not within the required one, retrain the neural network and obtain a new identification model, take this model as the prediction model of next time.

2. CONTROL SYSTEM DESIGN

The effect of each factor on the percentage of obtain is impossible to be described with the linear model, because the chemical actions in the ladle furnace are very complex. On the other hand, the neural network has the learning function, it can approximate any nonlinear function(Li, C. J.1995).The whole control structure used in this paper is shown as Fig1. This paper uses neural network to determine the nonlinear function among the percentage of obtaining the alloy, the slag and the profit from oxygen. Then using identification model predicts the change of the percentage of obtaining alloy during operation process of the current furnace. The optimal computing the quantity of adding the alloy at the minimum cost.

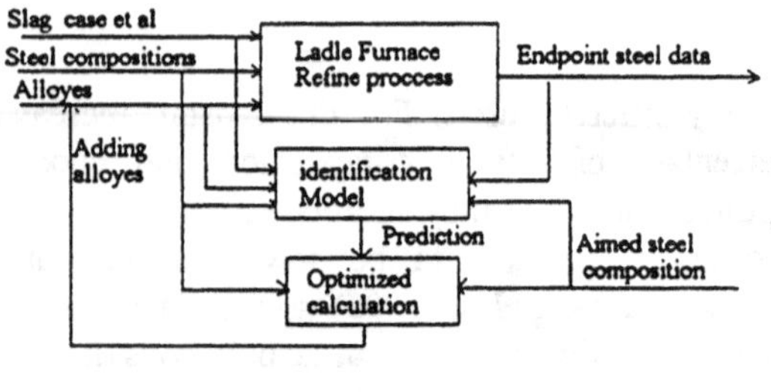

Fig1. control system structure diagram

2.1 predictive model

Self-learning model is constructed with three layers BP network with multi-input and single-output.

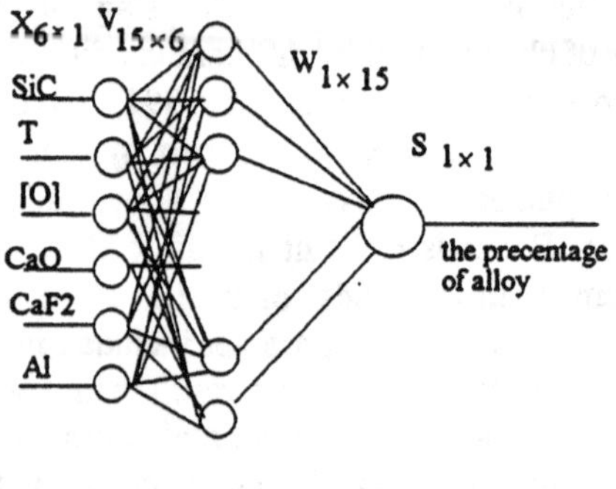

Fig2. the structure diagram of the neural network

Neural network input layer weight value V of the first kind of the percentage of obtaining alloy·the modify rules of hide layer weight value W and the output value of the neural network are presented in formula (1),(2)and (3) respectively.

$$V^i_{new} = V^i_{old} + \varepsilon[(W^{iT} \cdot e^i)\cdot(1 - f^{i2}(x))]\cdot X \qquad (1)$$

$$W^i_{new} = W^i_{old} + e \cdot e^i \cdot F^i(x)^T \qquad (2)$$

$$s^i = W^i \cdot F^i(x) \qquad (3)$$

Taking the cost is lowest as the target function of the optimization, the decision variable is alloy elements, limiting factor is upper limit and lower limit of the target compositions in the liquid steel as well as allow usage quantity of the alloy.. The optimal strategy is described as formula(4),(5), and (6).

2.2 Optimal calculation model

The n kinds of alloy are used when the composition of the liquid steel is adjusted, but the compositions of the liquid steel are controlled in m kinds of components of the liquid steel.

(1) The decision variables

Taking the burden quantity of ith kind of alloy material as the decision variable x_i, i=1,2, ··· n Because the burin quantity may not be less then zero, *i.e.* the decision variables meet the condition of non-negative, it can be described as

$$x_i \geq 0 \qquad i = 1,2,\cdots\cdots,n \qquad (4)$$

(2) The objective functions

Taking the minimum value as cost of the burden materials, the objective function will be constructed as follow

$$\min f = c_1x_1 + c_2x_2 + \cdots\cdots c_nx_n = \sum_{i=1}^{n} c_ix_i \qquad (5)$$

where c_i is the cost on one kilogram of ith alloy material

(3) The boundary conditions

The composition of the liquid steel will be changed when the n kinds of alloy are added into the ladle furnace, the result of composition changed, must make the m kinds of composition meet the requirement of composition analysis values *i.e.* it is not less than the lower value, but not more than the upper value of the technology index. The boundary of burden material compositions

$$\begin{cases} \sum_{i=1}^{n}(x_i \cdot a_{ij} \cdot s_{ij}) + W \cdot E_j \geq (W + \sum_{i=1}^{n} x_i \cdot s_i) \cdot E_j(min) \\ \sum_{i=1}^{n}(x_i \cdot a_{ij} \cdot s_{ij}) + W \cdot E_j \leq (W + \sum_{i=1}^{n} x_i \cdot s_i) \cdot E_j(max) \\ x_i \leq G_i \end{cases} \quad (6)$$

where Q---is the quantity of melten steel in Ladle furnace, before the alloy materials added into the Ladle furnace.

a_{ij}-- is the *j*th contents of composition in the *i*th alloy material (%)

S_{ij}--is the gains of the *j*th composition in *i*ht alloy material(%)

S_i--is the total gains of the *i*th alloy material (%)

E_j--is the present analysis value of the *j*th composition contents in melten steel, before alloy material added into the Ladle furnace (%)

$Emin_j$--is the low boundary controlled of the *j*th composition in the melten steel (%)

$Emax_j$--is the upper boundary controlled of the *j*th composition in melten steel (%)

G_i--is The allow quantity of the alloy in the refining period.

$\left(Q + \sum_{i=1}^{n} x_i \cdot s_i\right)$--is the melten steel quantity in the, Ladle furnace after the alloy material added into the Ladle furnace.

$\sum_{i=1}^{n}\left(x_i \cdot a_{ij} \cdot s_{ij}\right)$--is the actual quantity added into the Ladle furnace of the *j*th composition in melten steel

In order to prevent the other factors effect on the quality of the liquid steel when the compositions of the liquid steel is lower than the lower value of. the technology index, we can set up the high limit value is a little lower and the low limit value is a little higher than the technology index, when we make the high and low limit values of technology index according to each kind of steel number of the standard composition

The lowest cost *f(x)* can be acquired with single pure form strategy by the formula (4),(5) and(6).

3. THE RESULTS OF APPLICATION

The proposed method is made by us according to the 60T Ladle Furnace of Fushun special steel work. The main factors of effect on the percentage of obtain of alloy are T、 [O]、 SiC and Al basically of the slag. The four factors are taken as the input of the BP network, taking 12 hidden layer nodes and 124 groups historical data(metallized steel is 15CrMo), there are 62 groups data are used for training the network then taking learning rate as 0.0015 to learn 9000 times among the elements, the other 62 group data are used to predict, and with formula (7) to verify the correct of the identification model, that is Mn, Cr, Si and Mo, after that, we exchanges the two groups data and learns again 9000 times. Introduced evaluation function J:

$$J = \frac{1}{N}\sum_{i=1}^{N}\frac{|s_i - \hat{s}_i|}{|\hat{s}_i|} \times 100\% \quad (7)$$

where *N* is the quantity of the pair of data, s_i, $\hat{s}_i$ are the predictive value and the actual value of the model respectively. The identification and predictive precision of the model about the percentage of obtain on the four elements Mn, Cr, Si, Mo, is shown as table1.

Table 1 Identification and prediction accuracy

	Prediction data1	Prediction data2
Mn	0.89%	0.95%
Si	1.02%	1.18%
Cr	0.63%	0.57%
Mo	0.70%	0.63%

The conclusion from table 1 is that the predictive precision of the percentage of the alloy can meet the requirement of the technology, but the deviation about the percentage of the alloy of Si is much bigger, because the Si is easy to oxidize, it is the same with actual condition.

The optimal calculation on the added quantity of alloy is based on 15CrMo which is metallized (number of the furnace is 14759), alloy above 4 elements is acquired by the model predictive strategy shown as table 2

Table 2 predicted data alloy utilization ratio

	Mn	Si	Cr	Mo
Alloy utilization ratio	0.96	0.95	0.97	0.97

There are high content of Fe-Mn, middle content of Fe-Mn, high content of Fe-Cr, middle content of Fe-Cr, Fe-Si and Fe-Mo six kinds of alloy, it is given for the content of alloy, C content, cost and the analysis value of the composition for each kind of alloy material. The usage quantity of the alloy and the composition of the liquid steel are acquired by the formula (4),(5) and (6) is shown on table 3 and table 4

Table 3 Alloy used weight

	High Fe-Mn	Middle Fe-Mn	High Fe-Cr	Middle Fe-Cr	Fe-Si	Fe -Mo
Weight (Kg)	119.7	65.3	250.9	81.1	2203. 1	265. 0

Table 4 Trajectory and practical data

	Chemical Composition %				
Target	C	Mn	Si	Cr	Mo
Value	.12_.18	.40_.70	.17_.37	.80_1.1	.40_.55
Actual Value	0.15	0.41	0.21	0.82	0.43

It is shown that the chemical compositions are controlled at the lower limit of the target value from the table 4.

4. CONCLUSIONS

Some conclusions may be acquired from the theoretical analysis and the actual test:

(1) The nonlinear relationship between the percentage of the alloy and the liquid steel as well as the slag *etc* factors can be correctly shown with the neural network model, then the percentage of the alloy can be determined with predictive model.

(2) When the error between the predictive value and the actual value is over the allowing error of the technology index, the neural network model is trained again, hence, the correct of the percentage of the alloy is realized on line.

(3). The optimal calculation of the percentage of the alloy can be acquired by the single pure form method of linear programming. The cost of the steelmaking can be decreased because the compositions of the liquid steel is strictly controlled at the low limit of the technology index under the condition of the chemical composition of the liquid steel when it is in the range of the technology requirement.

Note: Although this controlling strategy is only used in the 60T ladle furnace , the input is effective the percentage of obtaining alloy can be adjusted so that the method may be extended to all kinds of the alloy burden in metallurgy industry.

REFERENCES

Zhang, J. M (1995). The proportional factor computing of the alloy additive of the steelmaking. *Jiangxi Metallurgy* (in Chinese) Vol.9, No.2, 37--41.

Zhao, B.T. (1993) Optimal control and its application software system. *Journal of Changqing university*,(in Chinese) Vol. 16, No.4 ,155--160.

Li, S.Q.&Li, W.l.and Liu, Y, G. (1995). *Modern steel making in Electric Arc Furnace*. 255-266. the Atomic Energy Publishing House, Beijing.

Fukunaga, M. (1994). Establishment of LF Automatic Operation. *CAMP-ISIJ* (in Japanese). Vol.7 ,205.

M. seike (1994). Deployment of LFV Guide control system Using the Expert system. *CAMP--ISIJ*(in Japanese). Vol.7, 1260.

Li, C. J.(1995) *Neural network System Theory*.(in Chinese) Xian Electronic Science and Technology University Publishing Mouse Xian. 57--89.

Qu Y. (1993). *Physical and Chemical Computing for Steelmaking Process* (in Chinese)Metallurgical Industry Publishing House Beijing, 31-57

MOLD LEVEL CONTROL IN CONTINUOUS CASTER VIA NONLINEAR CONTROL TECHNIQUE

YeongSeob Kueon and Seung-Yeol Yoo

Instrumentation and Control Research Team, POSCO Technical Research Lab., Pohang PO Box 36, 1- Koedong-Dong, Nam-Ku, Pohang-Shi, Kyungbuk, 790-785, Korea (E-mail: pc548332@smail.posco.co.kr)

Abstract: Mold level control system for continuous caster involves stick-slip friction in the hydraulic sliding gate, time-delay, non-linearity, and certain uncertain factors such as friction force variations between molten steel and the inner wall of mold. In this paper, nonlinear control technique was used to solve these complex control problems. The controller is then designed and implemented to the real continuous caster. Testing result shows that the new controller can decrease the fluctuant magnitude of the mold level and is superior to the previous PID controller. *Copyright © 1998 IFAC*

Keywords: Nonlinear control, Sliding mode control, Steel industry.

1. INTRODUCTION

The continuous caster is actually to produce continuous slab from molten steel and cuts into specified lengths to be supplied to the consequent processes or direct sale.

In the steel making industries, the mold level (or the molten steel level in mould) control plays an important role in assuring safety in the production of slab caster, improving the surface/internal quality of slab, increasing the yield rate of liquid steel, and reducing the production cost. The purpose of the control system is to decrease mold level fluctuation as much as possible since the fluctuation of the mold level is one of the most important factors determining quality of the slab. If the mold level fluctuation occurs severely, the oscillation mark is formed on the slab and mold powder and slag are infiltrated into the slab. (Graebe, et al., 1992).

In general, the main problems of control difficulties in the process were found to be the friction force variations between the inner wall of the mold and the molten steel in accordance with molten steel temperature and steel grade, the stick-slip friction in the sliding gate in the hydraulic system, and the clogging/erosion of submerged entry nozzle. They are very difficult to be included in the mathematical model and controlled properly.

The dynamic system to be controlled can be modeled by *1st order ordinary differential equation*, although it possesses unmodeled dynamics and wave effects. The mold level should be controlled to within a few *mm* in accordance with specifications. A simple PID controller is not robust enough to satisfy the specifications since the dynamic system to be controlled is changed during casting and possesses unmodeled dynamics.

Nonlinear control techniques were used to design the controller to overcome the above mentioned control difficulties and implemented to the real continuous caster since nonlinear control techniques provide consistent performance for the non-linear system in the face of modeling imprecision and parameter uncertainties. (Slotine, et al., 1991).

2. PROCESS AND MODELING

Fig. 1 depicts a schematic picture of a continuous caster of POSCO #2 continuous steel casting process, where the three subsystems which make up the whole process are the ladle, the tundish, and the mold. The molten steel is first poured from the ladle into the tundish which keeps approximately 50 tons and works as a secondary reservoir to hold a constant steel head. The molten steel flows from the tundish into the mold through the submerged entry nozzle then passes through the primary and the secondary cooling zones and is casted into continuous slab. The molten steel must flow into the mold at the same rate as the strand is withdrawn at casting speed controlled by pinch rolls. (Hesketh, et al, 1993; Kurokawa, et al, 1993).

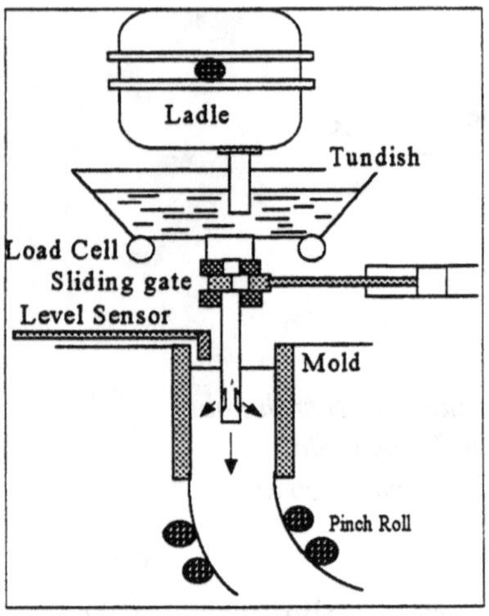

Figure 1: A Schematic picture of the process

In this paper, the speed of pinch roll or casting speed can be measured and is not controlled by the mold level controller, the mold level can only be controlled by adjusting the position of the sliding gate attached below the tundish.

The dynamic model of Fig. 2 is

$$\frac{d}{dt}\left[\int_0^{h(t)} A_m(z)dz\right] = M_{in}(t) - M_{out}(t) \tag{1}$$

where $A_m(h)$ is the cross sectional area of the mold, $M_{in}(t)$ is the inflow of molten steel which is defined the effective flow area times the height of molten steel, $l(t)$, and $M_{out}(t)$ is the outflow which is defined the mold size times the casting speed. If the initial level h_0 is quite different from the desired level h_d, the control of h involves a nonlinear regulation problem. (Graebe, et al, 1992).
The dynamics of (1) can be rewritten as

$$\begin{aligned} A_m(h)\dot{h} &= M_{in}(t) - M_{out}(t) \\ &= A_v c\sqrt{2gl(t)} - A_m V_{cs} \end{aligned} \tag{2}$$

where c is the coefficient of discharge, A_v is an effective flow area, and V_{cs} is the casting speed. Here, assuming that the outflow speed from the exit of the submerged entry nozzle (SEN) is equal to the passing speed of molten steel at the sliding gate, then $l(t)$ can be approximated as the molten steel head in the tundish. The main consideration of SEN is how to convert effective flow area A_v (mm^2), into the linear displacement of the sliding gate x_v(*mm*, stroke). Due to the geometry of the overlapping circular orifices in the sliding gate, the exact transformation is nonlinear as shown in (3) and Fig. 3.

$$\begin{aligned} A_v = 1600\pi + 3200\sin^{-1}\left(\frac{x_v - 120}{80}\right) \\ + \frac{x_v - 120}{2}\sqrt{6400 - (x_v - 120)^2} \end{aligned} \tag{3}$$

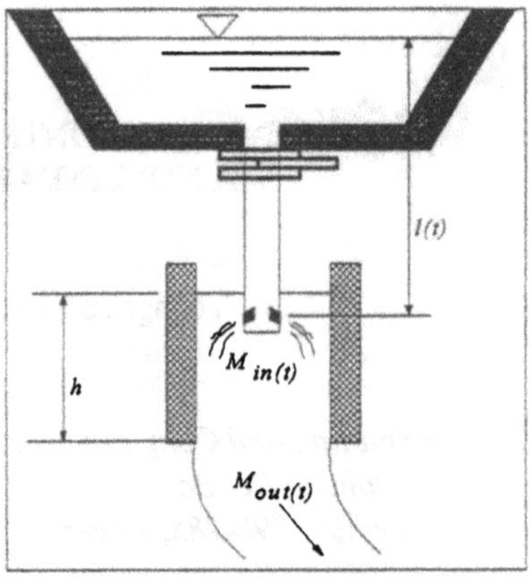

Fig. 2. Schematic picture of mold level control

To obtain the proper transformation, the effective flow area and the stroke value as shown in Figure 3 were chosen randomly and trained using an artificial neural networks. That is, the control algorithm computes the effective flow area according to the dynamic behavior of the plant, and then the stroke of the sliding gate is computed by using an artificial neural networks (ANN) as shown in Figure 4.

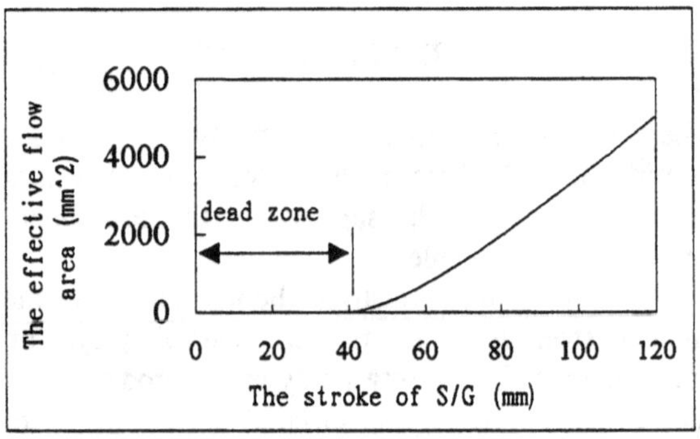

Fig. 3. The stroke Vs the effective flow area

Another main consideration is that the control system is time varying. It is the clogging/the erosion of the inner wall of SEN or the assembly between the tundish and the sliding gate. Because of the clogging/the erosion, the average stroke of the sliding gate differs up to 21 % of its initial one during casting. This problem was handled by using a proposed sliding mode control technique discussed in the following section.

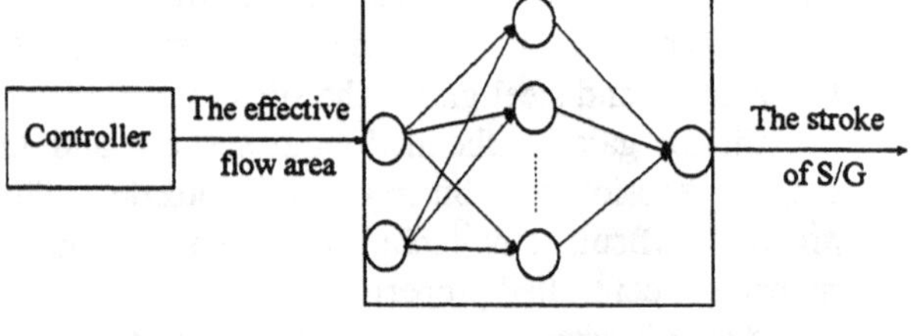

Fig. 4. The method of transformation using ANN

3. NONLINEAR CONTROL METHODS

Nonlinear control technique is an effective control method to handle parameter and modeling uncertainties for nonlinear control systems. In this paper, feedback linearization technique and sliding mode control method were considered to design and implement the mold level control system.

The single-input dynamic system is considered:

$$x^{(n)} = f(\mathbf{x},t) + b(\mathbf{x})u \tag{4}$$

where the state vector $\mathbf{x}(t)$ is $\left(x, \dot{x}, \ldots\ldots, x^{(n-1)}\right)^T$ and the scalar u is the control variable.

Furthermore, let $f(x,t)$ be a nonlinear function of the state vector x and of time t. The control problem is to obtain the state x for tracking a desired state x_d in the presence of model uncertainties and disturbances. The tracking error can be defined as follows:

$$e = \mathbf{x}(t) - \mathbf{x}_d(t) = (e, \dot{e}, \ldots\ldots, e^{(n-1)})^T \tag{5}$$

With feedback linearization technique, the dynamic system described by (4) can be controlled by the control input as follows:

$$u = \frac{1}{b(x)}(v - f(x)) \tag{6}$$

Then the nonlinearities can be canceled and the simple input-output relation can be obtained:

$$x^{(n)} = v \tag{7}$$

Thus, the control law

$$v = -k_0 x - k_1 \dot{x} - \ldots\ldots k_{n-1} x^{(n-1)} \tag{8}$$

with k_i chosen so that the dynamic system is stable and (6), (7), and (8) leads to

$$x^{(n)} + k_{n-1} x^{(n-1)} + \ldots + k_0 x = 0 \tag{9}$$

which implies that x(t) goes to 0 as t goes to infinity. For the tracking problem, equation (8) can be modified as follows

$$v = x_d^{(n)} - k_0 e - k_2 \dot{e} - \ldots - k_{n-1} e^{(n-1)} \tag{10}$$

and exponentially convergent tracking can be obtained.

Note that $b(x)$ should be invertible. (Slotine, et al., 1991). However, the system with unmodeled dynamics and time-varying parameters cannot be obtained the perfect tracking and linearization via feedback. In this case, equations (8) and (10) should be modified as follows

$$v = -k_0 x - k_1 \dot{x} - \ldots\ldots k_{n-1} x^{(n-1)} - k_{n-2} \int x.dt \tag{11}$$

$$v = x_d^{(n)} - k_0 e - k_2 \dot{e} - \ldots - k_{n-1} e^{(n-1)} - k_{n-2} \int e.dt \tag{12}$$

to perform the desired tracking and the desired linearization via feedback.

With sliding mode control technique, the control input can be designed as following procedures.
A sliding hyperplane

$$s(x,t) = e^{(n-1)} + k_1 e^{(n-2)} + \ldots + k_{n-1} e = 0.0 \tag{13}$$

is defined. Starting from the initial conditions

$$e(0) = 0 \tag{14}$$

the tracking problem $x{=}x_d$ can be considered as the state vector e remaining on the sliding surface $s(x,t)=0$ for all $t \geq 0$. The corresponding sliding condition is

$$s.\dot{s} \leq -\eta |s| \tag{15}$$

where η is a strictly positive constant. To achieve the sliding mode of equation (15), the following reaching law

$$\dot{s}(t) = -\alpha_1 s - \alpha_2 \int s.dt - \alpha_3 \frac{ds}{dt} - \alpha_4 .sign(s) \tag{16}$$

is used. Equation (16) is called the reaching law, which is the constant plus proportional plus integral plus derivative rate reaching law. The reaching law approach establishes the reaching condition and specifies the dynamic characteristics of the system during the reaching phase.

An undesirable disadvantage of this method is chattering which involves high control activity and further may excite high frequency dynamics neglected in the course of modeling. To handle the undesirable chattering the function sgn(s) in equation (16) is substituted into the followings: (Φ is boundary layer thickness)

$$-\alpha_4 .sat(\frac{s}{\Phi}) \; or - \alpha_4 .\tanh(s) \; or - \alpha_4 .\tanh(\frac{s}{\Phi}) \tag{17}$$

where $sat(x) = \begin{cases} x & if\ |x| < 1.0 \\ \mathrm{sgn}(x) & if\ |s| \geq 1.0. \end{cases}$

These substitutions correspond to the introduction of the control discontinuity in a thin boundary layer neighboring the switching surface. (Olgac, et al., 1995; Hwang, et al., 1992; Raymond, et al., 1988; Weibing, et al., 1993).

In this paper, sliding mode control technique was

used to control mold level in continuous caster, since feedback linearization technique cannot be guaranteed the system robustness in the presence of parameter uncertainty or unmodeled dynamics. The proposed feedback linearization technique described in equations (11) and (12) is under hot-test in POSCO #2 continuous casting process.

4. IMPLEMENTATION

The new type of sliding mode control technique with the proposed rate reaching law was implemented and tested in POSCO #2 continuous caster. The developed controller was installed in parallel with the existing PID controller. The sliding gate is adjusted by the selected controller as shown in Fig. 5.

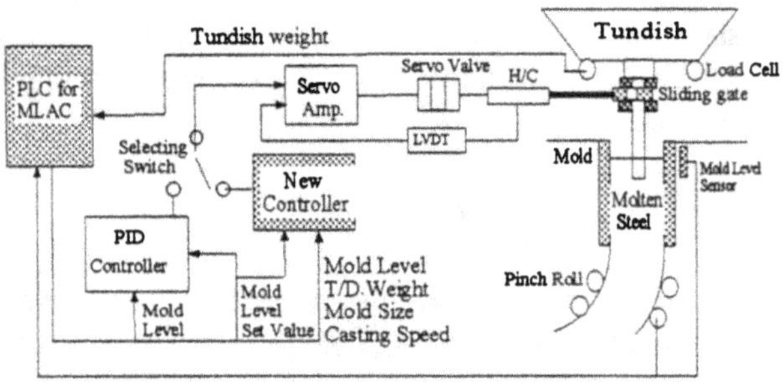

Fig. 5. New controller connections and signals

Before testing, control parameters were tuned on the computer simulation by using the data collected from the process. Then the hot-run-test was conducted on the real process at the last one of 8 charges of every sequence because the production should not be interrupted by the test.

Unwanted chattering was observed during operation when *sign* function was chosen for discontinuous control action and α_2 and α_3 were chosen as zero at the first hot run test. Immediately after observing undesirable chattering, the existing PID controller was selected and conducted the production to the last. The second hot run test was conducted on the condition of choosing *tanh(s)* function and non-zero α_2 and α_3 and the result is shows in figure 6.
As shown in Figs. 6 and 7, the control performance was very impressive, and the new controller was achieved much smoother control action than the existing PID controller. This means that *tanh* function makes the control action as smooth as a *saturation* function does. Many testing results tell that the proposed sliding mode controller shows the good performance even though the system is time-varying, which is the clogging/the erosion of the inner wall of SEN or the assembly between the tundish and the sliding gate because of integral action in equation (16).

5. CONCLUSION

The proposed sliding mode controller is successfully designed and implemented for the 1st orde time-varying dynamic system which is the mold level control system in continuous caster. The proposed sliding mode controller adopted constant plus proportional plus integral plus derivative rate reaching law, it is shown that good control performance can be obtained in the presence of stick-slip friction in the sliding gate, the friction force variations between the inner wall of the mold and molten steel according to molten steel temperature and steel grade, the tundish weight and the casting speed variations, and so on.

REFERENCES

S.F. Graebe, G. Elsey, and G. C. Goodwin. (1992). Accurate Control of Mould Level in Continuous Bloom Casting. **Control '92. Perth 2-4.**

Guang-Chyan Hwang and Shih-Chang Lin. (1992). A Stability Approach to Fuzzy Control Design for Nonlinear Systems. *Fuzzy Sets and Systems.* **48,** 79-287.

T. Hesketh, D. J. Clements, and R. Williams. (1993). Adaptive Mould Level Control for Continuous Steel Slab Casting. *Automatica,* **Vol. 29, No. 4.**

T. Kurokawa and T. Kondo. (1993). Development of Mold Level Control in Continuous Casting by H∞ Control Theory. *2nd IEEE Conference on Control Applications.* September 13-16, Vancouver, B.C.

N. Olgac and V.R. Iragavarapu. (1995). Sliding Mode Control with Backlash and Saturation Laws. *International Journal of Robotics and Automation.* **Vol. 10, No. 2.**

Raymond A. DECarlo, Stanislaw H. Zak, and Gregory P. Matthews. (1988). Variable Structure Control of Nonlinear Multivariable Systems: A Tutorial. *Proceedings of the IEEE.* **Vol.** 76, No. 3.

Jean-Jacques E. Slotine and Weiping Li. Applied Nonlinear Control. Prentice Hall, Englewood Cliffs, Jew Jersey.

Weibing Gao and James C. Hung. (1993). Variable Structure Control of Nonlinear Systems: A New Approach. IEEE Transactions on Industrial Electronics, **vol. 40, No. 1.**

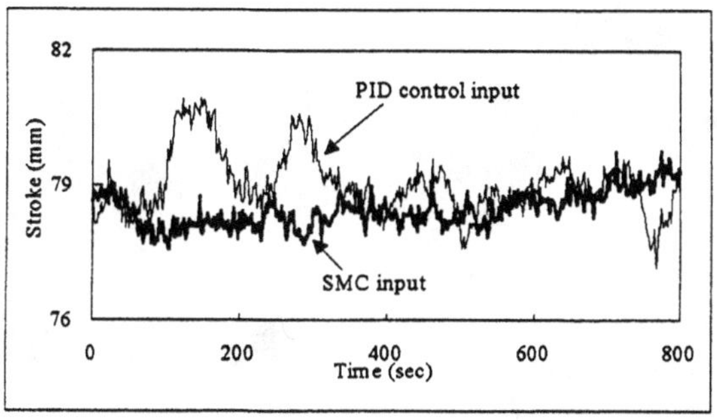

Fig. 6. Control inputs

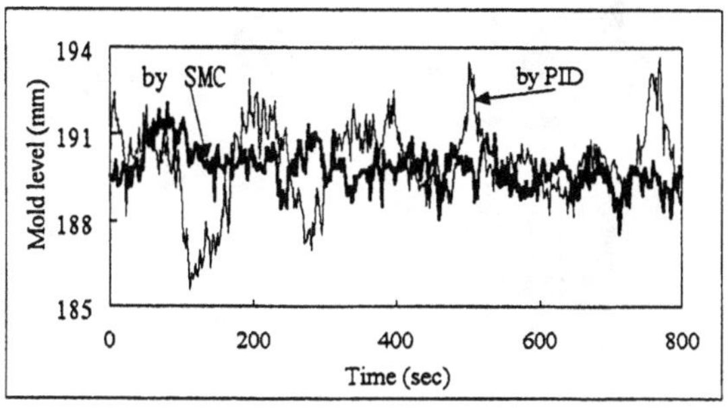

Fig. 7. Actual mold levels

OPTIMAL CONTROL SYSTEM FOR HOT STRIP FINISHING MILL

M.Okada*, K.Murayama*, A.Urano*, Y.Iwasaki*, A.Kawano*, H.Shiomi**

* Mizushima Works, Kawasaki Steel Corporation,
Kawasakidori 1-chome, Mizushima Kurashiki 712, Japan
phone : 086-447-2503
E-mail : m-okada@miz. kawasaki-steel. co. jp
** Chiba Works, Kawasaki Steel Corporation,
Kawasaki-cho 1-chome, Chuo-ku Chiba 260, Japan
phone : 043-264-9530

Abstract: This paper describes how to control strip gauge, looper angle, and strip tension for a hot strip finishing mill and the application result. This is based on the optimal servo theory and the model decoupling method. In finishing mill process, there exists mutual interaction among strip gauge , looper angle, and strip tension. Conventionally, each of them is controlled independently. To improve the control performance, the multivariable control considering this interaction is applied to all productions. The new control system is also used for the tension control of the automatic width control. *Copyright © 1998 IFAC*

Keywords : optimal control, autonomous control, multivariable feedback control, decoupling precompensators, steel industry

1. Introduction

In the hot strip mill process of the iron and steel industry, heated slabs are rolled to the required strip thickness and width. The finishing mill is the final sizing process in the hot strip mill, so the gauge control of the finishing mill is very important. Recently, more and more accurately sized hot strips are being demanded by customers, and it is necessary to improve the performance of the Automatic Gauge Control (AGC).

In order to improve the control performance, hydraulic screwdown systems were installed at the final three stands of the finishing mill, and they effected the gauge accuracy.

There exists mutual interaction among strip gauge, looper angle, and strip tension. But, conventionally, each of them is controlled independently, and this interaction is ignored.

When AGC works, the mass flow between stands is changed, then looper angle and strip tension are disturbed. That is, AGC is a disturbance of looper angle control and strip tension control. Then, if the response of AGC is improved by itself, this interaction can't be ignored.

The multi variable control, which is based on the finishing mill model considering this interaction, looks effective. But the finishing mill at Mizushima works consists of 7 stands and 6 loopers as shown in Fig.1, there exists too many variables for the multi variable control to be applied to.

In this paper, a model decoupling method by means of a similarity transformation of the state space is suggested. This model decoupling method makes it possible to treat the finishing mill model as a set of 6-order systems. Therefore the multi variable control can be applied. The result in the actual mill operation is also discussed.

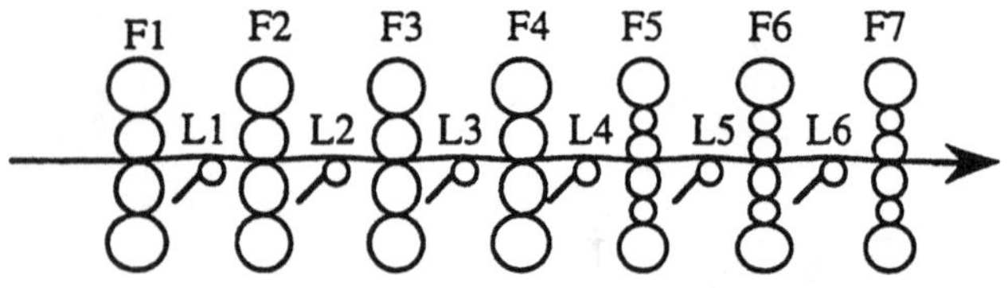

Fig.1 Schematic diagram of hot strip finishing mill

2. Problem of Conventional Control System

Fig.2 shows the conventional system of the strip tension control, the looper angle control and the AGC. The strip tension control is the feed forward control, which calculates the looper motor reference using the looper angle feedback. The looper angle control is the proportional integral control, which calculates the mill speed reference using the looper angle feedback. The AGC is the integral control, which calculates the roll gap reference using the strip gauge feedback.

There exists mutual interaction among the strip tension, the looper angle and the strip gauge. But the conventional controls do not consider the mutual interaction. So if the feedback gain is made bigger to improve the control performance, the plant will be unstable because of the mutual interaction.

Fig.3 shows the unstable looper control with the high gain AGC. If the looper control becomes unstable, the operator will not use the AGC.

To improve control performance, the new control system should consider the mutual interaction.

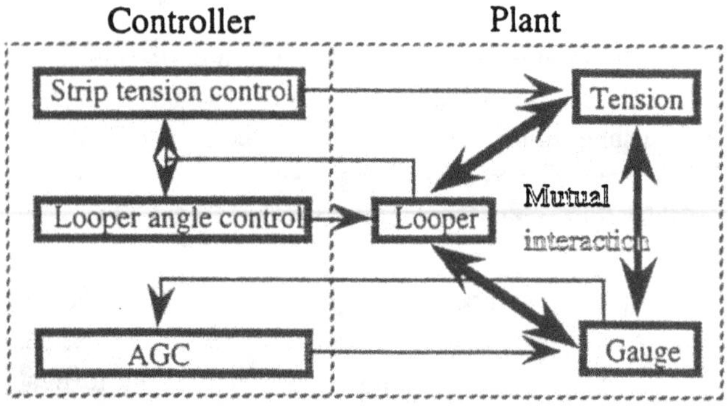

Fig.2 Outline of conventional control system

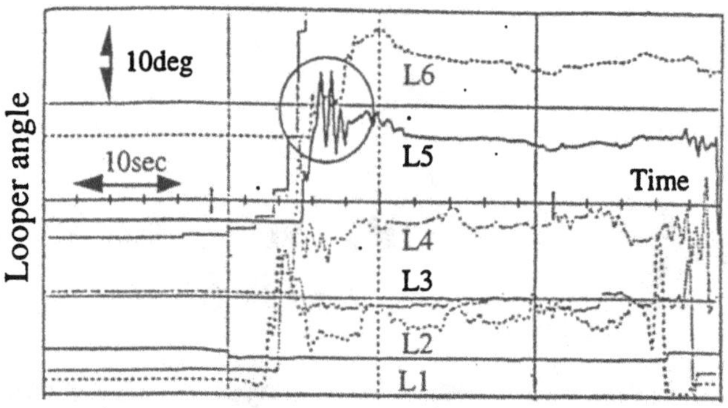

Fig.3 Example of unstable looper control

3. Modeling of Finishing Mill

The block diagram of the finishing mill model is shown in Fig.4. Where i is the stand number, Δ is the deviation from the starting point, V is the roll peripheral speed, Vref is the reference of the roll peripheral speed, Tv is the time constant of the automatic speed regulator, G is the looper torque, Gref is the reference of the looper torque, Tg is the time constant of the automatic current regulator, S is the roll gap, Sref is the reference of the roll gap, Ts is the time constant of the screwdown system, h is the strip thickness, M is the mill modulus, Q is the plasticity coefficient, f is the forward slip ratio, T is the strip tension, E is the Young's modulus of elasticity, L is the distance between stands, θ is the looper angle, ω is the looper angle speed, J is the looper inertia and D is the looper damping factor. This model is considering the mutual inter action.

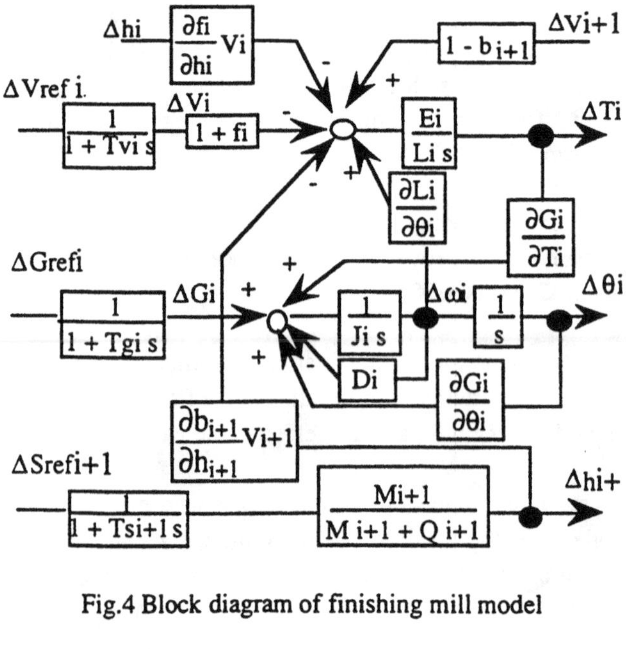

Fig.4 Block diagram of finishing mill model

From the block diagram shown in Fig.4, the finishing mill model by means of the state equation can be obtained as

$$\dot{\mathbf{x}}(t) = \mathbf{A}\,\mathbf{x}(t) + \mathbf{B}\,\mathbf{u}(t) \qquad (1)$$

$$\mathbf{y}(t) = \mathbf{C}\,\mathbf{x}(t) \qquad (2)$$

$$\mathbf{x}(t)^T = [\,\Delta V1,\ldots,\Delta V7,\ \Delta T1,\ldots,\Delta T6,\ \Delta G1,\ldots,\Delta G6,\ \Delta\omega 1,\ldots,\Delta\omega 6,\ \Delta\theta 1,\ldots,\Delta\theta 6,\ \Delta h1,\ldots,\Delta h7\,] \qquad (3)$$

$$\mathbf{u}(t)^T = [\,\Delta Vref\,1,\ldots,\Delta Vref\,7,\ \Delta Gref\,1,\ldots,\Delta Gref\,6,\ \Delta Sref\,1,\ldots,\Delta Sref\,7\,] \qquad (4)$$

where **x** is the state vector, **u** is the input vector, **y** is the output vector, **A** is 38×38 size matrix, **B** is 38×20 size matrix, and **C** is 20×38 size matrix.

4. Structure of the Finishing Mill Model

The finishing mill can be considered as a set of units which is constituted by a No.i looper, a Fi mill motor, and a Fi+1 screwdown. In this paper the unit is named as ' No.i unit '. Eqs.(1)-(4) can be rewritten as

$$\dot{\mathbf{x}}(t) = \mathbf{A}\,\mathbf{x}(t) + \mathbf{B}\,\mathbf{u}(t) \qquad (5)$$

$$\mathbf{y}(t) = \mathbf{C}\,\mathbf{x}(t) \qquad (6)$$

$$\mathbf{x}(t)^T = [\Delta h1,\ \mathbf{x1},\ldots\mathbf{x6},\ \Delta V7\,] \qquad (7)$$

$$\mathbf{u}(t)^T = [\,\Delta Sref\,1,\ \mathbf{u1},\ldots,\mathbf{u6},\ \Delta Vref\,7\,] \qquad (8)$$

$$\mathbf{xi} = [\Delta Vi,\ \Delta Ti,\ \Delta Gi,\ \Delta\omega i,\ \Delta\theta i,\ \Delta hi{+}1] \qquad (9)$$

$$\mathbf{ui} = [\Delta Vref\,i,\ \Delta Gref\,i,\ \Delta Sref\,i{+}1] \qquad (10)$$

$$\mathbf{A} = \begin{bmatrix} A0 & A01 & & & & \\ A10 & A1 & A12 & & \mathbf{0} & \\ & A21 & A2 & \ddots & & \\ & & \ddots & \ddots & \ddots & \\ & \mathbf{0} & & \ddots & A6 & A67 \\ & & & & A76 & A7 \end{bmatrix} \qquad (11)$$

$$\mathbf{B} = \begin{bmatrix} B0 & & \mathbf{0} \\ & B1 \ddots & \\ \mathbf{0} & & B7 \end{bmatrix} \qquad (12)$$

$$\mathbf{C} = \begin{bmatrix} C0 & & \mathbf{0} \\ & C1 \ddots & \\ \mathbf{0} & & C7 \end{bmatrix} \qquad (13)$$

Where **Aij** describes the state response from the change of the state in No.j unit to that in No.i. On the other hand, **Ai** describes the state response from the change of the state in No.i unit to that in No.i unit. In other words, **Ai** describes the response within the unit. **Aij** describes the response between units.

5. Model Decoupling Method

In this section, the model decoupling method by a similarity transformation is suggested.

There exists the invertible transfer matrices, **Tx Ty Tu**, as follows

$$\mathbf{x} = \mathbf{Tx}\ \mathbf{x}' \qquad (14)$$

$$\mathbf{y} = \mathbf{Ty}\ \mathbf{y}' \qquad (15)$$

$$\mathbf{u} = (1+\mathbf{Tu})\,\mathbf{u}' \qquad (16)$$

By the similarity transformation, the state equations, (5) and (6), are transformed as

$$\dot{\mathbf{x}}'(t) = \mathbf{Tx}^{-1}\mathbf{A}\ \mathbf{Tx}\ \mathbf{x}'(t) + \mathbf{Tx}^{-1}\mathbf{B}\,(1+\mathbf{Tu})\,\mathbf{u}'(t) \qquad (17)$$

$$\mathbf{y}'(t) = \mathbf{Ty}^{-1}\mathbf{C}\ \mathbf{Tx}\ \mathbf{x}'(t) \qquad (18)$$

If Ts1 = Ts2 = ... = Ts7 and Tv1 = Tv2 =...= Tv7 (This assumption is quite reasonable because each stand's controller is usually similar.), then the transfer matrices which satisfy following equations can be calculated algebraically.

$$\mathbf{A} = \begin{bmatrix} A0 & & \mathbf{0} \\ & A1 \ddots & \\ \mathbf{0} & & A7 \end{bmatrix} \qquad (19)$$

$$\mathbf{B} = \begin{bmatrix} B0 & & \mathbf{0} \\ & B1 \ddots & \\ \mathbf{0} & & B7 \end{bmatrix} \qquad (20)$$

$$\mathbf{C} = \begin{bmatrix} C0 & & \mathbf{0} \\ & C1 \ddots & \\ \mathbf{0} & & C7 \end{bmatrix} \qquad (21)$$

Therefore, with these transformations the mathematical model of the finishing mill can be divided into eight matrix equations, as follows

$$\Delta\dot{h}1(t) = A0\,\Delta h1(t) + B0\,\Delta Sref1(t)$$
$$\Delta h1(t) = C0\,\Delta h1(t)$$
$$\dot{x}1'(t) = A1\,x1'(t) + B1\,u1'(t)$$
$$y1'(t) = C1\,x1'(t)$$
$$\vdots$$
$$\dot{x}6'(t) = A6\,x6'(t) + B6\,u6'(t)$$
$$y6'(t) = C6\,x6'(t)$$
$$\Delta\dot{V}7(t) = A7\,\Delta V7(t) + B7\,\Delta Vref7(t)$$
$$\Delta V7(t) = C7\,\Delta V7(t) \qquad (22)$$

This result implies that the finishing mill can be regarded as a set of units that don't affect each other.

6. Compensator Design

In this section, the autonomous control of a hot strip finishing mill is introduced. This method is based on the model decoupling method, introduced earlier, and the optimal servo theory.

The block diagram of the control system is shown in Fig.5. Where, **Tx** and **Tu** are transfer matrices. **F** is the state feedback gain and **K** is the output feedback gain. **F** and **K** are calculated in the following way.

First, make the state equations as Eqs.(22).

Second, minimize the following performance index of No.i unit (i=0...7).

$$Ji = \int_0^{\infty} (ei^T\, Qi\, ei + vi^T\, Ri\, vi)\, dt \qquad (23)$$

$$ei = (y'i^{ref}(t) - yi'(t))$$

If $y'i^{ref}(t)$ is selected as

$$yi'^{ref}(t) = yi^{ref} - (yi(t) - yi'(t))$$

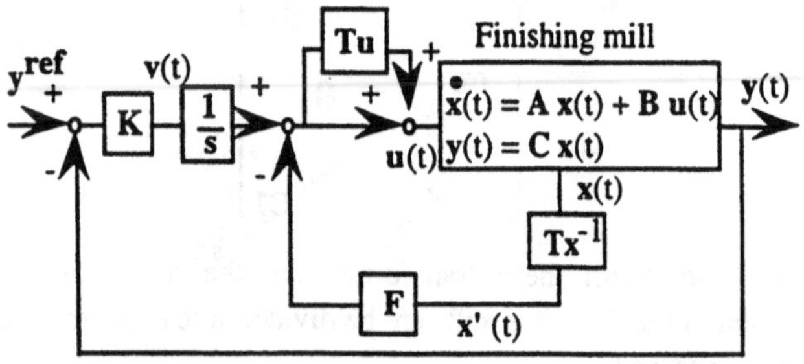

Fig.5 Block diagram of the autonomous control system

then

$$ei = (yi^{ref} - yi(t))$$

where **Qi** is 3×3 positive definite weighting matrices and **Ri** is 3×3 positive definite weighting matrices. The optimal feedback gain **Fi** and **Ki** are derived by

$$[\,Ki\;\; Fi\,] = -\,Ri^{-1}\,Bmi^{T}\,Pi \qquad (24)$$

$$Bmi = \begin{pmatrix} Bi \\ Zbi \end{pmatrix} \qquad (25)$$

where **Zbi** is 3×3 zero matrix and **Pi** is the positive definite solution of the following Riccati equation.

$$Pi\,Ami + Ami^T\,Pi - Pi\,Bmi\,Ri^{-1}\,Bmi^T\,Pi + Cmi^T\,Qi\,Cmi = 0 \qquad (26)$$

$$Ami = \begin{pmatrix} Ai & Za1i \\ -Ci & Za2i \end{pmatrix}$$

$$Bmi = \begin{pmatrix} Bi \\ Zbi \end{pmatrix}$$

$$Cmi = \begin{pmatrix} Zci & Uci \end{pmatrix}$$

where **Za1i** is 3×6 zero matrix, **Za2i** is 3×3 zero matrix, **Zbi** is 3×3 zero matrix, **Zci** is 3×6 zero matrix, and **Uci** is 3×3 unit matrix.

So, the multivariable autonomous controller which optimizes the No.i unit is derived as follows by using the state feedback matrix **Fi** and the output feedback matrix **Ki** .

$$ui(t) = Ki\int_0^{\tau} ei(t)\,dt - Fi\,(xi'(t) - xi'(0)) + ui(0) \qquad (27)$$

Third, put together the feedback gain **F** and **K** from **Fi** and **Ki** as

$$F = \begin{pmatrix} F0 & & & \\ & F1 & & 0 \\ & & \ddots & \\ & 0 & & F7 \end{pmatrix} \qquad (28)$$

$$K = \begin{pmatrix} K0 & & & \\ & K1 & & 0 \\ & & \ddots & \\ & 0 & & K7 \end{pmatrix} \qquad (29)$$

Last, add the similarity transfer matrices **Tu** and **Tx**. In this case, each unit is independent of other units, so **F** and **K** minimize the performance index J.

$$J = \int_0^{\infty} (e^T(t)\,Q\,e(t) + v^T(t)\,R\,v(t))\,dt \qquad (30)$$

where

$$Q = \begin{pmatrix} Q0 & & & \\ & Q1 & & 0 \\ & & \ddots & \\ 0 & & & Q7 \end{pmatrix} \tag{31}$$

$$R = \begin{pmatrix} R0 & & & \\ & R1 & & 0 \\ & & \ddots & \\ 0 & & & R7 \end{pmatrix} \tag{32}$$

Therefore, the multivariable controller which optimize the finishing mill is derived as follows.

$$\mathbf{u}(t) = \mathbf{u}(0) + (\mathbf{Un} + \mathbf{Tu})\,\mathbf{K}\left\{\int_0^t \mathbf{e}(t)\,dt - \mathbf{F}\,\mathbf{Tx}\,(\mathbf{x}(t) - \mathbf{x}(0))\right\} \tag{33}$$

where **Un** is 20×20 unit matrix.

This controller is named ' Local Autonomous Control '. Compared with the method based on the optimal servo theory only, this method has many benefits, because the finishing mill system can be considered as a set of independent units. First, the solution of the Riccati equation can be calculated more easily. Secondly, it isn't necessary to care for the system's total optimality, but it is sufficient to select the weighting matrices **Qi** and **Ri** so as to optimize each unit. Last, each unit's control can start independently without considering other units. The strip goes through from No.0 unit to No.7 unit, so the control of each unit will start sequentially.

7. Result of Application

The local autonomous control has been applied to all production at Mizushima works. The weighting matrices Q and R for Eq.(23) are selected by analysis of step response, frequency-singular value plots, and experimental results. The frequency-singular value plots were referred to for estimating the control system's robustness. The optimal gain for each strip is calculated by the process computer for mill set-up when the strip is in front of the finishing mill. It takes 2 seconds to solve the Riccati equation. The optimal inputs are calculated by the process computer for dynamic control every 20 milliseconds.

The experimental results in the production mill are shown in Fig.6. In this case, the material is low carbon, strip thickness is 1.8 millimeters, and strip width is 1322 millimeters. To compare with conventional control methods, Q and R were selected so as to fit the response of gauge control of the new method to that of conventional method. In the case of local autonomous control compared with the conventional control, No.4 , No.5, and No.6 looper are more stable and F4-5 strip tension , F5-6 strip tension, and F6-7 strip tension are tracking more exactly, without increasing F7 delivery thickness deviation.

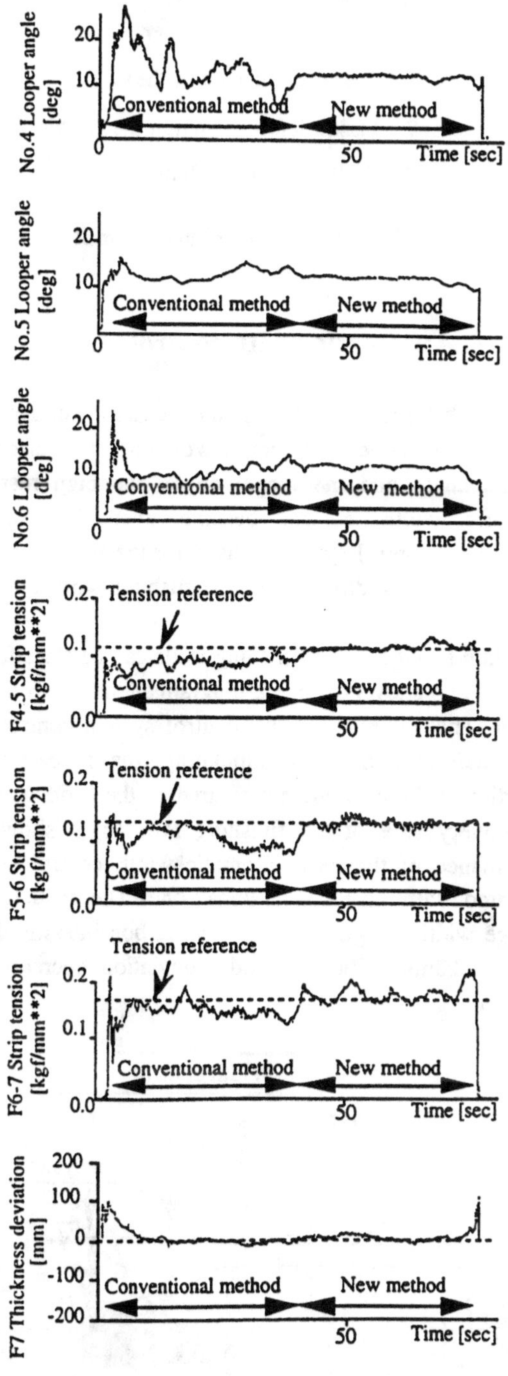

Fig.6 Experimental results in the actual production mill

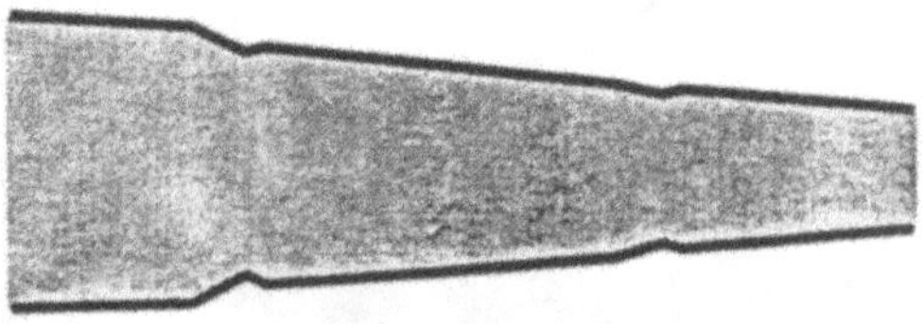

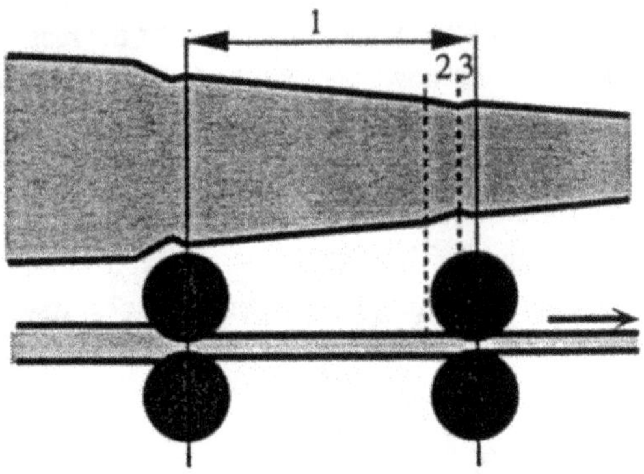

1 : Creep deformation by the tension

2 : Plastic deformation near the roll-bite

3 : Width spread by rolling

Fig.7 Width change at finishing mill

8. Automatic width control system

Because the performance of the tension control was improved and the tension control would no more affect the strip thickness and the looper angle, the high response automatic width control became available. So the autonomous control system is used for the tension control of the automatic width control at Mizushima works.

The width change in the finishing mills can be classified as Fig. 7. Fig.8 shows the block diagram of the width control system. The automatic width control system controls the creep deformation by manipulating the strip tension, according to the strip width measured by the width gauge at the delivery side of the finishing mill. Fig.9 shows the performance of the new automatic width control system compared with the conventional one. The reduction of average width margin for ultra low carbon interstitial free steel is 2.3mm. The standard deviation decreases by 0.29mm.

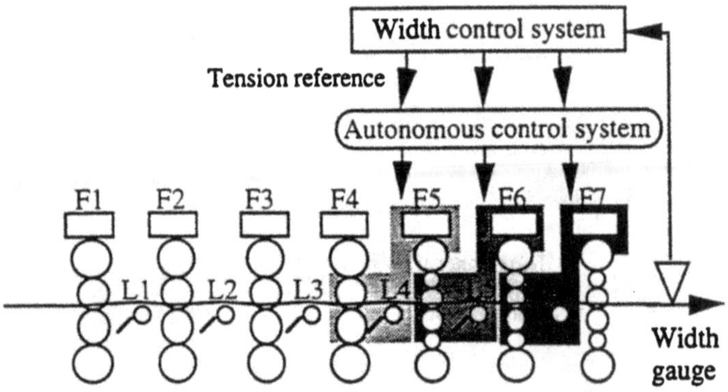

Fig.8 Block diagram of the width control system

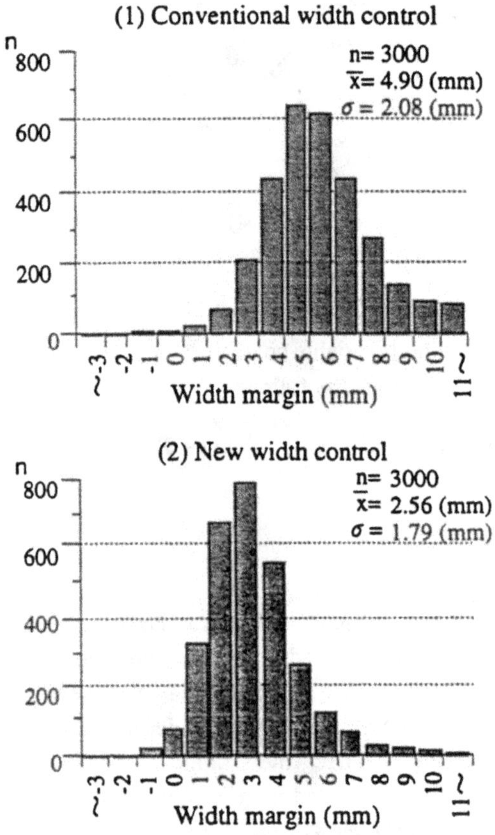

Fig.9 Result of new automatic width control

9. Conclusion

In this paper, the multivariable control method of the strip gauge, the looper angle and the strip tension, and the new automatic width control in the hot strip finishing mill was described. Experimental results show that the multivariable control and the new automatic width control are superior to the conventional methods.

References

[1] H.Miura, et al. (1993), "Gauge and tension control system for hot strip finishing mill", IEEE

[2] H. Yoshida, et al. (1982), "A mathematical model of rolling load estimation in hot strip mills", Journal of the Japan Society for Technology of Plasticity, vol. 23, No.252, 63-70.

[3]Y.Kotera, et al . (1981), "Multivariable control of hot strip mill looper", IFAC 8th World Congress-Kyoto.

[4]Y. Nishikawa, et al . (1986), "Advanced control in hot strip finishing mill", IFAC Symposium of Automation in Mining, Mineral and Metal Processing.

H^{∞} Speed Controller Design for Hot Rolling Mill Drives

Jong Hae Kim*, Hong Bae Park*, Tae Ho Um**,
Jin Yang Jeung***, Sang Ho Lee***, Jae Kyu You***, Kwang Ok Kim***

* : Kyungpook National University, Taegu, Korea
** : Kumi College, Kumi, Korea
*** : POSCO, Pohang, Korea

Abstract : An H^{∞} speed controller design method for hot rolling mill drives, #4 drive system in #2 hot rolling mill of POSCO, is proposed. A weighted mixed sensitivity minimization method is used for obtaining an H^{∞} speed controller, and loop shaping is used for improving speed tracking performance. The designed speed controller guarantees the robust stability of closed loop system against parameter variation and minimizes the effects of disturbance by change of load torques. And simulation results is given to validate the proposed design method.

Keywords : H^{∞} control, speed control, hot rolling mill drive, mixed sensitivity problem, loop shaping

1. Introduction

The increasing demands for higher quality steel strips and higher strip thickness precision in the steel industry have led to better performance requirements. To meet these requirements, many factors must be controlled accurately in the hot rolling process. Speed is one of the these major factors. Until now, POSCO(Pohang Iron & Steel Company) has used PI controller to control the speed of DC drive system in the hot rolling mill. In practical application, the hot rolling mill drive system operates under wide range of load characteristics and the parameters of system vary substantially. Unfortunately, PI speed controller cannot exhibit good control capability under these situations. Thus to ensure a specified dynamic response independent of variations of parameters, an H^{∞} speed controller scheme is required. So many researchers have tried to apply robust control theory to steel process(Liaw *et al.*, 1994; Dhaouadi *et al.*, 1993; Umeno *et al.*, 1991; Iwasaki *et al.*, 1993), including speed control, and good results on some field applications have been reported(Iwasaki *et al.*, 1993; Dhaouadi *et al.*, 1993).

In this paper, an H^{∞} speed controller design method for hot rolling mill drives is proposed. For designing the speed controller, #4 drive system in #2 hot rolling mill of POSCO is used. The plant model includes the models of DC motor, thyristor converter, sensors, compensators and etc. A weighted mixed sensitivity minimization method with sensitivity minimization and complementary sensitivity minimization is used for guaranteeing the robust stability of closed loop system with parameter variations and load torque disturbances. And loop shaping technique is used for improving tracking

This work was supported by POSCO, Korea.

performance. The principal idea of loop shaping is that the maximum singular values of closed loop transfer functions can be directly determined over appropriate frequency ranges by the singular values of the corresponding open loop transfer functions.

The control performance from the proposed speed controller is compared with that from the existing PI speed controller through the computer simulation. The reference speed signals of step and pulse signals, load torque disturbances of step and pulse signals, and parameter variations of inertia constant and current sensor gain are treated in simulation. Then the characteristics of bandwidth, overshoot, and steady state error are investigated.

To achieve quality improvement of hot strips using the proposed speed control method, the speed signals of hot rolling mill drives must be linked and regulated properly.

2. Drive System Modeling and Control System Formulation

High performance DC motors are used extensively as the actuator of drive system in hot rolling mill. The accurate modeling of such devices is important, not only to analyze the control system, but also to aid controller design and quality control in production. Pasek(1974) introduced a measurement technique by which all the parameters of the linear model of a DC motor can be determined by observing a single current response of the motor to a step input of armature voltage. Fig. 1 shows a typical test circuit for obtaining current response. The linear model of DC motor is given as

$$\begin{aligned} V &= K_e\omega + R_a i + L_a\frac{di}{dt} \\ T &= J\frac{d\omega}{dt} + B\omega + T_f \end{aligned} \qquad (1)$$

where, V is magnitude of the applied armature step-voltage, K_e is voltage constant of the DC motor, ω is angular velocity of the motor shaft, R_a is armature resistance effect, i is motor armature current, L_a is armature inductance effect, T is developed torque, J is inertial effect, B is viscous damping effect, and T_f is friction effect, respectively.

This technique is tested under zero load condition. The DC motor model in drive system is obtained as

$$\frac{\omega(s)}{V(s)} = \frac{6.74}{s^2+53s+601.4}. \qquad (2)$$

through Pasek's method. The nominal plant is constructed from the full modeling of stand drive system except speed controller part. Fig. 2 shows the hot rolling mill drive system. The drive system can be changed to the simplified system configuration in Fig. 3. The H^∞ speed controller will be designed on the basis of the system.

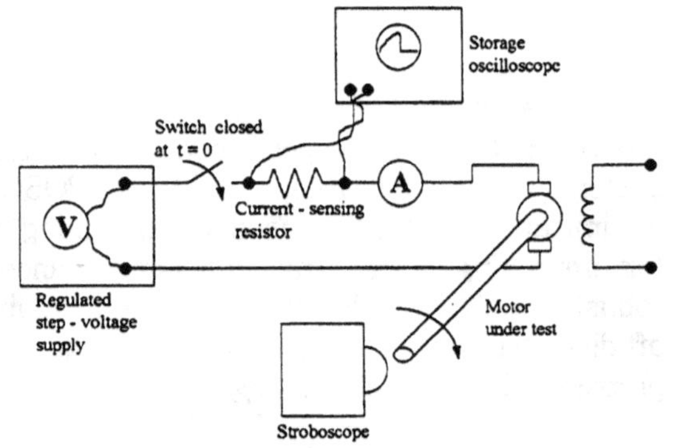

Fig. 1 The test circuit for Pasek's method.

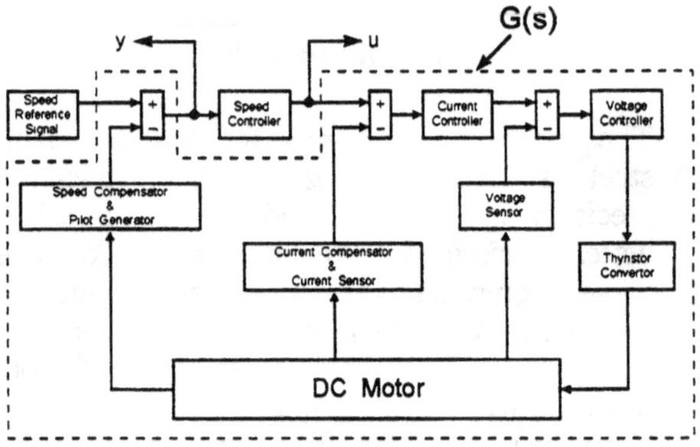

Fig. 2 Drive system block diagram including speed controller.

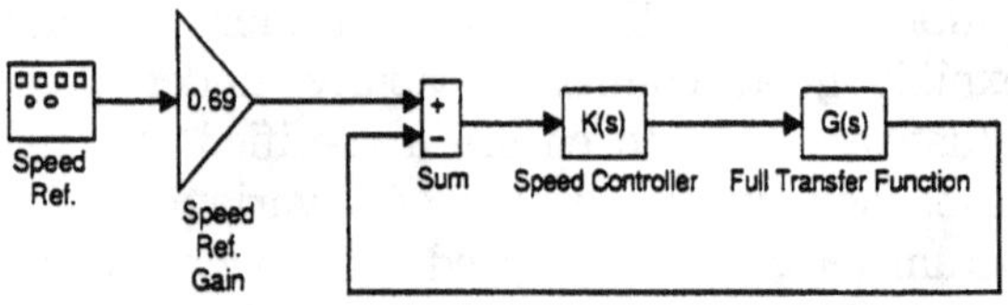

Fig. 3 Simplified drive system block diagram.

3. H^{∞} Speed Controller Design

In this section, an H^{∞} speed controller design method using Glover and Doyle's H^{∞} controller design algorithm(1988) and loop shaping technique(McFarlane *et al.*, 1990) is presented. The basic objective in controller design is to guarantee robust stability and good performance. Loop shaping is to incorporate the simple performance and robustness trade-off obtained in loop shaping, with the guaranteed stability properties of H^{∞} design methods. It is resonable to treat performance problem in low frequency and to consider robust stability in high frequency. The loop shaping design procedure is shown in Fig. 4 and summarized as:

① Using a pre compensator, W_1, and a post compensator, W_2, the singular values of the nominal plant, G, and shaped functions, W_1, W_2 are combined to form the shaped plant, G_s where $G_s = W_2 G W_1$.

② A feedback controller, K_{∞}, for G_s is obtained using H^{∞} controller design algorithm.

③ The final feedback controller, K, is then constructed by combining the H^{∞} controller, K_{∞}, with the shaping functions, W_1 and W_2 such that $K = W_1 K_{\infty} W_2$.

The sensitivity minimization is just focused on disturbance rejection. In fact, speed controller should be satisfied with sensitivity minimization and robust stability simultaneously. Therefore, it is necessary to formulate mixed sensitivity H^{∞} control problem. The closed loop system for mixed sensitivity minimization problem is depicted in Fig. 5, where G is nominal plant, K is controller, $\alpha(s)$ is weighting function for loop shaping, $W_s(s)$ is weighting function for sensitivity minimization, $W_t(s)$ is weighting function for complementary sensitivity minimization, w is exogenous input, y is measured output, and r is reference speed input. The complementary sensitivity function from w to z_1 is represented by

$$\frac{z_1}{w} = W_t G K_{\alpha}(I + GK_{\alpha})^{-1} \tag{3}$$

and the sensitivity function from w to z_2 is described as

$$\frac{z_2}{w} = W_s(I + GK_{\alpha})^{-1} \tag{4}$$

where $K_{\alpha} = \alpha(s)K(s)$. Mixed sensitivity minimization problem is written as

$$\min_{K(s)} \left\| \begin{matrix} W_s(I+GK)^{-1} \\ W_t GK(I+GK)^{-1} \end{matrix} \right\|_{\infty}. \tag{5}$$

The standard plant for using the H^{∞} controller design algorithm(Glover *et al.* 1988) is expressed as

$$P = \begin{bmatrix} W_s & -W_s G \\ 0 & W_t G \\ I & -G \end{bmatrix} \tag{6}$$

and state space representation of (6) can be transformed into

$$P = \begin{bmatrix} a_g & 0 & 0 & 0 & -b_g \\ b_{ws}c_g & a_{ws} & 0 & b_{ws} & -b_{ws}d_g \\ 0 & 0 & a_{wtg} & 0 & b_{wtg} \\ d_{ws}c_g & c_{ws} & 0 & d_{ws} & -d_{ws}d_g \\ 0 & 0 & c_{wtg} & 0 & d_{wtg} \\ c_g & 0 & 0 & I & -d_g \end{bmatrix} \tag{7}$$

where, G_s, $W_t G$, and W_s is defined as

$$G_s = \begin{bmatrix} a_g & b_g \\ c_g & d_g \end{bmatrix} \tag{8}$$

$$W_t G = \begin{bmatrix} a_{wtg} & b_{wtg} \\ c_{wtg} & d_{wtg} \end{bmatrix} \tag{9}$$

$$W_s = \begin{bmatrix} a_{ws} & b_{ws} \\ c_{ws} & d_{ws} \end{bmatrix}. \tag{10}$$

In Fig. 5, the weighting functions chosen by trial and errors are

$$\alpha(s) = 0.6s$$

$$W_s(s) = \frac{1}{0.26(s+0.001)} \tag{11}$$

$$W_t(s) = \frac{s^2 + 200s + 10^4}{3 \times 10^8}.$$

Using the H^{∞} controller design algorithm, the H^{∞} speed controller for #4 drive system in #2 hot rolling mill is obtained by

$$K(s)=\frac{\begin{array}{c}1468s^9+(3.444\,E6)s^8+(3.143\,E9)s^7\\+(1.471\,E12)s^6+(3.881\,E14)s^5+(5.433\,E16)s^4\\+(1.781\,E18)s^3+(1.885\,E19)s^2+(6.093\,E19)s\end{array}}{\begin{array}{c}s^9+2572s^8+(2.694\,E6)s^7+(1.542\,E9)s^6\\+(5.426\,E11)s^5+(1.179\,E14)s^4+(1.249\,E16)s^3\\+(1.394\,E17)s^2+(4.111\,E17)s+(4.109\,E14)\end{array}} \qquad (12)$$

The simulation results of the existing PI speed controller and the designed H^∞ speed controller are shown in fig. 6 ~ fig. 8. For step reference speed signal and pulse-wave load torque, the designed H^∞ speed controller has better effects of disturbance rejection and overshoot minimization than the PI speed controller in fig. 6. Also, for pulse-wave reference speed signal and step change load torque, the designed H^∞ speed controller shows good performance in the sense of tracking and disturbance rejection in fig. 7. When current sensor gain changes -80% from original constant value, the simulation results are shown in fig. 8. In the case of PI speed controller, the system is diverged but the designed H^∞ speed controller make the system converged.

Therefore, the proposed H^∞ speed controller can guarantee good performance and robust stability against parameter variations and disturbances.

4. Conclusions

The H^∞ speed controller design method for #4 drive system in #2 hot rolling mill of POSCO was proposed. For designing an H^∞ speed controller, DC motor was modeled by Pasek's method and full control system was constructed from the hot rolling mill drive system except speed controller. Using weighted mixed sensitivity H^∞ control problem, the drive system with designed H^∞ speed controller can be satisfied with sensitivity minimization and robust stability. Through various simulation conditions, it was confirmed that the proposed H^∞ speed controller guaranteed the robust stability of closed loop system against parameter variation and minimized the effects of load torque disturbances. Moreover, the designed H^∞ speed controller had better performance than existing PI speed controller in the sense of good tracking, wide bandwidth, and small overshoot.

To achieve good performance in field application to hot rolling mill drives using the proposed speed control method, the speed signals from all of the hot rolling mill drives must be linked and regulated properly.

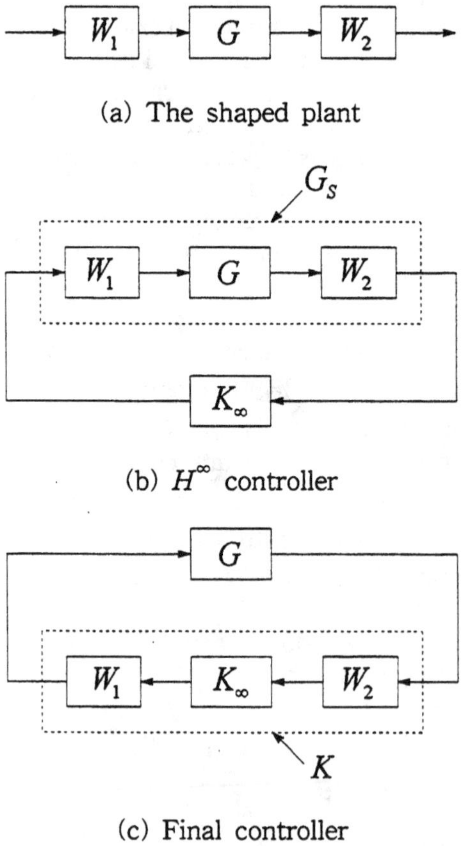

Fig. 4 The loop shaping design procedure.

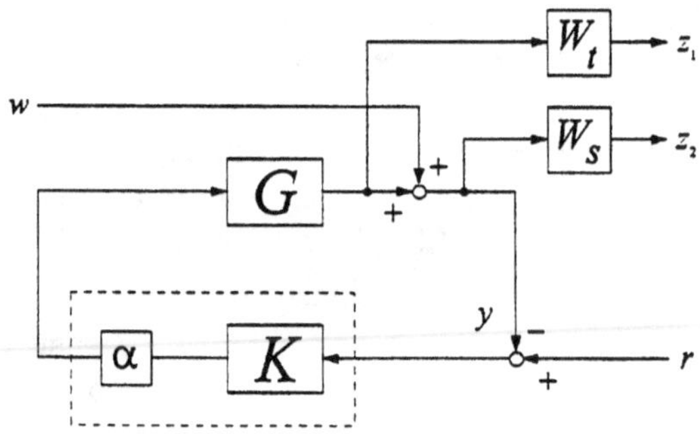

Fig. 5 The closed loop system for mixed sensitivity problem.

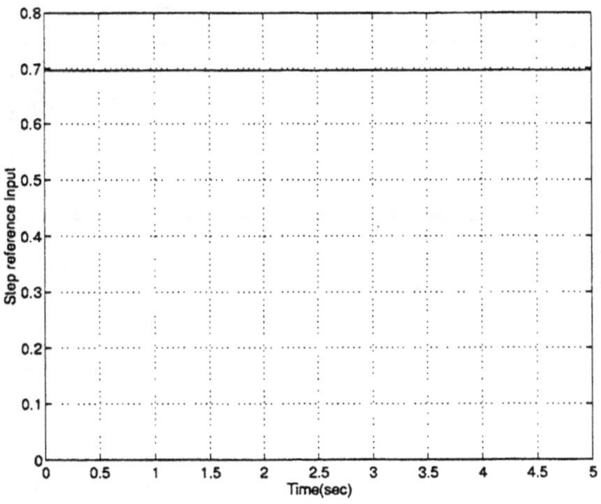

(a) Step reference speed signal

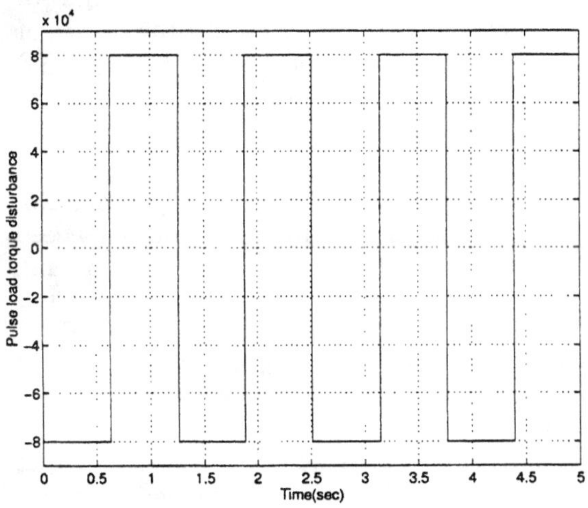

(b) Pulse-wave load torque

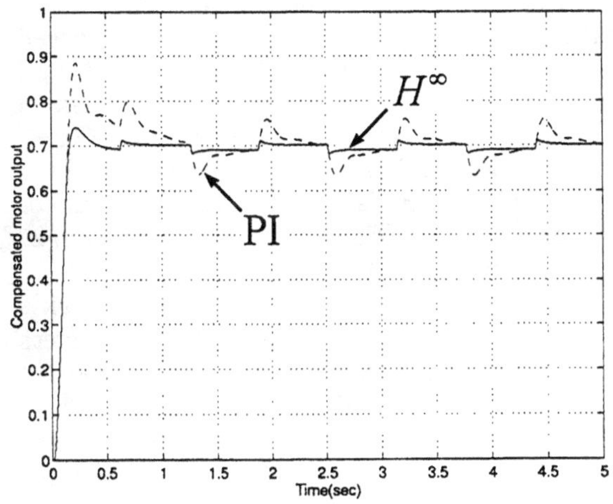

(c) Motor response of PI and H^{∞} speed controller

Fig. 6. The output response of motor in the case of step reference signal and pulse-wave load torque.

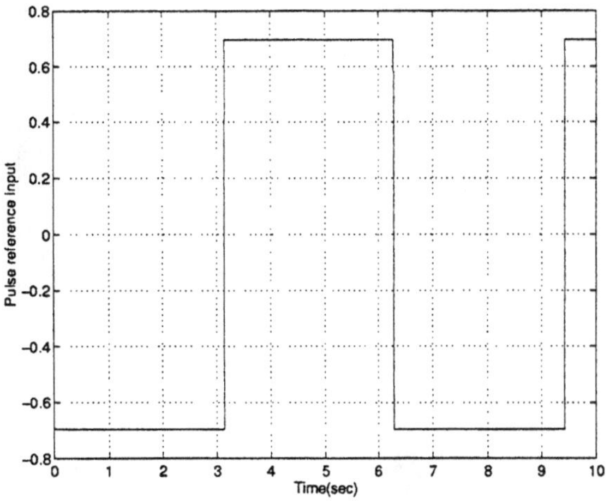

(a) Step-wave reference speed signal

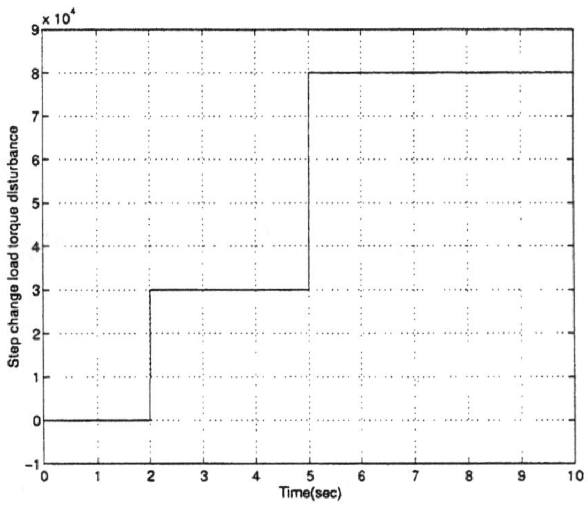

(b) Step change load torque

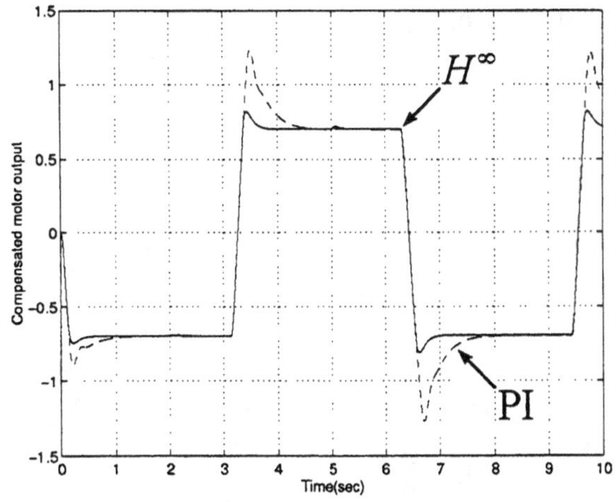

(c) Motor response of PI and H^{∞} speed controller

Fig. 7. The output response of motor in the case of pulse-wave reference signal and step change load torque.

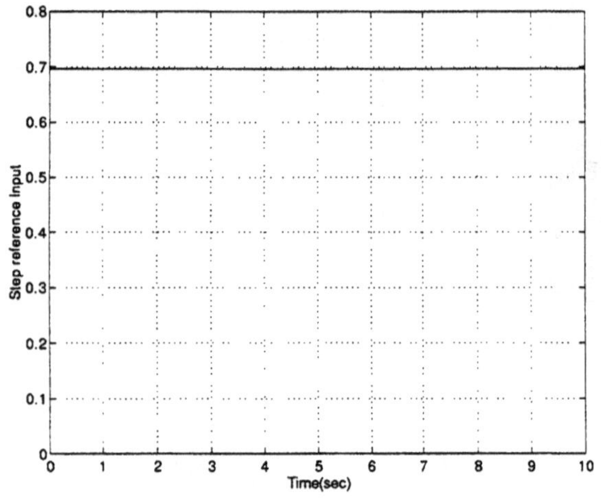

(a) Step reference speed signal

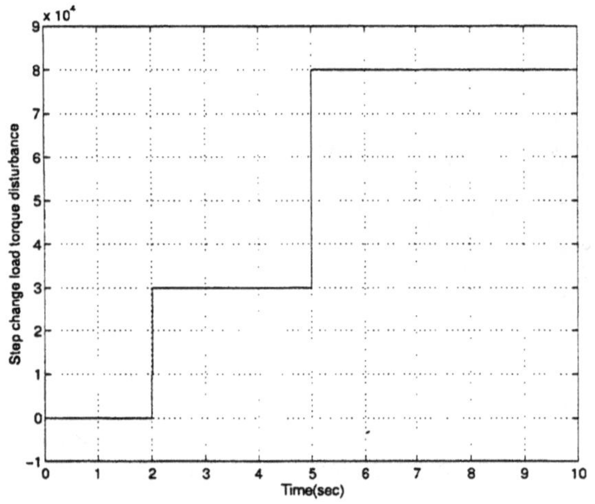

(b) Step change load torque disturbance

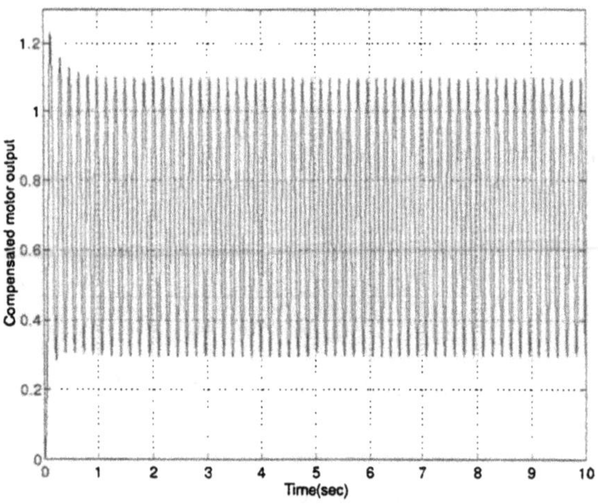

(c) PI case

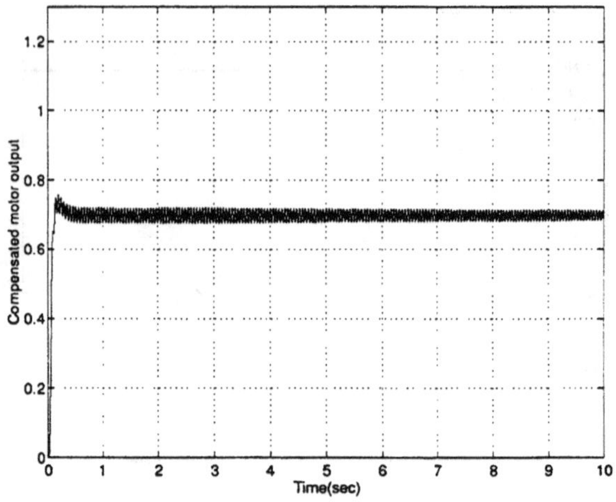

(d) H^{∞} case

Fig. 8 The motor response of change of current sensor gain in the case of pulse-wave reference signal and step change load torque.

References

Dhaouadi, R., K. Kubo and M. Tobise(1993), Two degree of freedom robust speed controller for high performance rolling mill drives, *IEEE Trans. Industrial Application*, vol. 29, no. 5, pp. 919-926.

Francis, B. A.(1987), *A Course in H^{∞} Control theory*, pp. 15-47, Springer-Verlag.

Glover, K. and J. C. Doyle(1988), State-space formula for all stabilizing controllers that satisfy an H^{∞} norm bound and relations to risk sensitivity, *Syst. & Contr. Lett.*, vol. 11, pp. 167-172.

Green, M. and D. J. N. Limebeer(1995), *Linear Robust Control,* Prentice Hall International.

Iwasaki, M. and N. Matsui(1993), Robust speed control of IM with torque feedforward control, *IEEE Trans. Industrial Electronics,* vol. 40, no. 6, pp. 553-560.

Liaw, G. M. and F. J. Lin(1994), A robust speed controller for induction motor drives, *IEEE Trans. Industrial Electronics*, vol. 41, no. 3, pp. 308-315.

Lord, W. and J. H. Hwang(1974), Pasek's technique for determining the parameters of high performance DC motors, *Report.*

McFarlane D. and K. Glover(1990), *Robust Controller Design using Normalized Coprime Factor Plant Descriptions, Lecture Notes in Control and Information Sciences*, Springer-Verlag.

Umeno, T. and Y. Hori(1991), Robust speed control of DC servomotors using modern two degrees of freedom controller design, *IEEE Trans. Industrial Electronics,* vol. 38, no. 5, pp. 363-368.

IMPLEMENTABLE RECEDING HORIZON CONTROL FOR CONSTRAINED SYSTEMS

Jae-Won Lee, Hyeong Jun Lim, and Wook Hyun Kwon *

* *ERC-ACI, School of Electrical Eng. Seoul National University*
Tel : +82-2-880-7314 and e-mail : ljw @cisl.snu.ac.kr

Abstract: In this paper, new state feedback and output feedback receding horizon controllers, based on a finite input and state horizon cost with a finite terminal weighting matrix, are proposed for discrete linear systems with input and state constraints. We propose matrix inequality conditions on the terminal weighting matrix under which closed loop exponential stability is guaranteed for both cases of state feedback and output feedback. We show that such a terminal weighting matrix can be obtained by solving an LMI (Linear Matrix Inequality). An artificial ***invariant ellipsoid*** constraint is introduced in order to relax the conventional terminal equality constraint and to handle constraints. Using the invariant ellipsoid constraints, a feasibility condition of the optimization problem is presented and a region of attraction is characterized for constrained systems with the proposed receding horizon control. A couple of illustrative examples are included. *Copyright © 1998 IFAC*

Keywords: Receding horizon control, Constraints, Stability

1. INTRODUCTION

Most practical systems have some constraints on inputs, states, and outputs. For example, actuators, valves, pumps, and compressors which have been widely used as input devices have their limits in the operating region. Drum pressure, temperature, and rpm (revolution per minutes) which may be expressed as outputs or states for boiler-turbine control systems must not exceeded their upper bounds for safety.

The most widely used methods to overcome such constraints are Anti-Windup Bumpless Transfer (AWBT) control and Receding Horizon Control. The history of AWBT control precedes that of RHC. Originally, AWBT method was studied for PID control and has been studied for single input and single output (SISO) systems. And it has been widely known that it works very efficiently for SISO systems (Åström and Rundqwist, 1989; Hanus and Kinnaert, 1989). However, it has some drawbacks. In (Doyle *et al.*, 1987), it is shown that AWBT cannot be utilized for multi-input and multi-output (MIMO) systems, because the saturation may cause a change in the input direction which may result in dangerous aspects. Moreover, there is no result on AWBT control for states (outputs) constrained systems.

Recently, the receding horizon control has been emerged as an alternative and powerful strategy for such constrained systems with limitations on inputs, states, and outputs (Richalet *et al.*, 1978; Richalet, 1993; Rawlings and Muske, 1993; Zheng and Morari, 1995). Because it is based on the state-space frameworks, it is easy to handle MIMO systems with both input and state (or output) constraints. However the stability property of the receding horizon control has been shown very recently (Rawlings and Muske, 1993; Zheng and Morari, 1995; Lee *et al.*, 1997).

There are two approaches in obtaining the stabilizing receding horizon control for constrained systems, according to the cost to be optimized.

The first approach is to adopt an infinite state horizon and finite input horizon. In this approach, if the optimization problem is feasible at some time step, then the feasibility at the next time is also guaranteed, which makes the stability proof simple. Hence, this approach has been used widely in order to guarantee stability for I/S constrained systems (Rawlings and Muske, 1993; Zheng and Morari, 1995). In the case of unstable systems, this approach needs a geometric method in order to partition the system matrix into stable and unstable modes. In order to guarantee stability, this approach needs the artificial terminal equality constraint that unstable modes should be zero within a finite horizon. This terminal equality constraint, may reduce the feasibility region and the flexibility of the controller. In this approach, there is also a restriction on the input horizon size, which should be larger than or equal to the number of unstable modes, and there is no explicit feasibility condition of the optimization problem in the case of I/S constrained unstable systems.

The second approach is to adopt a finite input and state horizon with *a finite terminal weighting matrix* (Lee *et al.*, 1997). In this approach, a so called *invariant ellipsoid constraint* is introduced in order to relax terminal equality constraint and guarantee stability. The basic concept of this artificial constraint is to put the terminal state into the ellipsoid which is invariant for the constrained system using a linear state feedback gain. In this approach, the uniform stabilizability condition, rather than controllability condition, is sufficient for the closed loop stability and the cost horizon size can be selected more freely than that of the first approach. The existing receding horizon controller stemming from the infinite state horizon cost (Rawlings and Muske, 1993; Zheng and Morari, 1995) can be obtained as a special case of the second approach.

In this paper, we propose new receding horizon controllers for constrained systems based on the result in (Lee *et al.*, 1997) for the both cases of state feedback and output feedback. And a receding horizon tracking controller will be proposed, based on the proposed new receding horizon controllers which guarantee exponential stability. A couple of illustrative examples will be followed, where inverted pendulum model and position tracking model are adopted.

2. STABILIZING RECEDING HORIZON CONTROL FOR CONSTRAINED SYSTEMS

In this paper, we will consider the following time-invariant discrete linear system with input and state constraints :

$$\begin{cases} x_{k+1} &= Ax_k + Bu_k \\ y_k &= Cx_k \end{cases} \tag{1}$$

subject to

$$\begin{cases} -u_{lim} \le u_k \le u_{lim}, & k = 0, 1, \cdots, \infty \\ -g_{lim} \le Gx_k \le g_{lim}, & k = 0, 1, \cdots, \infty \end{cases} \tag{2}$$

where $u_{lim} \in R^m, G \in R^{n_g \times n}$, and $g \in R^{n_g}$. It is assumed that $u_k = 0$ and $Gx_k = 0$ satisfy the constraint (2). Note that output constraint $-y_{lim} \le y_k \le y_{lim}$ can be expressed as state constraint in (2), because generally output equation is expressed as $y_k = C_k x_k$. We denote $\mathcal{U}$ as the feasible set for the above input constraint and $\mathcal{X}$ as the feasible set for the above state constraint. Since the constraint (2) is linear, $\mathcal{U}$ and $\mathcal{X}$ are polyhedra and therefore convex.

Since the constrained systems are nonlinear, careful arguments for stability should be addressed. Moreover, global stability is not possible for unstable systems with bounded input constraints. In this section, we propose a new receding horizon control for the linear system (1) with constraints (2).

Before stating main result, we introduce so called *invariant ellipsoid* property which can be interpreted in terms of quadratic stability (Boyd *et al.*, 1994). Suppose that there exists a $K \in R^{m \times n}$ and a positive definite matrix $P \in R^{n \times n}$ such that

$$(A + BK)'P(A + BK) - P \le 0. \tag{3}$$

and define an ellipsoid $\mathcal{E}_P$ centered at the origin $\mathcal{E}_P = \{\xi \in R^n | \xi' P \xi \le 1\}$. Then, for every initial state $x_0 \in \mathcal{E}_P$, the state trajectory $x_k (\forall k > 0)$ with the state feedback control $u_k = Kx_k$ remains in the ellipsoid $\mathcal{E}_P$. Based on this property, we introduce the following lemma for the stability of the system (1) while satisfying the constraint (2) with a state feedback controller.

Lemma 1. Suppose that there exists a $P > 0$ satisfying both (3) and the following LMIs :

$$\begin{bmatrix} Z & G \\ G' & I \end{bmatrix} \ge 0, Z_{ii} \le g_{j,lim}^2, \quad j = 1, 2, \cdots, n_g \tag{4}$$

$$\begin{bmatrix} Z & Y \\ Y' & X \end{bmatrix} \ge 0, Z_{ii} \le u_{j,lim}^2, \quad j = 1, 2, \cdots, m \tag{5}$$

where $X = P^{-1}, Y = KX$, and $g_{j,lim}$ and $u_{j,lim}$ are the jth elements of g_{lim} and u_{lim}, respectively. Then the state feedback controller $u_k = Kx_k$ exponentially stabilizes the system for all $x_0 \in \mathcal{E}_P$ while satisfying the constraint (2). And the resultant state trajectory x_k always remains in the region $\mathcal{E}_P$.

Proof: This lemma is an extension of the result in (Boyd *et al.*, 1994) which is based on continuous time systems. ■

Now using the property of the invariant ellipsoid, we propose a new receding horizon controller guaranteeing the constraint on input and state. Consider the following performance index with finite horizon and a finite terminal weighting matrix : $J(k) = \sum_{i=0}^{N-1}(||x_{k+i|k}||_Q^2 + ||u_{k+j|k}||_R^2)$ $+ ||x_{k+N|k}||_\Psi^2$.

The basic idea for guaranteed stability of receding horizon control for constrained system is to put the terminal state ($x_{k+N|k}$) into the invariant ellipsoid. Hence, we introduce the following optimization problem with an artificial invariant ellipsoid constraint at the terminal step.

$$\underset{u_{k|k},\cdots,u_{k+N-1|k}}{\text{Minimize}} \quad J(x_k, k) \tag{6}$$

subject to

$$\begin{cases} -u_{lim} \le u_{k+i|k} \le u_{lim}, & i = 0, 1, \cdots N-1 \\ -g_{lim} \le Gx_{k+i|k} \le g_{lim}, & i = 0, 1, \cdots, N \\ x'_{k+N|k}\Psi x_{k+N|k} \le 1 & \end{cases} \tag{7}$$

where we assume that the terminal weighting matrix Ψ and H satisfy the following assumption.

Assumption *1.* The finite terminal weighting matrix Ψ satisfies the following LMIs :

$$\begin{bmatrix} X & F' & (Q^{1/2}X)' & (R^{1/2}Y)' \\ F & X & 0 & 0 \\ Q^{1/2}X & 0 & I & 0 \\ R^{1/2}Y & 0 & 0 & I \end{bmatrix} > 0, \tag{8}$$

$$\begin{bmatrix} Z & G \\ G' & I \end{bmatrix} \ge 0, Z_{ii} \le g_{j,lim}^2, \begin{bmatrix} Z & Y \\ Y' & X \end{bmatrix} \ge 0, Z_{ii} \le u_{j,lim}^2, \tag{9}$$

where $F = AX + BY$, $X = \Psi^{-1} > 0$ and $Y = R^{1/2}HX$.

Now we define a new state feedback receding horizon controller as the first control move of the optimal solution which minimizes the above constrained performance index with the terminal weighting matrix satisfying the assumption 1 .

Theorem 2. Suppose that the terminal weighting matrix Ψ satisfies the assumption 1. Then the optimization problem minimizing $J(x_k, k)$ subject to the constraint (7), is always feasible for all $k \ge 0$ and for all initial states $x_0 \in \mathcal{E}_\Psi$ where $\mathcal{E}_\Psi = \{\xi \in R^n | \xi'\Psi\xi \le 1\}$. Also, $x_k = 0$ is the exponential stable equilibrium of the closed loop system with the state feedback receding horizon controller stemming from the optimization problem (6), for all initial states $x_0 \in \mathcal{E}_\Psi$.

Proof: See (Lee *et al.*, 1997). ■

Remark 3. In conventional results (Rawlings and Muske, 1993; Zheng and Morari, 1995) where an infinite state horizon cost is adopted, an additional terminal equality constraint that unstable modes should go to zero in some finite time should be included if the system is unstable. We can observe that the invariant ellipsoid constraint $x'_{k+N|k}\Psi x_{k+N|k} \le 1$ is less restrictive than the terminal equality constraint.

In the above theorem, we observe that $\mathcal{E}_\Psi$ specifies the region of attraction for the proposed receding horizon controller. The following LMI problem will give us a way to enlarge the region of attraction :

$$\begin{aligned} \text{Minimize} \quad & \textit{log det } X^{-1} \\ \text{subject to} \quad & (8) \text{ and } (9). \end{aligned}$$

This is possible because there are some flexibilities in choosing the finite terminal weighting matrix.

Now we will propose an output feedback receding horizon controller, which is based on the above result on state feedback case. The state is estimated by an asymptotic observer as follows :

$$\bar{x}_{k+1} = A\bar{x}_k + Bu_k + L\{C[A\bar{x}_k + Bu_k] - y_{k+1}\}, \tag{10}$$

where $\bar{x}_k$ denotes the state estimated at the time k.

Then, the dynamics of the error $e_k = \bar{x}_k - x_k$ becomes

$$e_{k+1} = (I + LC)Ae_k. \tag{11}$$

Given the estimated state $\bar{x}_k$, the performance index is defined by

$$\begin{aligned} J_o(\bar{x}_k) = & \sum_{i=0}^{N-1}(||\bar{x}_{k+i|k}||_Q^2 + ||u_{k+i|k}||_R^2) \\ & + ||\bar{x}_{k+N|k}||_\Psi^2 \end{aligned} \tag{12}$$

and the optimization problem for output feedback receding horizon controller is defined by

$$\underset{u_{k|k},\cdots,u_{k+N-1|k}}{\text{Minimize}} \quad J_o(\bar{x}_k, k) \tag{13}$$

subject to

$$\begin{cases} \bar{x}_{k+i+1|k} = A\bar{x}_{k+i|k} + Bu_{k+i|k}, & i = 0, 1, \cdots, \infty \\ -u_{lim} \le u_{k+i|k} \le u_{lim}, & i = 0, 1, \cdots, N-1 \\ -g_{lim}Gx_{k+i|k} \le g_{lim}, & i = 0, 1, \cdots, N-1 \\ \bar{x}'_{k+N|k}\Psi\bar{x}_{k+N|k} \le 1 & \end{cases} \tag{14}$$

where $\bar{x}_{k|k} = \bar{x}_k$.

Theorem 4. Suppose that the terminal weighting matrix Ψ satisfies the assumption 1 and $(I + LC)A$ is stable. Then, the closed loop system with the state feedback receding horizon controller stemming from the optimization problem (13), is exponentially stable.

To establish the exponential stability of the closed loop system with output feedback, consider the following Lyapunov function candidate : $V(\hat{x}_k, e_k) = J_o^*(\hat{x}_k) + \sum_{i=0}^{\infty} ||e_{k+i|k}||_O^2 = J_o^*(\hat{x}_k) + ||e_k||_{\bar{O}}^2$ where $J_o^*(\cdot)$ is the optimal value of $J_o(\cdot)$, and O is a positive definite $n \times n$ matrix and $\bar{O}$ is a unique positive definite solution of the following Lyapunov equation : $\bar{O} = A'(I + LC)'\bar{O}(I + LC)A + O$. From the property of the cost function $J_o(\cdot)$,

$$V(\hat{x}_k, e_k) \geq ||\hat{x}_k||_Q^2 + ||e_k||_{\bar{O}}^2$$
$$\geq \lambda_{min}(Q)|\hat{x}_k|^2 + \lambda_{min}(\bar{O})|e_k|^2 \quad (15)$$

And, let the control law $u_{k+i|k} = 0, k = 0, \cdots, N-1$. Then $\hat{x}_{k+i+1|k} = A\hat{x}_{k+i|k}$. We can find $\bar{A}$ such that $A^i \leq \bar{A}, i = 0, 1, \cdots, N$. Then

$$V(\hat{x}_k, e_k) \leq \sum_{i=0}^{N-1} ||\hat{x}_{k+i|k}||_Q^2 + ||\hat{x}_{k+N|k}||_\Psi^2 + ||e_k||_{\bar{O}}^2$$
$$\leq (N\lambda_{max}(Q) + \lambda_{max}(\Psi))|\bar{A}||\hat{x}_k|^2 + \lambda_{max}(\bar{O})|e_k|^2 (16)$$

Next, we assume that the control sequences of the cost $J_o(\hat{x}_{k+1})$ are $u_{k+i|k+1} = u^*_{k+i|k}$, $i = 1, 2, \cdots, N-1$ and $u_{k+N|k+1} = H_{k+N}x_{k+N|k+1}$. Then

$$J_o^*(\hat{x}_k) = J_o(\hat{x}_{k+1}) + ||\hat{x}_k||_Q^2 + ||u_k||_R^2 + M_k, \quad (17)$$

where

$$M_k = ||\hat{x}^*_{k+N|k}||_\Psi^2 - ||\hat{x}_{k+N|k+1}||_Q^2 - ||u_{k+N|k+1}||_R^2$$
$$-||\hat{x}_{k+N+1|k+1}||_\Psi^2. \quad (18)$$

Under the assumption on the control sequence, we can obtain $\hat{x}_{k+N|k+1} = \hat{x}^*_{k+N|k} + \xi_{k+N|k+1}$ and $\hat{x}_{k+N+1|k+1} = (A + BH)(\hat{x}^*_{k+N|k} + \xi_{k+N|k+1})$. From (18) and the condition on the terminal weighting matrix Ψ, it follows that $M_k = -2||\hat{x}^*_{k+N|k}||_\Psi^2 - ||\xi_{k+N|k+1}||_\Psi^2$, where $z = \hat{x}^*_{k+N|k} + \xi_{k+N|k+1}$. From the fact that, for $\delta > 0$, $-2a'\Psi b \geq -\delta a'\Psi a - \frac{1}{\delta}b'\Psi b$, we obtain

$$M_k \geq -\delta||\hat{x}^*_{k+N|k}||_\Psi^2 - (1 + \frac{1}{\delta})||\xi_{k+N|k+1}||_\Psi^2$$
$$\geq -\delta\lambda_{min}(\Psi)|\hat{x}^*_{k+N|k}|^2 - (1 + \frac{1}{\delta})\lambda_{min}(\Psi)|\xi_{k+N|k+1}|^2$$

where the value of δ will be determined later. Since there exists a K such that $|\hat{x}^*_{k+N|k}| \leq (|A| + |B|K)^N|\hat{x}_k|$ and $\xi_{k+i|k+1} = A^{i-1}LCAe_k$, we obtain $|\xi_{k+N|k+1}| = |A^{N-1}LCA||e_k|$. Hence, the equation (17) becomes

$$J_o^*(\hat{x}_k) \geq J_o^*(\hat{x}_{k+1}) + \lambda_{min}(Q)|\hat{x}_k|^2 + \lambda_{min}(R)|e_k|^2$$
$$-\delta(|A| + |B|K)^{2N}|\hat{x}_k|^2 - (1 + \frac{1}{\delta})\lambda_{min}(\Psi)|A^{N-1}LCA|^2|e_k|^2$$

Therefore, we obtain $\Delta J^*(\hat{x}_k) \leq -\alpha|\hat{x}_k|^2 + \beta|e_k|^2$ where $\alpha = \lambda_{min}(Q) - \delta(|A| + |B|K)^{2N}$, $\beta = \lambda_{min}(R) + (1 + \frac{1}{\delta})\lambda_{min}(\Psi)|A^{N-1}LCA|^2$. Hence, from this inequality, (15), and (16), we obtain exponential stability. ■

3. RECEDING HORIZON TRACKING CONTROL FOR CONSTRAINED SYSTEMS

One of the advantages of receding horizon control is that it can presents good tracking performances, since it can utilize given tracking commands for a finite future time. However, there is no results on receding horizon tracking control for constrained while guaranteeing stability. In this section, based on the results in the previous section, we propose a new receding horizon tracking control where the tracking commands for a finite future time are utilized.

Consider the following reference model where the plant output should follow the output signal of reference model :

$$z_{k+1} = A_r z_k,$$
$$\tilde{y}_k = C_r z_k$$

We also consider the following performance index :

$$J_r(k) = \sum_{i=0}^{N-1} (||y_{k+i|k} - \tilde{y}_{k+i}||_{Q_y}^2 + ||u_{k+j|k}||_R^2)$$
$$+||y_{k+N|k} - \tilde{y}_{k+N}||_{\Psi_y}^2, \quad (19)$$

and the augmented state $\hat{x}_k = [x_k' \; z_k']'$. Then the system equation for augmented state $\hat{x}_k$ is written by

$$\hat{x}_{k+1} = \begin{bmatrix} A & 0 \\ 0 & A_r \end{bmatrix} + \begin{bmatrix} B \\ 0 \end{bmatrix} u_k$$
$$= \hat{A}\hat{x}_k + \hat{B}u_k$$

The performance index J_r is represented as $J_r(k) = \sum_{i=0}^{N-1}(||\hat{x}_{k+i|k}||_{\hat{Q}}^2 + ||u_{k+j|k}||_R^2) + ||\hat{x}_{k+N|k}||_{\hat{\Psi}}^2$, where

$$\hat{Q} = \begin{bmatrix} C'Q_yC & -C'Q_yC_r \\ -C_r'Q_yC & C_r'Q_yC_r \end{bmatrix} \hat{\Psi} = \begin{bmatrix} C'\Psi_yC & -C'\Psi_yC_r \\ -C_r'\Psi_yC & C_r'\Psi_yC_r \end{bmatrix}.$$

This method is well known one in LQ tracking control theory. Now we are trying to construct the reference model using given tracking commands for a finite future time. Assume that the tracking commands $r_0, r_1, \cdots, r_N$ are given. Then the system parameters for the reference model can be obtained as follows :

$$A_r = \begin{bmatrix} a_0 & a_1 & \cdots & a_N \\ 1 & 0 & \cdots & 0 \\ 0 & 1 & \cdots & 0 \\ \vdots & \ddots & \ddots & \vdots \\ 0 & \cdots & 1 & 0 \end{bmatrix}, z_0 = \begin{bmatrix} 1 \\ 0 \\ \vdots \\ 0 \end{bmatrix}$$

$$C_r' = \begin{bmatrix} a_0 \\ a_1 \\ \vdots \\ a_N \end{bmatrix} = \begin{bmatrix} 1 & 0 & \cdots & 0 \\ r_0 & 1 & \cdots & 0 \\ \vdots & \vdots & \ddots & \vdots \\ r_{N-1} & r_{N-2} & \cdots & 1 \end{bmatrix}^{-1} \begin{bmatrix} r_0 \\ r_1 \\ \vdots \\ r_N \end{bmatrix}$$

4. ILLUSTRATIVE EXAMPLE

In this section, we present two examples which illustrate the implementation of the proposed receding horizon control.

In the first example, we adopt an inverted pendulum model and make a comparison between he proposed state feedback receding horizon control and the conventional infinite-horizon receding horizon control (Rawlings and Muske, 1993; Zheng and Morari, 1995). The discretized dynamic equation of the pendulum is given by

$$x_{k+1} = \begin{bmatrix} 1.0018 & 0.01 & 0 \\ 0.3595 & 1.0018 & 0 \\ -0.0098 & 0 & 1 \end{bmatrix} x_k + \begin{bmatrix} -0.0002 \\ -0.334 \\ 0.01 \end{bmatrix} u_k,$$

where $x_k = [\theta\ \dot{\theta}\ \dot{d}]'$, θ denotes the angular of the rod, $\dot{d}$ denotes the velocity of the cart, and control input u_k denotes the applied force to the cart. The input constraint is given by $-5 \leq u_k \leq 5$. In this example, we choose the design parameters as $Q = I$, $R = 0.00001$. For the proposed controller, the minimum horizon for feasibility is obtained as $N_{min} = 14$. In case of infinite-horizon RHC, we must partition the Jordan form of the system matrix into stable and unstable parts $A = TJT^{-1} = \begin{bmatrix} T_1 & T_2 \end{bmatrix} \begin{bmatrix} J_1 & 0 \\ 0 & J_2 \end{bmatrix} \begin{bmatrix} \tilde{T}_1 \\ \tilde{T}_2 \end{bmatrix}$, where J_1's eigenvalues are unstable, and need the terminal equality constraint which brings the unstable modes, $z_k^1 = \tilde{T}_1 x_k$ to be zero at $k = N$. Because of such a restrictive condition, the minimum horizon size for feasibility could be larger, especially in case that the unstable modes z_k^1 are very far from zero even if actual states x_k are not the case. This example illustrates such a case and the minimum horizon for feasibility of infinite-horizon RHC is obtained as $N_{min} = 21$, which is much larger horizon compared with the proposed RHC. Besides, such a constraint of infinite-horizon RHC may result a larger optimized cost value compared with our method.

The initial optimal cost, i.e., $J^*(0) = \sum_{i=0}^{\infty}(||x_{i|0}||_Q^2 + ||u_{i|0}||_R^2)$ for these two controllers with $x_0 = [0.1\ 0\ 0]'$ is plotted in Fig. 1. $J^*(0) = 7.9218$ with $N = 14$, $J^*(0) = 6.3007$ with $N = 21$ for our controller, but $J^*(0) = 14.8264$ with $N = 21$ for infinite-horizon RHC. The actual closed-loop cost from $k = 0$ to $k = 100$ $(\sum_{i=0}^{100}(x_i'Qx_i + u_i'Ru_i))$ with $N = 21$ for our method is also smaller. The cost for our method is 5.7430, but the cost for infinite-horizon RHC is 7.9095.

In the second example, we present the simulation results for the receding horizon tracking control using a classical angular positioning system. The system consists of a rotating antenna driven by an electric motor. The control problem is to use input voltage to the motor to rotate the antenna so that it always points in the direction of a moving object

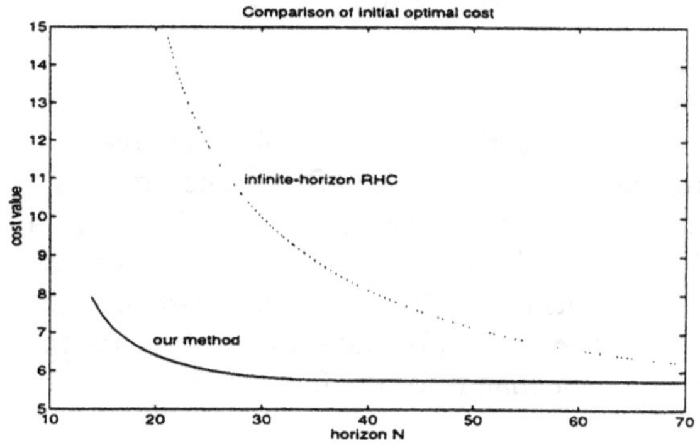

Fig. 1. Comparison with infinite-horizon RHC

in the plane. The discrete-time equations for this system are

$$x_{k+1} = \begin{bmatrix} \theta_{k+1} \\ \dot{\theta}_{k+1} \end{bmatrix} = \begin{bmatrix} 1 & 0.1 \\ 0 & 1 - 0.1\alpha \end{bmatrix} x_k + \begin{bmatrix} 0 \\ 0.1\kappa \end{bmatrix} u(k)$$

$$y_k = \begin{bmatrix} 1 & 0 \end{bmatrix} x_k \qquad (20)$$

with $\kappa = 0.787$, $\alpha = 1$, and $Q = I$, $R = 0.00002$. Using the model following method described in the previous section, the simulation results for the proposed receding horizon tracking controller with horizon $N = 6$ is plotted in Fig.2.

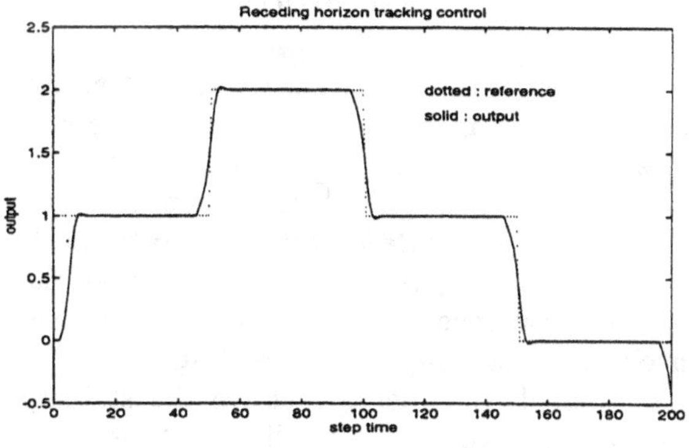

Fig. 2. Proposed receding horizon tracking control

In the infinite-horizon RHC, the above model following method is difficult to apply, because it seems that the terminal equality constraints could not be included to such a method. Therefore, constant set-point tracking method is somewhat easy to apply in the infinite-horizon RHC. But, in this method, the controller couldn't see the future reference values in advance. This implies that this tracking method has a disadvantageous property with the comparison to the above model following method. If we denote the equilibrium points as x_s, u_s, y_s, we can describe the transformed states, input and output as follows : $\hat{x}_k = x_k - x_s$, $\hat{u}_k = u_k - u_s$, and $\hat{y}_k = y_k - y_s$.

Then, the equations for the transformed variables are as follows

$$\hat{x}_{k+1} = A\hat{x}_k + B\hat{u}_k$$
$$\hat{y}_k = C\hat{x}_k. \qquad (21)$$

With the equation, we can solve the tracking problem for infinite-horizon RHC. The simulation result for the infinite-horizon RHC with horizon $N = 6$ is plotted in Fig. 3. This shows the inferior tracking performance compared with the proposed controller, because it cannot utilize the tracking commands for the future time.

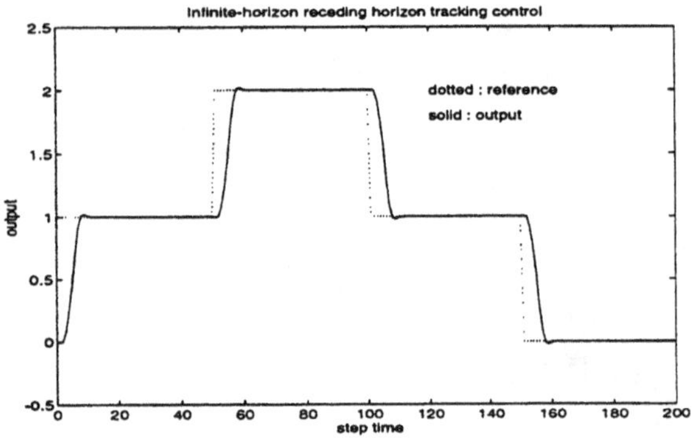

Fig. 3. Infinite-horizon receding horizon tracking control

5. CONCLUSION

In this paper, we proposed a new state and output feedback receding horizon controller for discrete linear systems with input and state constraints. In the proposed scheme, we adopted an optimization problem with *the finite horizon cost* and *the finite terminal weighting matrix*. It is noted that the zero terminal constraint, which has been widely involved for stability in the existing results, is not required. In order to achieve the stability property, we proposed a condition on the finite terminal weighting matrix, which has some free parameters and can be converted to an LMI. The stability conditions for the proposed receding horizon controller are relaxed compared with the conventional results. The uniform controllability condition, which is a necessary condition for the stability of the receding horizon control with zero terminal equality constraint, is relaxed to the uniform stabilizability condition. Under the proposed receding horizon control scheme, the cost horizon size can be more freely selected than that of the conventional results. In this paper, we introduced an artificial *invariant ellipsoid constraints* in order to relax the terminal equality constraint, and then derived an explicit feasibility condition for the optimization problem. We also characterized the region of attraction for I/S constrained systems with the proposed controller. It is noted that a constrained system is nonlinear and hence global stabilization is difficult. In this paper, we also propose a receding horizon tracking control for constrained systems. One of the advantages of the receding horizon control is that it can present good tracking performances by using tracking commands for a finite future time. The proposed tracking controller is based on the model following controller which has been widely used for LQ tracking control. In order to utilize the model following controller, we propose a reference model, where the output is exactly same as the given tracking commands for a finite future time. Through the illustrative examples, we verified the stability and tracking property of the proposed controller and made some comparisons with the existing controllers.

6. REFERENCES

Åström, K.J. and L. Rundqwist (1989). Integrator windup and how to avoid it. In: *Proceedings of the 1989 American Control Conference.* pp. 1693–1698.

Boyd, S., L.E. Ghaoui, E. Feron and V. Balakrishnan (1994). *Linear Matrix Inequalities in System and Control Theory* . Vol. 15. SIAM. Philadelphia, PA.

Doyle, J.C., R.S. Smith and D.F. Enns (1987). Control of plants with input saturation nonlinearities. In: *Proceedings of the 1987 American Control Conference.* Minneapolis, MN. pp. 1034–1039.

Hanus, R. and M. Kinnaert (1989). Control of constrained multivariable systems using the conditioning technique. In: *Proceedings of the 1989 American Control Conference.* pp. 1711–1718.

Kwon, W.H. and A.E. Pearson (1978). On feedback stabilization of time-varying discrete linear systems. *IEEE Trans. Automat. Contr.* **23**(3), 479 – 481.

Lee, J.-W., W.H. Kwon and J. Choi (1997). On Stability of Constrained Receding Horizon Control with Finite Terminal Weighting Matirx . *Submitted to Automatica.*

Rawlings, J.B. and K.R. Muske (1993). The stability of constrained receding horizon control . *IEEE Trans. Automat. Contr.* **38**, 1512 – 1516.

Richalet, J. (1993). Industrial applications of model based predictive control. *Automatica* **29**(5), 1251–1274.

Richalet, J., A. Rault, J. L. Testud and J. Papon (1978). Model predictive heuristic control: applications to industrial processes. *Automatica* **14**(5), 413–428.

Zheng, A. and M. Morari (1995). Stability of model predictive control with mixed constraints . *IEEE Trans. Automat. Contr.* **40**, 1818 – 1823.

A MATHEMATICAL MODEL OF THE HEAT LOSS OF STEEL IN A METALLURGICAL LADLE

Tom P. Fredman* J. Torrkulla* H. Saxén*

* *Heat Engineering Laboratory, Åbo Akademi University*
Biskopsgatan 8, FIN-20500 Åbo, Finland
E-mail: <tfredman/jtorrkul/hsaxen>@abo.fi

Abstract:
A mathematical model suitable as a tool for improving temperature control in the steel plant is presented. With this tool, a number of variables such as holding time, material choice and refractory layer thicknesses can be studied with regard to their influence on the steel temperature evolution during casting. In addition, the model can be used as a decision support and forms a basis for automation of temperature control. *Copyright © 1998 IFAC*

Keywords: Mathematical models, Process simulators, Temperature control, Heat flows, Steel industry

1. BACKGROUND

In modern steelmaking, after the introduction of the continuous casting process and new refractory materials, steel temperature control within the narrow bounds called for by quality requirements has become increasingly demanding. As holding times for steel in the ladle have increased and process logistics often overrule thermal- and energy-efficiency aspects, improved strategies for heat loss estimation for the steel are necessary. Contemporary refractory materials with improved durability and longer campaign life are thermally inferior to traditional brick material, i.e. their heat conductivity and specific heat are larger, resulting in increased heat losses from the steel. Over multiple ladle cycles this can result in unacceptably high shell temperatures for the ladle, introducing problems with shell buckling and lining separation.

2. INTRODUCTION

Over the years, a number of thermal models for ladle systems have been presented, ranging in complexity from simple correlations for steel temperature drop variation with time, specific to a certain plant (Olika and Björkman, 1993) to more involved models (Austin *et al.*, 1992*b*) accounting for stratification phenomena in the steel (Verhoog *et al.*, 1974), (Egerton *et al.*, 1979), (Perkins *et al.*, 1986), (Ilegbusi and Szekely, 1987), (Austin *et al.*, 1992*a*) and (Neifer *et al.*, 1993) and effects of stirring and ladle additions. Early contributions feature simulation of heat losses and refractory temperature profiles on an analog computer (Paschkis, 1956), (Paschkis and Hlinka, 1957). In order to describe steel temperature variation with time it is necessary to formulate an energy balance equation for the steel in the ladle and integrate it at the desired time instant. Included in the balance are the heat loss terms due to radiation and convection from the slag surface and due to conduction to the refractory. Since there is a two-way coupling between the energy balance and the temperature profile in the refractory these, strictly speaking, have to be solved for simultaneously. Hence, an accurate heat balance for the steel requires good estimation of loss terms of which an essential component is the calculation of the heat loss to the ladle refractory. In a number of works this was done by numerical integration of the heat conduction equation for the ladle wall, see (Alberny and Leclercq, 1973), (Morrow and Russell, 1985), (Hoppmann *et al.*, 1989), (Verhoog *et al.*, 1974), (Austin *et al.*, 1992*b*) and (Zoryk and Reid, 1993). Due to the intricate boundary condition changes for the equation as the ladle cycles through various stages of secondary metallurgy, one-dimensional solutions at different height positions in the refractory have been adopted as a sufficient means for steel temperature prediction (Pfeifer *et al.*, 1984), (Zoryk and Reid, 1993). Although this approach is appropriate for on-line simulation, more powerful modeling and simulation tools are required to achieve deeper understanding of thermal phenomena in the ladle cycle e.g. through parametric studies.

Another application of thermal modeling of metallurgical ladles is design of refractory configurations, see (Pfeifer *et al.*, 1983) and (Barber *et al.*, 1994). This has been a growing field with the advent of new materials capable to withstand longer campaigns and higher temperatures, but unfortunately having less advantageous thermal properties. Previously, engineering of ladle refractories was done by checking with steady-state calculations that refractory and shell temperature stayed within permissible temperature bounds. Although this mostly produced acceptable lining material assemblies, since the steady-state would give substantially overestimated temperatures, it was hard to foresee long-term effects over multiple ladle cycles. In addition the designs were prone to overspecification, resulting in higher labor and material cost than necessary. Possible negative end-effects in the joint area between the ladle bottom and side wall and in the upper side wall of the ladle were also hard to predict.

In this work, a mathematical model of the thermal state of a system consisting of molten steel in a refractory lined metallurgical ladle was developed. The molten steel was assumed to be perfectly mixed and homogeneous in temperature and composition. The overall energy balance for the steel, including loss terms due to conduction of heat through the refractory and radiation from the surface, was formulated. As the variation of the steel level in the ladle can be considered through alteration of model boundary conditions, the thermal status of the system can be simulated throughout the ladle cycle. The heat balance for the steel is discretized with respect to the storage term and the loss terms are recomputed at each time step for the steel temperature. At teeming, the rate of descent of the steel surface is calculated from the teeming or casting time prescribed by the user prior to starting the simulation. The ladle was modeled as an axisymmetric system, reducing the geometry to two dimensions. As a starting point a commercial computer code, for solution of partial differential equations in two-dimensional geometries, was chosen to solve for the heat flow due to conduction to the refractory. The solution is obtained by the finite element method, which is especially suitable for irregular boundary shapes. As a complement to this program, a number of routines for calculation of radiation losses, consideration of tapping/teeming situations and steel heat content were created. This approach brought programming time down to a minimum and ensured that the necessary routines for the numerical mathematics were efficient and well tested. Moreover, a user-friendly interface was constructed to allow fast specification of geometry, materials, initial thermal state and ladle cycle schedule. Together with pertinent visualization routines, a self-consistent program package suitable for ladle refractory design and cycle schedule analysis was formed.

In this paper the tool is illustrated by two case studies. First, a ladle lining specification situation is examined in which the advantages of including auxiliary insulation on steel temperature evolution and ladle shell temperature are demonstrated. Second, the impact of ladle cycle irregularities on the steel temperature is investigated by simulation.

The modeling offers an aid to gain further insight into how the thermal state of the ladle affects the evolution of the steel temperature during continuous casting. Furthermore, the applicability of different refractory materials can be readily evaluated.

3. MATHEMATICAL MODEL

The energy balance equation for the steel in the ladle can be expressed as

$$\frac{\mathrm{d}}{\mathrm{d}t}\left[C_p(t)T_{\mathtt{m}}(t)\right] + \int_A q_{\mathtt{m}}(r,z,t)drdz = 0, \quad (1)$$

$$A = (\mathtt{m} \cap \mathtt{l}_1) \cup (\mathtt{m} \cap \mathtt{s}), \quad (2)$$

where r and z are the two spatial variables, the radius measured from the axis of symmetry and the height measured from the bottom shell of the ladle, and t is time. The steel temperature, $T_{\mathtt{m}}$, is assumed to depend only on time and not on position within the molten steel. Subscript $\mathtt{m}$ denotes steel, $\mathtt{s}$ slag and $\mathtt{l}_j$, $j = 1, \ldots, N$ the different refractory lining layers $\mathtt{l}_1$ being the working lining and $\mathtt{l}_N$ the ladle shell. Heat capacity of the steel is denoted by C_p and heat flux outward from the molten steel by $q_{\mathtt{m}}$. The integration in (1) is carried out over two surfaces, the boundary between steel and slag, $\mathtt{m} \cap \mathtt{s}$, and the hot face of the ladle refractory, $\mathtt{m} \cap \mathtt{l}_1$. Using the same notation, heat conduction in the jth ladle lining layer with *constant* thermal diffusivity α_{1_j} is governed by

$$\begin{aligned}\alpha_{1_j}\nabla^2 T_{1_j}(r,z,t) &= \dot{T}_{1_j}(r,z,t)\\ T_{1_j}(r,z,0) &= f_{1_j}(r,z).\end{aligned} \tag{3}$$

The dot on the symbol is used for partial differentiation with respect to time and the spatial differential operator is defined as

$$\nabla^2 = \frac{1}{r}\frac{\partial}{\partial r}\left[r\frac{\partial}{\partial r}\right] + \frac{\partial^2}{\partial z^2}. \tag{4}$$

The boundary conditions for (3) have different forms at a boundary between two lining materials and at a boundary to steel ($\mathtt{m}$), slag ($\mathtt{s}$) or ambient air ($\mathtt{a}$). In the former case, denoting the outward surface normal derivative by $\frac{\partial}{\partial \mathbf{n}}$ and thermal conductivity by k, their general form is

$$\begin{aligned}k_{1_j}\frac{\partial T_{1_j}(r,z,t)}{\partial \mathbf{n}_{1_j}} &= k_{1_{j+1}}\frac{\partial T_{1_{j+1}}(r,z,t)}{\partial \mathbf{n}_{1_j}},\\ (r,z) \in 1_j \cap 1_{j+1}, j &= 1,\ldots,N-1.\end{aligned} \tag{5}$$

Consequently, the heat flow across material boundaries is continuous. In the latter case, on a surface of the ladle wall in contact with either metal, slag or ambient air, $\mathtt{i} = \mathtt{m},\mathtt{s},\mathtt{a}$;

$$\begin{aligned}&q_{\mathtt{i}}(r,z,t) =\\ &\quad h_{1_j}(r,z,t)\left[T_{\mathtt{i}}(t) - T_{1_j}(r,z,t)\right]\\ &\quad j = 1, N.\end{aligned} \tag{6}$$

For $\mathtt{i} = \mathtt{m}$ it is seen how (1) and (6) are coupled both through the heat flux and the steel temperature. As (6) is a boundary condition in the conduction problem (3) the heat conduction problem has to be solved simultaneously to the integration of the heat balance. The variation of the steel-slag level is incorporated into the storage term of the energy balance and modeled after specification of empty, full and casting times of the ladle cycle in progress. By assuming a constant casting mass outflow the corresponding steel level can be obtained by straightforward calculation on the basis of ladle geometry. Convection and radiation heat transfer at a surface exposed to the ambient air is described with the coefficient $h_{1_j}(r,z,t)$, which is highly nonlinear (roughly proportional to the third power of the temperatures), see (Bird *et al.*, 1960) and hence must be updated at each time step in the solution of (1). The coefficient can be decomposed into a part describing convection and a part describing radiation, as

$$\begin{aligned}h_{1_j}(r,z,t) &= h_{1_j}^{conv}(r,z,t) \;+\; \epsilon(r,z,t) \;\cdot\\ &\left[T_{1_j}^2(r,z,t) + T_{\mathtt{i}}^2(t)\right]\left[T_{1_j}(r,z,t) + T_{\mathtt{i}}(t)\right]\\ &j = 1, N.\end{aligned} \tag{7}$$

When evaluating the effective emissivity, $\epsilon(r,z,t)$, radiation from surrounding surfaces was considered by calculation of relevant view factors and the net heat flux due to radiation at each position.

4. IMPLEMENTATION

The mathematical model outlined was implemented in the MATLAB^{TM} environment. This approach was initiated when a MATLAB application toolbox for solution of partial differential equations was released. The routines use the finite element method and provide a graphical user interface for convenient geometry and equation parameter specification. Since these routines can be supported by the user's own macros and graphical user interfaces, the heat conduction problem could be solved using toolbox routines and supplemented by routines tailored to solve the energy balance equation for the steel (1), evaluate loss terms (6) and boundary conditions (7). For the energy balance, a simple Eulerian discretization method was adopted and in each time step the heat conduction subproblem was solved. These computations can run uninterruptedly throughout the ladle cycle, and the boundary conditions and steel level are updated regardless of which phase the ladle is in. At casting, descent of the steel level is accounted for in the boundary conditions, tapping, however, is modeled as an instantaneous event. Graphical user interfaces for specification of ladle cycle times, tapping temperatures and physical dimensions were created as well as routines for formatting the data into a form suitable for input to the MATLAB routines. Possible ladle geometries and lining configurations were studied in order to construct a generic physical structure, from which geometry specifications can be entered easily. A graphical user interface was then designed for this purpose. In case a more complicated geometry must be analyzed, the toolbox routines routines can readily be used in the specification stage. Lining materials are handled using trade names, while thermal and physical properties are retrieved from a data base which can be updated at the user's convenience. It should be noted that thermal properties of the lining materials are approximated as constants within each material layer.

The model output is available in a variety of representations thanks to the powerful graphics functions of MATLAB, such as steel temperature trajectory from tapping to end of casting, temperature profiles and color-field plots in the ladle wall and temperature trajectories for different points in the ladle wall. From a temperature control point of view the steel temperature evolution is the most interesting, enabling the user to experiment with different refractory assemblies, times etc. and to compare the outcome with

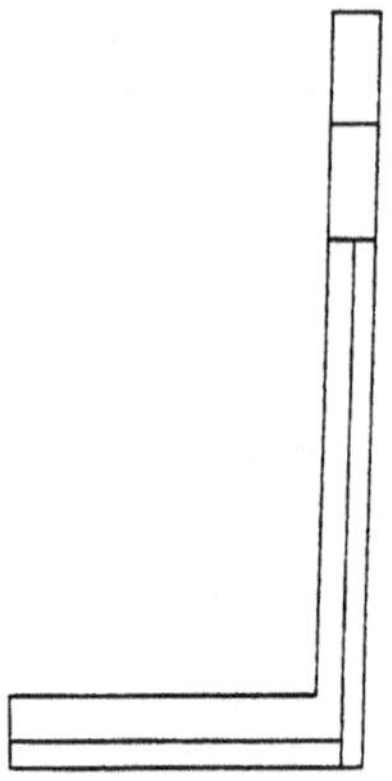

Fig. 1. Ladle lining sections, as handled in the axisymmetric model

respect to steel temperature at the end of casting. However, the capability of extracting trajectories at different positions in the lining is almost equally important, e.g., in order to compare with thermocouple measurements of lining temperatures. Here, it is possible to assess model performance and when necessary identify poorly known model parameters.

5. EXAMPLES

5.1 *The effect of an insulating lining layer*

The first example illustrates the impact of an insulating layer at the cold face of the lining.

The ladle geometry, as depicted in figure 1, is further specified with the graphical menu shown in figure 2.

In the case used as a reference for the study, the outermost lining layer is the steel shell. Even though an insulating layer is usually used inside the shell, the study carried out here simply replaced the shell by the insulation. After specifying times and boundary conditions of the ladle cycle (cf. figure 3), the model was executed through a sufficient number of cycles to reach a quasi steady state. (Typically, the lining is thermally soaked after three or four cycles.) Figure 4 presents the material properties of the layers, as well as the temperature distribution in the lining at the end of a cycle (i.e., prior to the subsequent tapping). The insulating layer is seen to yield elevated temperatures throughout the lining, except for at the cold surface, where lower temperatures occur (because of smaller heat losses). Figure 5 illustrates important information for the thermal control, i.e., the time evolution of the steel temperature in the ladle. The 85-min period depicted consists of a holding time of 35 min followed by a casting time of 50 min (cf. figure 3). Use of an insulating layer is observed to lead to a 10°C higher temperature of the steel in the ladle at the end of the cast.

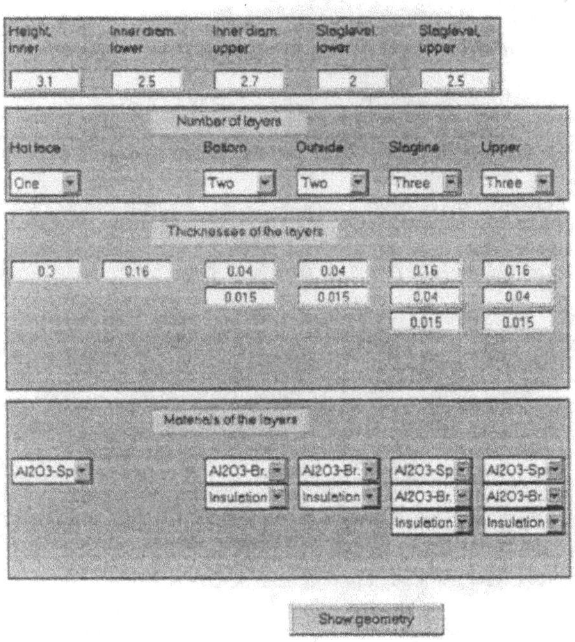

Fig. 2. Menu for specification of ladle physical dimensions and lining layers

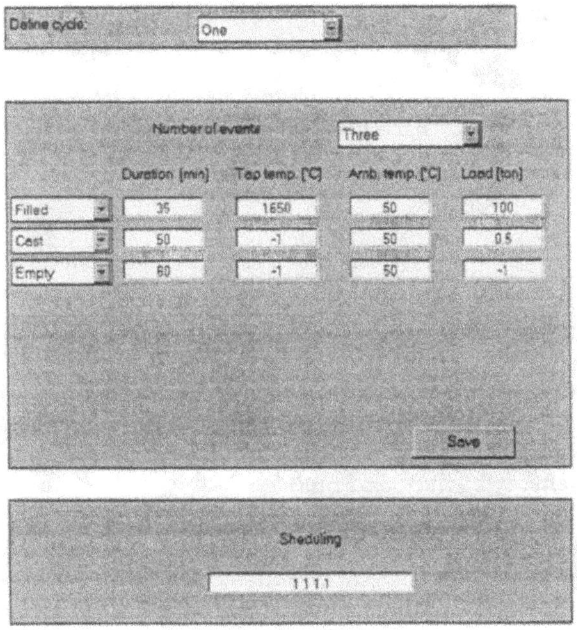

Fig. 3. Menu for specification of operation parameters for ladle cycle and scheduling

This observation agrees quite well with practical findings in the steel plant.

5.2 *Irregularities in the ladle cycle*

In the steel plant, ladle maintenance measures and scheduling problems may result in delayed arrival of the ladles to the converter. The present example illustrates the simulated effect of a delay, where the idle time for one cycle was increased from 60 min to 180 min. Materials, geometry and

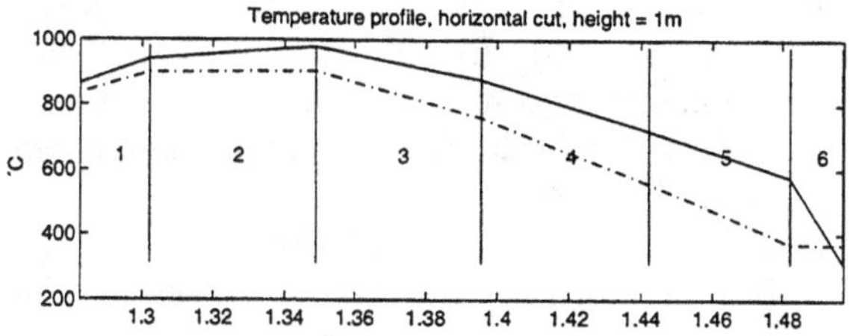

Material	[J/kgK]	[W/mK]	[kg/m^3]
1. Al2O3-sp.	1350	2.15	3100
2. Al2O3-sp.	1350	2.15	3100
3. Al2O3-sp.	1350	2.15	3100
4. Al2O3-sp.	1350	2.15	3100
5. Al2O3-Br.	1000	2.05	2700
6. Ins./Shell	1000/780	0.34/50	1000/7800

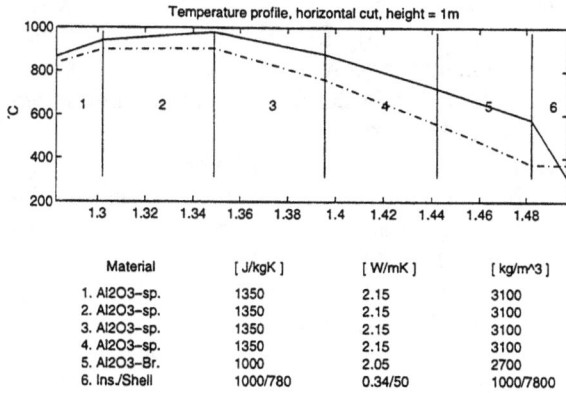

Fig. 4. Temperature profiles in the ladle lining, at the end of a cycle, in the two cases of the first example

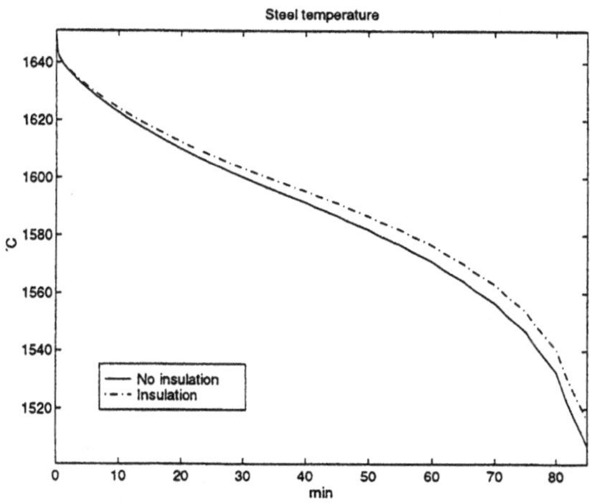

Fig. 5. Steel temperature trajectories for the two cases of the first example

boundary conditions are exactly the same as in the reference case (ladle with steel shell) of subsection 5.1. The temperature evolution at three points in the lining located 1 m above the bottom has been depicted in figure 6. The large variation in the lining temperature 5 cm from the hot surface as well as the damping and delaying action of the lining can be readily observed. As for the effect on steel temperature, figure 7, the longer idle time is seen to have resulted in a 15°C drop in temperature at the end of the casting.

6. CONCLUSIONS

Thermal control of molten steel is of considerable importance for the operation of modern steel plants. The steel temperature is influenced by the thermal state of the lining of the ladle, in which the steel is transported from the converter to the casting machine. In ladle design it is important to consider both the physical and the thermal resistances of the refractory material. This paper has presented a two-dimensional dynamic model of the ladle lining, which has been applied to simulate the state of the lining and the temperature

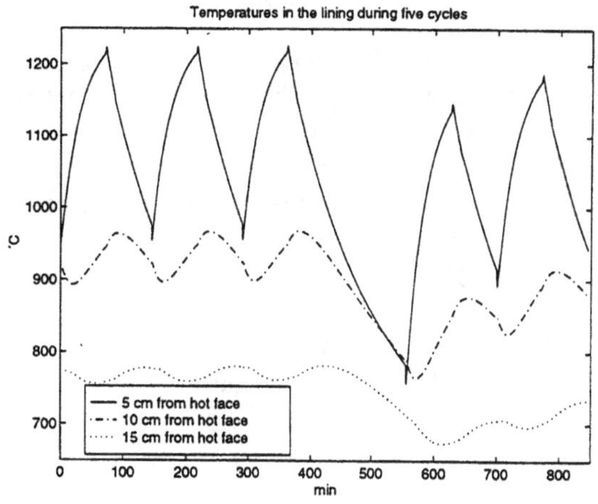

Fig. 6. Temperature trajectories over several cycles at different positions in the ladle lining, for the second example

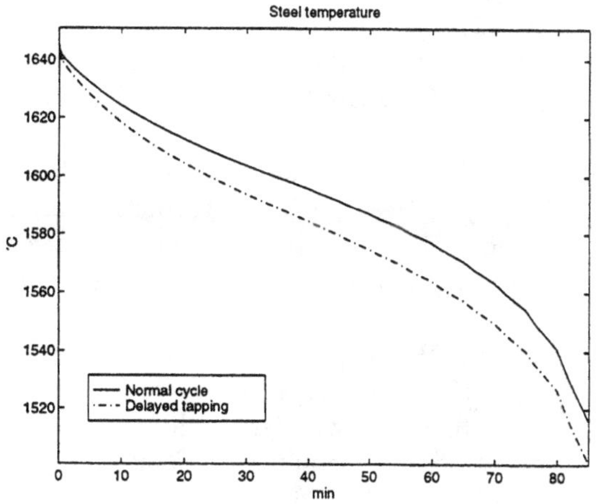

Fig. 7. Steel temperature trajectories for the second example

of the steel in the ladle during the ladle cycle. The model has been implemented using a commercial program for solution of partial differential equations by the finite element method. To the PDE solver have been added a graphical user interface and routines for setting up the geometry and computation of boundary conditions during the ladle cycle, the steel temperature, and a large number of auxiliary variables required in the simulations.

The model can be used to evaluate different ladle designs, e.g., for analysis of the expected performance of novel materials in the lining layers. The impact of the extent of different stages in the ladle cycle can also be analyzed, which may be helpful if scheduling problems are studied. Two brief examples have been included in this paper to illustrate how the model can be used. The results of the model have been found to agree reasonably well with practical observations and with measurement values obtained from thermocouples in the ladle lining during test campaigns.

7. REFERENCES

Alberny, R. and A. Leclercq (1973). Heat losses from liquid steel in the ladle and in the tundish of a continuous-casting installation. In: *Mathematical Process Models in Iron and Steelmaking, Proceedings Amsterdam*. The Metals Society. pp. 151–156.

Austin, P. R., J. M. Camplin, J. Herbertson and I. J. Taggart (1992*a*). Mathematical modelling of thermal stratification and drainage of steel ladles. *ISIJ International* **32**(2), 196–202.

Austin, P. R., S. L. O'Rourke, Q. L. He and A. J. Rex (1992*b*). Thermal modelling of steel ladles. In: *75th Steelmaking Conference Proceedings*. Vol. 75. Iron & Steel Society. pp. 317–323.

Barber, B., G. Watson and L. Bowden (1994). Optimum ladle design for heat retention during continous casting. *Ironmaking and Steelmaking* **21**(2), 150–153.

Bird, R. B., W. E. Stewart and E. N. Lightfoot (1960). *Transport Phenomena*. John Wiley & Sons. New York.

Egerton, P., J. A. Howarth, G. Poots and S. Taylor-Reed (1979). A theoretical investigation of heat transfer in a ladle of molten steel during pouring. *Int. J. Heat Mass Transfer* **22**, 1525–1532.

Hoppmann, W., F. N. Fett, G. Hsu and L. Fiege (1989). Konzept eines On-line-Modells zur Überwachung der Stahltemperatur in der Sekundärmetallurgie. *Stahl und Eisen* **109**(23), 1177–1186.

Ilegbusi, O. J. and J. Szekely (1987). Melt stratification in ladles. *Transactions ISIJ* **27**, 563–569.

Morrow, G. D. and R. O. Russell (1985). Thermal modeling in melt shop applications. *Am. Ceram. Soc. Bull.* **64**(7), 1007–1012.

Neifer, M., S. Rödl and D. Sucker (1993). Investigations on the fluiddynamic and thermal process control in ladles. *Steel Research* **64**(1), 54–62.

Olika, B. and B. Björkman (1993). Prediction of steel temperature in ladle through time/temperature simulation. *Scandinavian Journal of Metallurgy* **22**, 213–219.

Paschkis, V. (1956). Temperature drop in pouring ladles, part one. *AFS Transactions* **64**, 565–576.

Paschkis, V. and J. W. Hlinka (1957). Temperature drop in pouring ladles, part two. *AFS Transactions* **65**, 276–281.

Perkins, A., T. Robertson and D. Smith (1986). Improvements to liquid steel temperature control in the ladle and tundish. *Fachberichte Hüttenpraxis Metallweiterverarbeitung* **24**(8), 649–656.

Pfeifer, H., F. N. Fett, H. Schäfer and K.-H. Heinen (1983). Der Einfluss veränderlicher Pfannenbordgeometrien auf den Wärmeverlust der Stahlschmelze. *Stahl und Eisen* **103**, 1321–1326.

Pfeifer, H., F. N. Fett, H. Schäfer and K.-H. Heinen (1984). Modell zur thermischen Simulation von Stahlgiesspfannen. *Stahl und Eisen* **104**, 1279–1287.

Verhoog, H. M., S. Rosier, H. W. den Hartog, A. B. Snoeyer and P. J. Kreyger (1974). Heat balance and temperature stratification of liquid steel in ladles. *ESTEL-Berichte aus Forschung und Entwicklung unserer Werke* **3**, 114–120.

Zoryk, A. and P. M. Reid (1993). On-line liquid steel temperature control. *Iron & Steelmaker* **20**(6), 21–27.

ADAPTIVE INVERSE CONTROL FOR NONLINEAR PLANTS

Dong-Yeol Seok* , Jin S. Lee**

* *Maintenance Technology Department,Pohang Iron & Steel Co., Ltd.,5 Dongchon-dong, Nam-gu, Pohang-shi Kyungbuk,790-360,Korea*
** *Department of Electrical Engineering, Pohang University of Science and Technology, San 31, Hyoja-dong, Nam-gu, Pohang-shi, Kyungbuk, 790-784, Korea*

Abstract : This paper describes the concept of an adaptive inverse control system, the adaptive digital filters as the structures of the plant model and the controller, and LMS algorithm for adaptation. For the control of nonlinear plants, this control system is extended to the nonlinear adpative inverse control system and the fuzzy logic system is adopted. The cascading error backpropagation algorithm is formulated for the adaptive fuzzy logic system and applied to the adaptation of the controller.

Key words : adaptive inverse control, adaptive filtering theory, LMS adaptation algorithm, fuzzy logic system, cascading error backpropagation algorithm

1. INTRODUCTION

Most plants in the industrial processes are actually nonlinear systems. How can we control these nonlinear plants effectively? Many pioneering engineers and researchers have studied about this topic. Several methods have been developed and applied to the industrial applications. Among them, this paper describes the concept of adaptive inverse control(Widrow and Walach,1996) for nonlinear plant introduced by Widrow(1971). This adaptive inverse control is based on the adaptive filtering techniques and the least mean square(LMS) algorithm(Haykin,1986;Widrow, *et al.*,1976). The basic idea of an adaptive inverse control system is that the controller becomes an inverse of the plant. In this paper, an adaptive inverse control system is formulated by using the fuzzy logic systems. In modeling of the plant and the controller for uncertain system by using fuzzy logic, the linguistic information from experts can be exploited to generate the rules. The adaptive inverse control system developed in this paper, is applied to both minimum phase plants and nonminimum phase plants. Therefore, the objective of this paper adopts a concept of adaptive inverse control and implements the identifier and controller by using fuzzy logic systems. Also, the cascationg error backpropagation is formulated and used for the adaptation of the controller. This paper is organized as follows. In section 2, the concept of an adaptive inverse control is introduced. Section 3 presents the identifications of nonlinear plants and an adaptation of nonlinear inverse controller. Section 4 deals with a nonlinear adptive inverse control by using fuzzy logic systems. Section 5 shows the simulation results. Conclusion is given in section 6.

2. CONCEPT OF THE ADAPTIVE INVERSE CONTROL

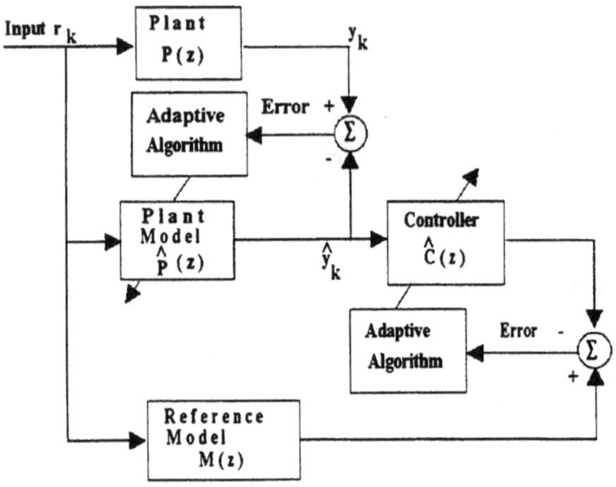

Fig. 1. An online adaptive inverse control system

In the adaptive inverse control systems, the adaptive plant inverses are used as controllers in the feedforward loop of the control system configuration. The structure of the adaptive inverse control system is shown in Fig. 1. The unknown plant is identified and the controller in a cascade with the identifier is updated by the output error. Here, the identifier and the controller adopt the same architectures which are in the form of adaptive filters.

The theory of adaptive filtering is fundamental to adaptive inverse control. This adaptive filters are used for modeling the unknown plant and the plant inverse.The form of an adaptive digital filter to be considered here comprises a tapped delay line and variable weights with causality constraint. Then the adaptive process automatically seeks for the optimal impluse response by adjusting the weights. Fig. 2 shows a modeling structure of an unknown system by discrete adaptive filtering concept. The input signal is delayed along the tapped delay lines and is used to adjust the variable weights. These weighted signals are adjusted by the error between the output of adaptive filter and the desired response. The LMS algorithm was derived by Widrow and Hoff in the late 1950s. The weights of the adaptive filter are adjusted by an automatic algorithm to minimize the mean square error.

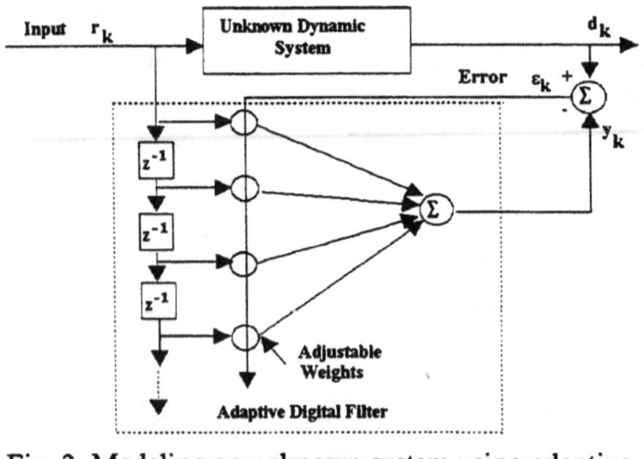

Fig. 2. Modeling an unknown system using adaptive digital filter

When the k-th input vector is represented as

$$X_k = [x_{1k}, x_{2k}, x_{3k}, \ldots, x_{nk}]^T \quad (1)$$

and the weight vector as

$$W^T = [w_1, w_2, w_3, \ldots, w_n]\,, \quad (2)$$

Then the k-th output signal becomes

$$y_k = \sum_{i=1}^{n} w_i x_{ik} = W^T X_k = X_k^T W \quad (3)$$

Denoting th desired response as d_k, we have the error equation at the k-th time as

$$\varepsilon_k^2 = (d_k - y_k)^2 = d_k^2 - 2d_k X_k^T W + W^T X_k X_k^T \quad (4)$$

and the mean square error(MSE), the expected value of ε_k^2 becomes

$$MSE = E[\varepsilon_k^2] = E[d_k^2] - 2E[d_k X_k^T]W + W^T E[X_k X_k^T]$$
$$= E[d_k^2] - 2P^T W + W^T R W\,, \quad (5)$$

where the cross correlation vector between the input signals and the desired response is $P = E[d_k X_k]$ and input correlation matrix is $R = E[X_k X_k^T]$. Note that R is symmetric and positive semidefinite. Now, the weight vector W_k is updated as

$$W_{k+1} = W_k + \mu(-\Delta_k) = W_k + 2\mu\varepsilon_k X_k, \quad (6)$$

where the gradient Δ_k of the error function is obtained by differentiating equation(5) and μ is a convergence factor that controls stability and the rate of adaptation.

3. NONLINEAR ADAPTIVE INVERSE CONTROL

Unlike the linear case, the nonlinear adaptive inverse control(NAIC) uses the filtered - ε LMS algorithm for the adaptation of the controller. The inverse controller for a linear plant has a transfer function which is a close approxination to the reciprocal of the plant transfer function. But in the nonlinear case, the transfer function doesn't exist. Strictly speaking, nonlinear systems does not even have inverses. Nevertheless, the inverse control can be performed by making use of nonlinear adaptive filter(Fig.3). The adaptive model for the nonlinear plant is the same as the linear plant case except that the nonlinear adaptive filter is used in the structure instead of the linear adaptive filter.

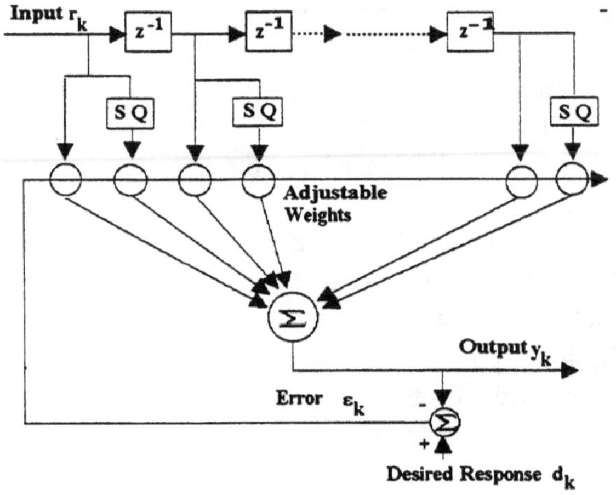

Fig. 3. Nonlinear adaptive filter

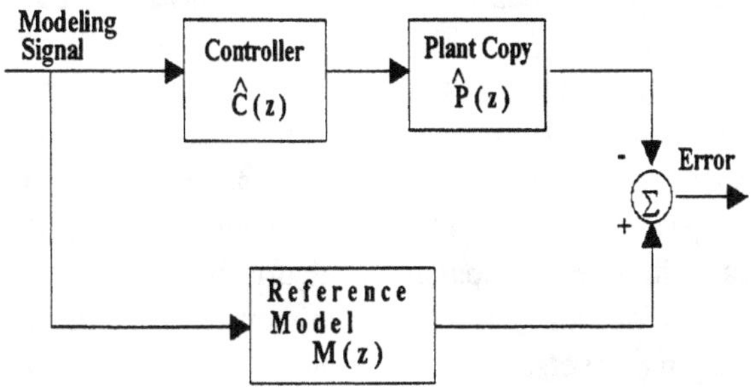

Fig. 4. Modeling of an adaptive controller for a nonlinear plant

But the method of obtaining the adaptive controller is different from the linear case. The modeling of the adaptive controller for the nonlinear plant is shown in Fig. 4. In this figure, $\hat{C}$ and $\hat{P}$ cannot be commuted since they are nonlinear. Hence, the LMS algorithm cannot be used because error at the output of $\hat{C}$ is not availbale. So, the filtered - ε LMS algorithm is used to find $\hat{C}$. This filtered - ε LMS algorithm includes modeling of a delayed inverse of $\hat{P}$, $\hat{P}_{\Delta}^{-1}$ and is used to filter the overall system error. The filtered error is used directly to update $\hat{C}$ instead of ideal error signal. The nonlinear adaptive control system using the filtered - ε LMS algorithm is shown in Fig. 5. The adaptive inverse controller is obtained by following procedures.

The adaptive plant model(or plant identification) is first found, and then it is copied and used in order to find the delayed plant inverse. The delayed plant inverse is then used to obtain the overall system error which is then used to update $\hat{C}$. Finally, the delayed plant inverse is copied and is used to filter the output of the plant and the output of the reference. The filtered error, the difference between two filtered outputs, updates the adjustable weights of the controller $\hat{C}$.

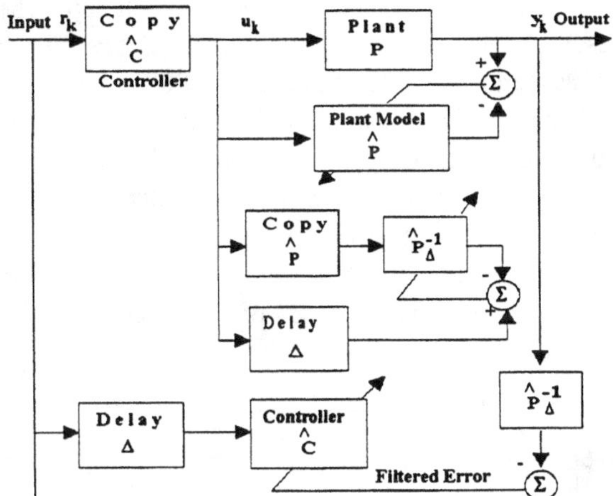

Fig. 5. Nonlinear adaptive control system the filtered-ε LMS algorithm

4. NONLINEAR ADAPTIVE INVERSE CONTROL BY USING FLS

4.1 FLS and cascading error backpropagation

The term *fuzzy* was first introduced by Zadeh(1965). Fuzzy *If-Then* rules were first applied to the fuzzy control by Mamdani and Assilian. Takagi and Sugeno developed another fuzzy logic systems and devoted to this field largely (Takagi and Sugeno, 1985; Sugeno and Kang,1986,1988; Sugeno and Tanaka, 1991,1992). Recently, fuzzy control has become a popular research topic in the control engineering area. Fuzzy logic system tries to emulate the software of human brain and it is capable of incorporating the linguistic information from operators. Fuzzy logic controllers are universal approximators capable of approxima- tion any nonlinear function(Buckley,1992,1993) to arbitrary error boundary. In fuzzy logic system, the error backpropagation algorithm is also applied and introduced by Wang(1992). A fuzzy rule base is composed of the following form:

$R^{(l)}$: If x_1 is F_1^l and ... and x_n is F_n^l, Then y is G^l (7)

where F_i^l and G^l are fuzzy sets and $\mathbf{x} = (x_1,...,x_n)^T$ and l = 1,2,...,M is the number of *IF-THEN* rules. The **x** and y are the input and output to the fuzzy logic system, respectively. The fuzzy logic system which is applied is

$$f_o(\mathbf{x}) = \frac{\sum_{l=1}^{M} \bar{y}_o^l [\prod_{o=1}^{n} a_o^l \exp(-((x_o-\bar{x}_o^l)/\sigma_o^l)^2)]}{\sum_{l=1}^{M} [\prod_{o=1}^{n} a_o^l \exp(-((x_o-\bar{x}_o^l)/\sigma_o^l)^2)]} \quad (8)$$

where the subscripts of *i,m,*and *o* denote input,midde, and output fuzzy logic systems and $\bar{y}$ is the center of the fuzzy set G^l. This form of fuzzy logic system uses the center average defuzzifier, product inference rule, singleton fuzzifier, and Gaussian membership function. In the equation(8), assume that $a_o^l = 1$ and M is given. The error function is $\varepsilon^p = 1/2\ [\ f_o(\mathbf{x}^p) - d^p\ ]^2$. Therefore, $\bar{y}_o^l$, $\bar{x}_o^l$, σ_o^l must be trained to minimize ε^p. In Fig. 6, ε, f_o, and d denote ε^p, $f_o(\mathbf{x}^p)$, and d^p.

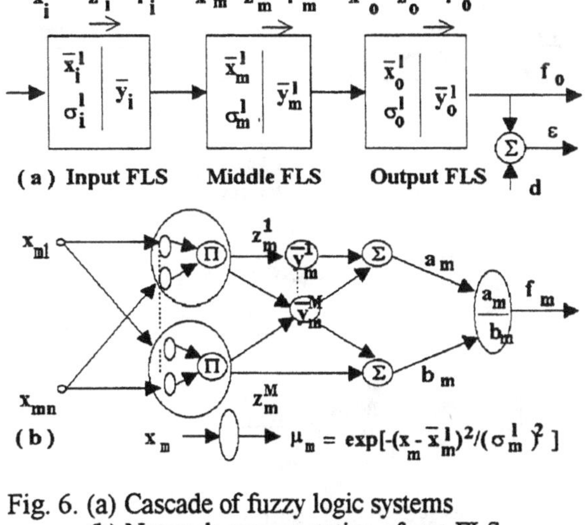

Fig. 6. (a) Cascade of fuzzy logic systems
(b) Network representation of one FLS

The training algorithm for one fuzzy logic system is described in Wang(1994), whose results are summarized as follows

The training algorithm for $\bar{y}_o^l$ is

$$\bar{y}_o^l(k+1) = \bar{y}_o^l(k)+\Delta\bar{y}_o^l(k) = \bar{y}_o^l(k)+\alpha\left(-\frac{\partial \varepsilon}{\partial \bar{y}_o^l}\Big|_k\right)$$

$$= \bar{y}_o^l(k)+\left(-\alpha\frac{f_o - d}{b_o}z_o^l\right) \quad (9)$$

where $l = 1,2,\ldots,M, k = 0,1,2,\ldots$, α is a learning ratio, $f_o = a_o/b_o$, $a_o = \sum_{l=1}^{M}(\bar{y}_o^l z_o^l)$, $b_o = \sum_{l=1}^{M} z_o^l$,

and $z_o^l = \prod_{o=1}^{n}\exp[-(\frac{x_o-\bar{x}_o^l}{\sigma_o^l})^2]$.

The training algorithm for $\bar{x}_o^l$ is

$$\bar{x}_o^l(k+1) = \bar{x}_o^l(k)+\Delta\bar{x}_o^l(k)$$

$$= \bar{x}_o^l+\left(-\alpha\frac{f_o-d}{b_o}z_o^l(\bar{y}_o^l-f_o)\frac{2(x_o-\bar{x}_o^l)}{(\sigma_o^l)^2}\right) \quad (10)$$

and the training algorithm for σ_o^l is

$$\sigma_o^l(k+1) = \sigma_o^l(k)+\Delta\sigma_o^l(k)$$

$$= \sigma_o^l(k)+\left(-\alpha\frac{f_o-d}{b_o}z_o^l(\bar{y}_o^l-f_o)\frac{2(x_o-\bar{x}_o^l)^2}{(\sigma_o^l)^3}\right) \quad (11)$$

In order to train parameters $\bar{y}_m^l$, $\bar{x}_m^l$ and σ_m^l for the fuzzy logic systems which is in the middle, the error is backpropagated from the output fuzzy logic syste m. As is common in neural networks, we use the chain rule to the previous fuzzy logic system the parameters in the input fuzzy logic system is trained as follows:

$$\Delta\bar{x}_i^l = -\alpha\frac{\partial\varepsilon}{\partial\bar{x}_i^l} \quad (12)$$

$$= -\alpha\left(\frac{\partial\varepsilon}{\partial f_o}\right)\left(\frac{\partial f_o}{\partial z_o^l}\right)\left(\frac{\partial z_o^l}{\partial x_o}\right)\left(\frac{\partial f_m}{\partial z_m^l}\right)\left(\frac{\partial z_m^l}{\partial x_m}\right)\left(\frac{\partial f_i}{\partial z_i^l}\right)\left(\frac{\partial z_i^l}{\partial \bar{x}_i^l}\right)$$

$$\Delta\sigma_i^l = -\alpha\frac{\partial\varepsilon}{\partial\sigma_i^l} \quad (13)$$

$$= -\alpha\left(\frac{\partial\varepsilon}{\partial f_o}\right)\left(\frac{\partial f_o}{\partial z_o^l}\right)\left(\frac{\partial z_o^l}{\partial x_o}\right)\left(\frac{\partial f_m}{\partial z_m^l}\right)\left(\frac{\partial z_m^l}{\partial x_m}\right)\left(\frac{\partial f_i}{\partial z_i^l}\right)\left(\frac{\partial z_i^l}{\partial \sigma_i^l}\right)$$

$$\Delta\bar{y}_i^l = -\alpha\frac{\partial\varepsilon}{\partial\bar{y}_i^l} \quad (14)$$

$$= -\alpha\left(\frac{\partial\varepsilon}{\partial f_o}\right)\left(\frac{\partial f_o}{\partial z_o^l}\right)\left(\frac{\partial z_o^l}{\partial x_o}\right)\left(\frac{\partial f_m}{\partial z_m^l}\right)\left(\frac{\partial z_m^l}{\partial x_i}\right)\left(\frac{\partial f_i}{\partial \bar{y}_i^l}\right)$$

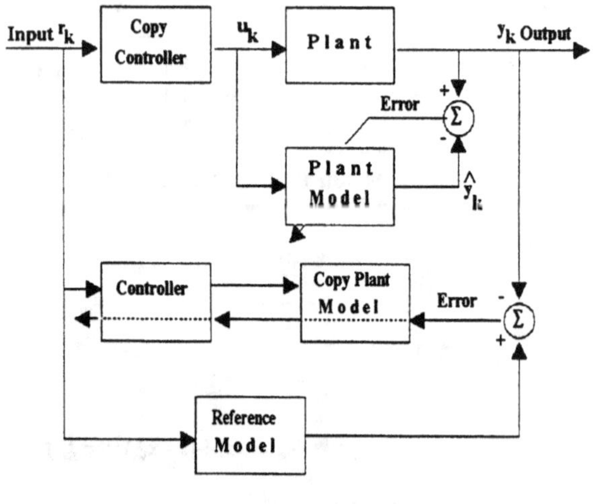

Fig. 7. Adaptive inverse control system by using FLS

4.2 Nonlinear adaptive inverse control by using FLS

A fuzzy controller was proved to be an universal appoximator by many researchers. In this paper, the nonlinear adaptive filters with tapped-delays are adopted as the architecture of controller and the cascading backpropagation algorithm is used for adaptation. Fig. 7 is a block diagram of the nonlinear adaptive inverse control system by using the fuzzy logic systems. The copy of the plant model and the controller to be updated are cascaded. And then, the controller can be directly updated by the cascading backpropagation even if the plant is nonlinear. Also, it is not necessary to implement the plant inverse for obtaining the controller since it is possible to find the controller by using the error backpropagation of the cascaded systems. In order to prevent the system from diverging, the universe of discourses for the identifier and the controller must be considered carefully.

5. SIMULATION

In this section, the algorithms are simulated for are modeling the unknown plants and finding the controllers for the given nonlinear systems. In each example, the nonlinear adaptive inverse control systems are implemented by using the filtered-ε LMS algorithm and the fuzzy logic system.

Example 1 : The simulated nonlinear plant is shown in Fig. 8. This nonlinear plant is composed of two linear digital functions and a hyperbolic tangent function for simulation. In simulating the filtered-ε method, the digital filter has 24 weights and the input vector has 24 elements, but it has 12 unit delays. The input vector is composed of unit delayed inputs and their squared inputs at each tap. The result of the simulation are shown in Fig. 9. Fig. 9 displays the last 500 samples after 5,000 iterations in example 1. The next simulation is performed by nonlinear adaptive inverse control by using the fuzzy logic systems. The initial parameters, $\bar{x}^l$, $\bar{y}^l$ and σ_i^l are obtained the initial parameter choosing method and the linguistic information is not used. The input vector is composed of 24 elements and the applied rules are 125 rules.

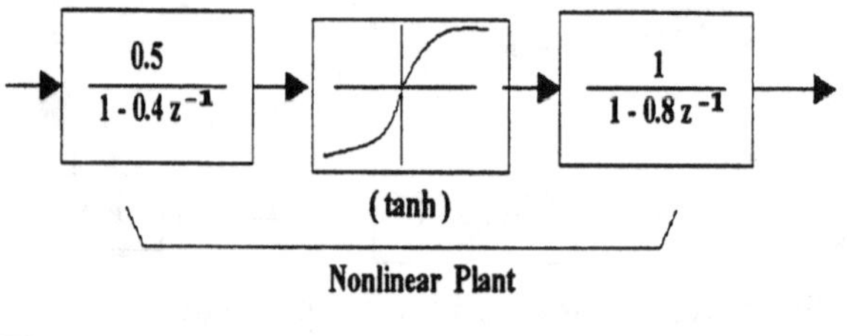

Fig. 8. Example 1: Nonlinear plant for simulation

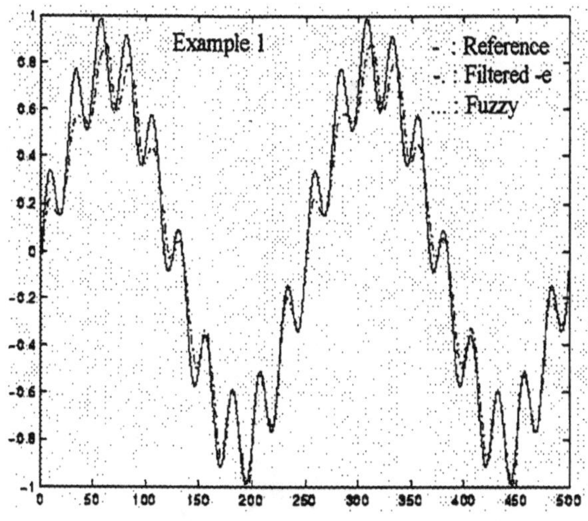

Fig. 9. Comparision with control outputs for example 1 : last 500 samples after 5,000 iterations

Example 2 : The nonlinear plant is given by

$$y(k+1) = g[y(k),y(k-1),y(k-2),u(k),u(k-1)] \quad (15)$$

where unknown function g has the form :

$$g(x_1,x_2,x_3,x_4,x_5) = \frac{x_1x_2x_3x_5(x_3 - 1) + x_4}{1 + x_3^2 + x_2^2} \quad (16)$$

The simulation is performed for the nonlinear plant in example 2 under the following conditions. In case of the filtered - ε method, the architecture of adaptive filter, number of adjustable weights, and number of delays are the same as in example 1. In using the fuzzy logic system, the input vector consists of 5 unit delayed inputs and the applied rules are 60. In plant identification, 5 linguistic rules are used. Fig.10 shows is the results for modeling the plant and Fig. 11 shows the control outputs for example 2.

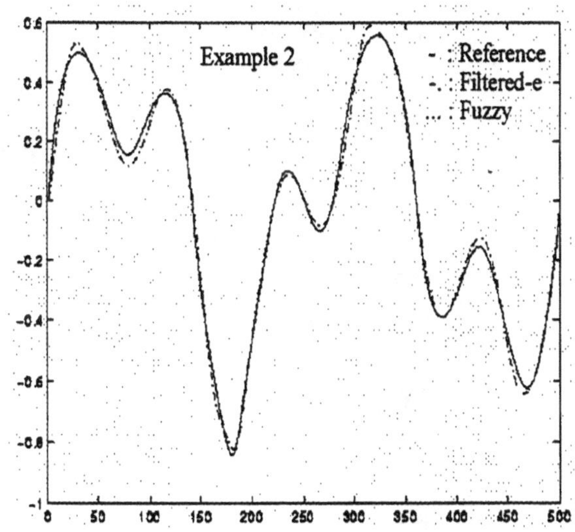

Fig. 10. Comparision with plant identifiers for example 2 : last 500 samples after 10,000 iterations

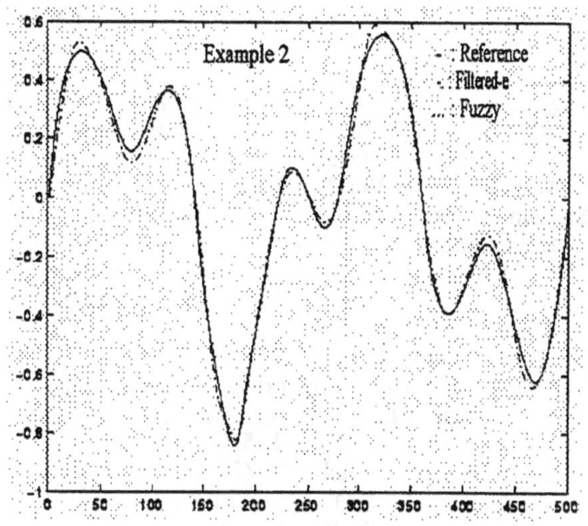

Fig. 11. Comparision with ontrol outputs for example 2 : last 500 samples after 10,000 iterations

6. CONCLUSION

The nonlinear adaptive inverse control method is an extension of a linear adaptive inverse control method , which is based on the adaptive filtering theory and LMS adaptation algorithm. This paper extended the linear adaptive inverse control to the nonlinear adap tive inverse control by use of fuzzy logic systems. For this extension, in the cascaded fuzzy logic systems, the error backpropagation is formulated. Therefore, the adaptation of the controller is done directly by this algorithm. The simulation results show that the performance of the identifiers is improved. The fuzzy controller achieves a good performance and fast convergence because fuzzy logic systems is able to use the initial parameter choosing method and the linguistic information from human operators. The nonlinear adaptive fuzzy inverse control can be applied to the control for the nonlinear plants in the industrial processes.

REFERENCE

Å strom,K.J. and Wittenmark,B.(1990). *Computer-Controlled Systems:Theory and Design.* Prentice Hall.

Buckley, J. J. (1992). Universal Fuzzy Controllers. *Automatica.*, **Vol. 28,** no. 6, pp. 1245-1248.

Buckley, J. J. (1993). Sugeno Type Controllers are Universal Controllers. *Fuzzy Sets and Systems.*, **Vol. 53,** pp. 299-303.

Haykin, S. (1986). *Adaptive Filter Theory.* Prentice - Hall.

Hoskins, D.A. *et al.*(1992). Iterative Inversion Neural Networks and Its Application to Adaptive Control. *IEEE Trans. on Neural Networks.*,**Vol. 3**, no. 2.

Horikawa, S. (1992). On Fuzzy Modeling using Fuzzy Neural Networks with the backpropagation Algorithm. *IEEE Transations on Neural Networks.* **Vol. 3**, no. 5.
Sugeno, M. and Takagi, T. (1985). Fuzzy Identification of Systems and its Application to Modeling and Control, *IEEE Trans. on Systems, Man, and Cybe.*, **Vol. SMC-15**, no. 1.
Sugeno, M. and Kang, G.T.(1986). Fuzzy Modelling and Control of Multilayer Incinerator. *Fuzzy Sets and Systems.* **Vol. 18**, pp. 329-346.
Sugeno, M. and Kang, G. T. (1988). Structure Identification of fuzzy model., *Fuzzy Sets and Systems.* **Vol. 28**, pp 15-33
Sugeno, M and Tanaka, K. (1992). Stability analysis and design of fuzzy control system.,*Fuzzy Sets and Systems.* **Vol. 45**, pp 135-156.
Wang, L.X. and Mendel, J.M. (1992a). Fuzzy Basis Functions, Universal Approximation, and Orthogonal Least-Squares Learning. *IEEE Trans. Neural Networks.*, **Vol. 3**, no. 5, pp. 807-814.
Wang, L. X. and Mendel, J. M. (1992b). Generating Fuzzy Rules by Learning from Examples., *IEEE Trans. on Systems, Man, and Cybe.*, **Vol. 22**, no. 6, pp. 1414-1427.
Wang, L. X. (1994). *Adaptive Fuzzy Systems and Control : Design and Stability Analysis.* Prentice - Hall.
Wang, L. X. (1995). Design and Analysis of Fuzzy Identifiers of Nonlinear Dynamic Systems. *IEEE Trans. on Automatic Control,* **Vol. 40**, no. 1.
Widrow,B.et al.(1976). Stationary and Nonstationary Learning Characteristics of the LMS Adaptive Filter. *Proceedings of the IEEE.*, **Vol. 64**, no. 8.
Widrow, B. and Walach, E.(1996). *Adaptive Inverse Control.* Prentice - Hall.
Li, H. X. and Gatland, H. B.(1995). A New Methodology for Designing a Fuzzy Logic Controller. *IEEE Trans. on Systems, Man, and Cybernetics.*, **Vol. 25**, no. 3, pp. 505-512.
Yager,R.R. and Filev, D.P.(1993). Unified Structure and Parameter Identification of Fuzzy Models. *IEEE Trans. on Sys.,Man,and Cybe.*,**Vol. 23**, no.4.

Copyright © IFAC Automation in the Steel Industry,
Kyongju, Korea, 1997

ON-LINE PROCESS MODELING AND OPTIMISATION FOR A 20HIGH COLD ROLLING MILL

A. Schneider [2], R. Werners [1] and J. Heidepriem [2]
[1] Mannesmann Demag Hüttentechnik MDS Walzwerktechnik
Daniel-Goldbach-Straße 17-19, D-40880 Ratingen
[2] Institute of Automation Technique, University of Wuppertal
Gaußstraße 20, D-42097 Wuppertal

Abstract : The paper describes the basic principles of a model formation used for the optimisation of mill utilisation and presetting of actuators in a 20high cold rolling mill. Besides the special requirements placed to the force and torque calculation model in a Sendzimir mill arrangement [1], the paper in particular describes the model effort for the description of the mill stand behaviour and the interaction with the strip. Finally, the chosen concept for the on-line implementation in a process control computer is shown.

Introduction

Product quality such as the final strip thickness, the strip flatness and surface as well as the mill utilisation and operation flexibility are ever increasing demands placed on cold rolling mill operators. However in case of complex mill arrangements such as the Sendzimir mill discussed herein, the adequate presetting of the available actuators to achieve the desired product quality represents an extremely difficult task. This results from the great number of variable process parameter and actuators to be considered.

Figure 1 gives an overview about the arrangement and the main technological actuators in a Sendzimir mill.

The Sendzimir mill arrangement provides high quality strip surface results due to its small work roll diameters, which are in contact with the strip. Those however force the use of a variety of intermediate and backup rolls in order to provide a stable and safe rolling process.

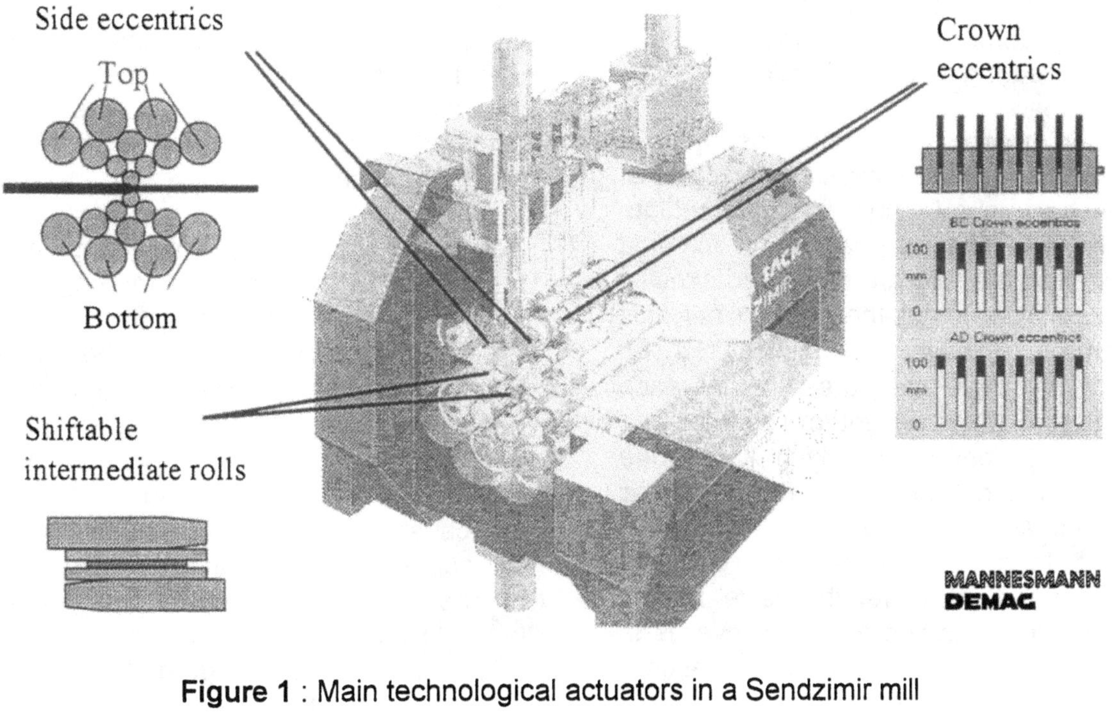

Figure 1 : Main technological actuators in a Sendzimir mill

So-called side eccentrics at each of the outer backup rolls are used to adjust the position of the corresponding roll axis in a wide range and by this indirectly serve for adjusting the roll gap geometry overall. The side eccentrics at the lower backup rolls mainly serve for maintaining the so-called mill pass line whereas those eccentrics at the upper backup rolls are used to achieve the desired strip thickness and also serve for compensation of changes during rolling. Side eccentrics appear to be mechanically or electrically coupled, which needs to be considered while performing model based pre-set calculations.

Crown eccentrics, which are available at several locations over the barrel length on all upper backup rolls, serve for specific backup roll axis contours. This affects the roll bite geometry respectively and is of particular importance for matching the roll gap contour to the profile of the strip entering the roll gap and for achievement of a certain desired strip flatness. The shiftable first intermediate rolls represent further contour actuators, which however mainly serve the modification of the roll gap contour in strip edge area. Crown eccentrics and shiftable first intermediate rolls need to be used to compensate for any kind of elastic deformation and thermal grow-up appearing in the roll stack.

Measurements of the geometrical relations in the roll stack are available at the backup rolls and first intermediate rolls. The shifting positions of the first intermediate rolls are measured directly. Information about the backup roll axis contour however is only based on rotation angle measurements at each side and crown eccentric. Together with the mill spring, for which knowledge about the roll separating force and the load share in the stand is needed, this then leads to the position of each backup roll axis over the barrel length. The roll separating force however is in most applications only measured indirectly through the adjustment pressure needed for certain eccentrics. Apart from hysteresis effects also the relation to the geometry related load share in the mill stand causes this indirect measurement to be usually insufficient.

Due to the complex mill arrangement of the Sendzimir mill, the reproducibility of the final product quality and the optimum usage of available mill resources to increase productivity represents an extremely difficult task, which can only be fulfilled with a comprehensive model approach, which takes all relevant mill and process parameter into consideration. Due to the fact, that neither direct geometrical information nor accurate roll separating force measurements exist, a variety of individual mathematical models is needed to describe the complex elastic mill stand behaviour and the elastic/plastic characteristic of the material to be rolled.

Roll gap analysis for force, power and slip and representation of friction

The roll separating force, roll torque and required drive power brought about by forming of the material in the roll gap are some of the most important process information. Only if exact knowledge about these process values exists, accurate further presetting of the mill actuators and optimum mill utilisation can be obtained.

Because reliable direct roll separating force measurements are missing in most Sendzimir mills, special analysis effort is required to describe the different stress components in the roll gap. Those can be broken down into vertical and tangential stresses acting on the work rolls. The sum of the vertical stress components leads to the roll separating force, whereas the sum of the tangential stress components leads to the roll torque and thus to the main drive power. Figure 2 shows exemplary the conditions in the roll bite including those tangential and vertical stress components.

A model approach, which simultaneously provides accurate information about the vertical and tangential stress components in the roll gap opens up capabilities of detecting or at least evaluating forces based on the main drive power, providing that friction and drive efficiency are known. This in particular enables material yield stress adaption even in case of missing or unreliable roll separating force measurements. Material yield stress adaption is required in any case, where high flexibility due to a great alloy spectrum to be rolled is needed.

slip, a further improvement of the strip surface can be obtained. The slippage present in the roll bite is correlated to the position of the neutral point (refer to figure 2), which is affected to a certain extent by the entry and exit tension applied. Entry and exit tension thus in small ranges serve as an actuator for the optimisation of the strip surface. To provide accurate presetting, the gap analysis model has also to deliver accurate information about the neutral point position.

The mathematical approach chosen here is a strip fibre model (Lippmann, Mahrenholtz [2]), the basics theory of which has been described by v. Karman [3] and Siebel [4,5]. The strip in

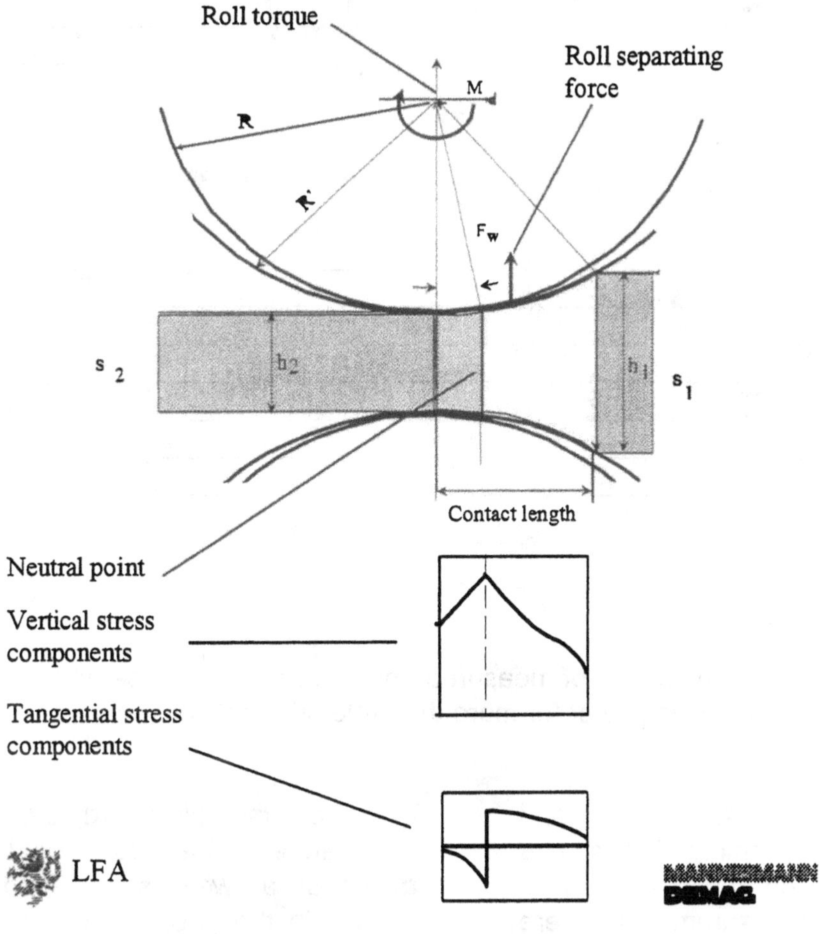

Figure 2 : Stress components in the roll gap

In particular the demand for optimisation of the strip surface requires an indication about the slip present in the roll gap. By minimising the forward

the roll gap is divided into several fibres for which the individual forming parameter are determined. The yield stress of the material is herein described with the aid of an equation introduced by Spittel [6]. The coefficients necessary for this have been determined by torsion bar trials and were collected to an extensive material data base for steel [7]. This then leads to detailed information about the vertical and tangential stress components and in particular also about the slip present.

The impacts of forming parameter to the yield stress are well known and have been determined in comprehensive trials off-line. Minor variations can thus be tuned on-line. However inaccuracies in the friction may also cause in parts significant deviations in the predicted roll force, power and torque. Therefore the various impacts to the friction need to be described precisely too.

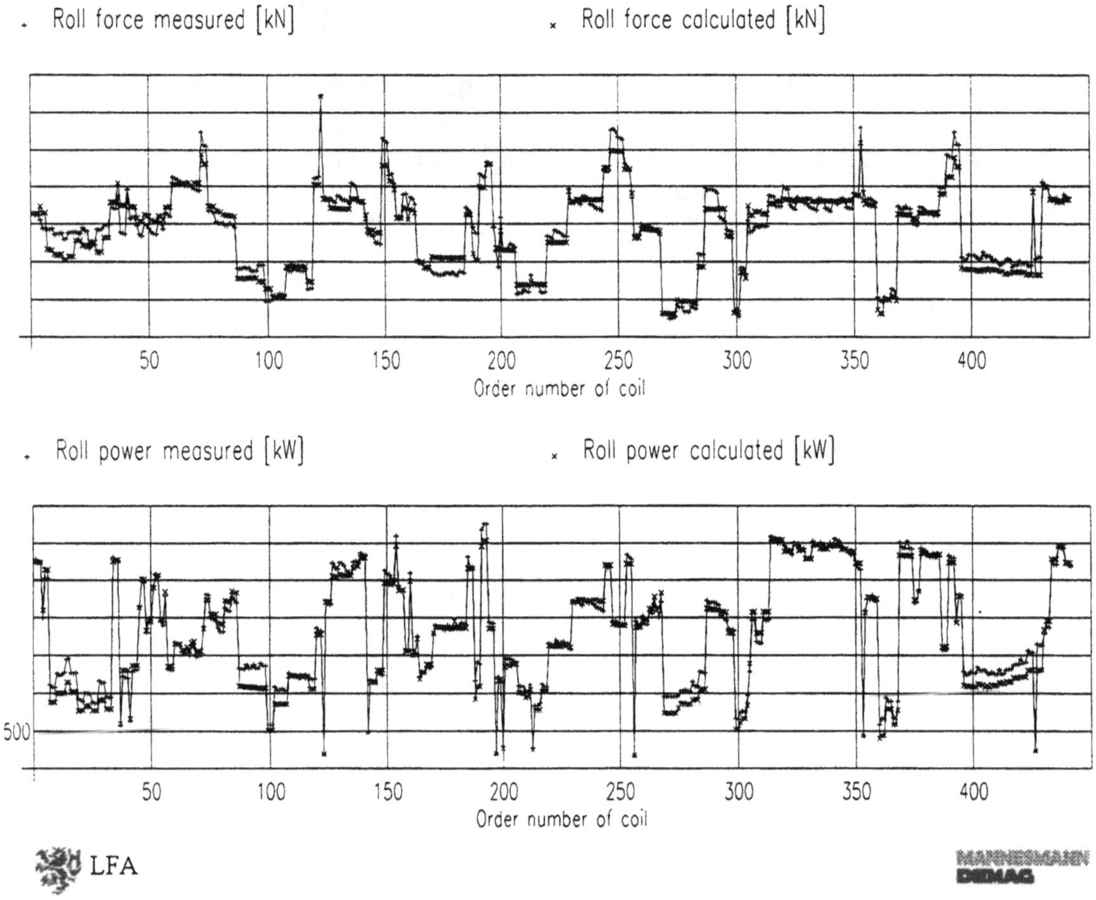

Figure 3 : Comparison of measured and calculated roll force and main drive power for more than 400 different coils

A comparison of measured and calculated roll separating forces and drive powers taken in a 4high mill, where direct measurements were available, is shown in figure 3. The comparison performed for a great number of different coils with different geometry and chemistry shows the simultaneously high accuracy of force and power calculation of the model developed.

Apart from the roll speed, the coolant temperature, the roll and strip roughness as well as the temperature balance in the gap, further impacts to the friction must be considered. Detailed friction trials appear to be time consuming and cost intensive so that here an approach has been chosen on the basis of artificial neural networks (Fig. 4).

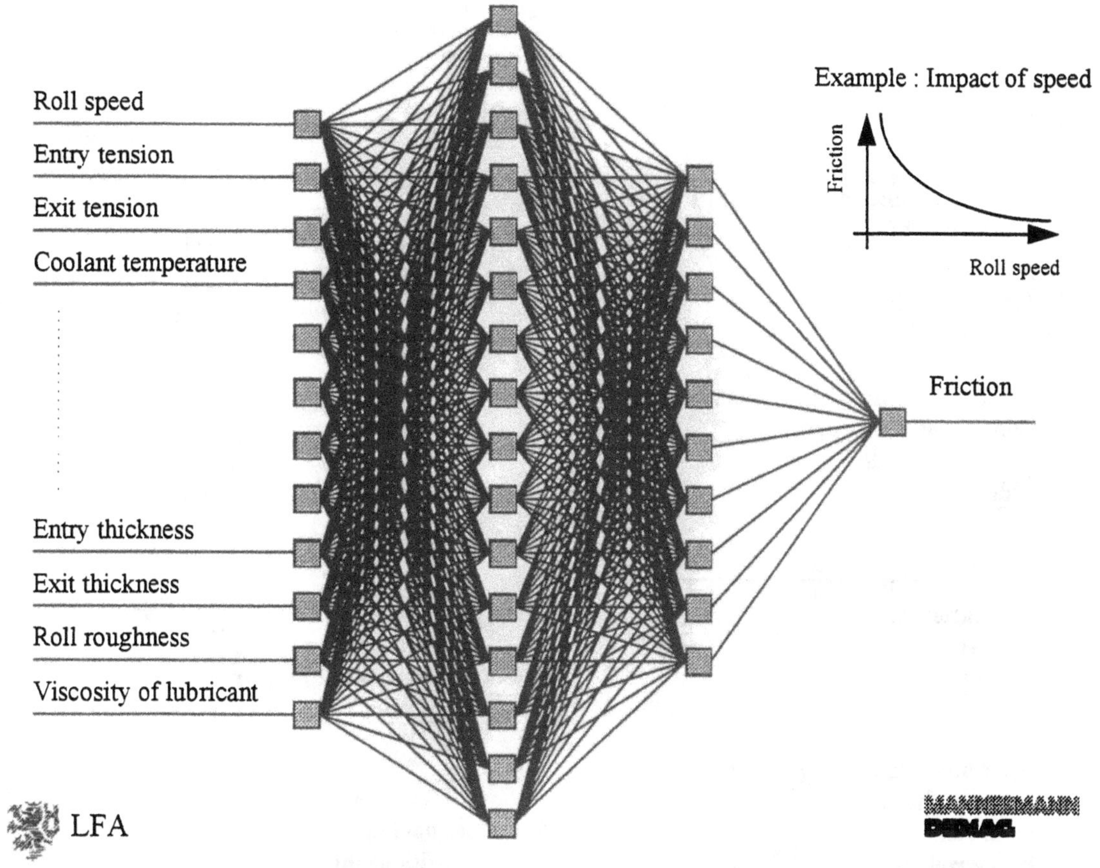

Figure 4 : Representation of friction through an artificial neural network

The entry layer of the neural network is arranged in that way, that transformed process information are provided in pure noisy nature. Those information are processed through the multi-layer perceptron feed-forward network and finally lead to the friction coefficient. Extensive studies showed, that with a representative number of information provided to the entry layer of the network even physical dependencies of the friction coefficient to the entries can be represented accordingly.

Model for description of elastic mill stand behaviour

It is of importance to describe all the several effects present in the mill stand in their entirety in order to allow propagation from the measured eccentric adjustments down to the roll bite contour, which is the target for further optimisation steps.

One effect considered here is the mill spring, which is equivalent to a position change of the saddle segments. Furthermore the flattening between the several rolls and likewise between the strip and the work rolls as well as the deflection of the several rolls is considered.

Another relevant effect is the thermal expansion of the work rolls brought about by the forming energy. A model has been implemented, which is a combination of an analytical approach in radial roll direction and a finite difference approach over the barrel of the roll [8]. The results of the thermal crown model are taken into consideration in the elastic mill stand model as offsets to the contour of the work rolls.

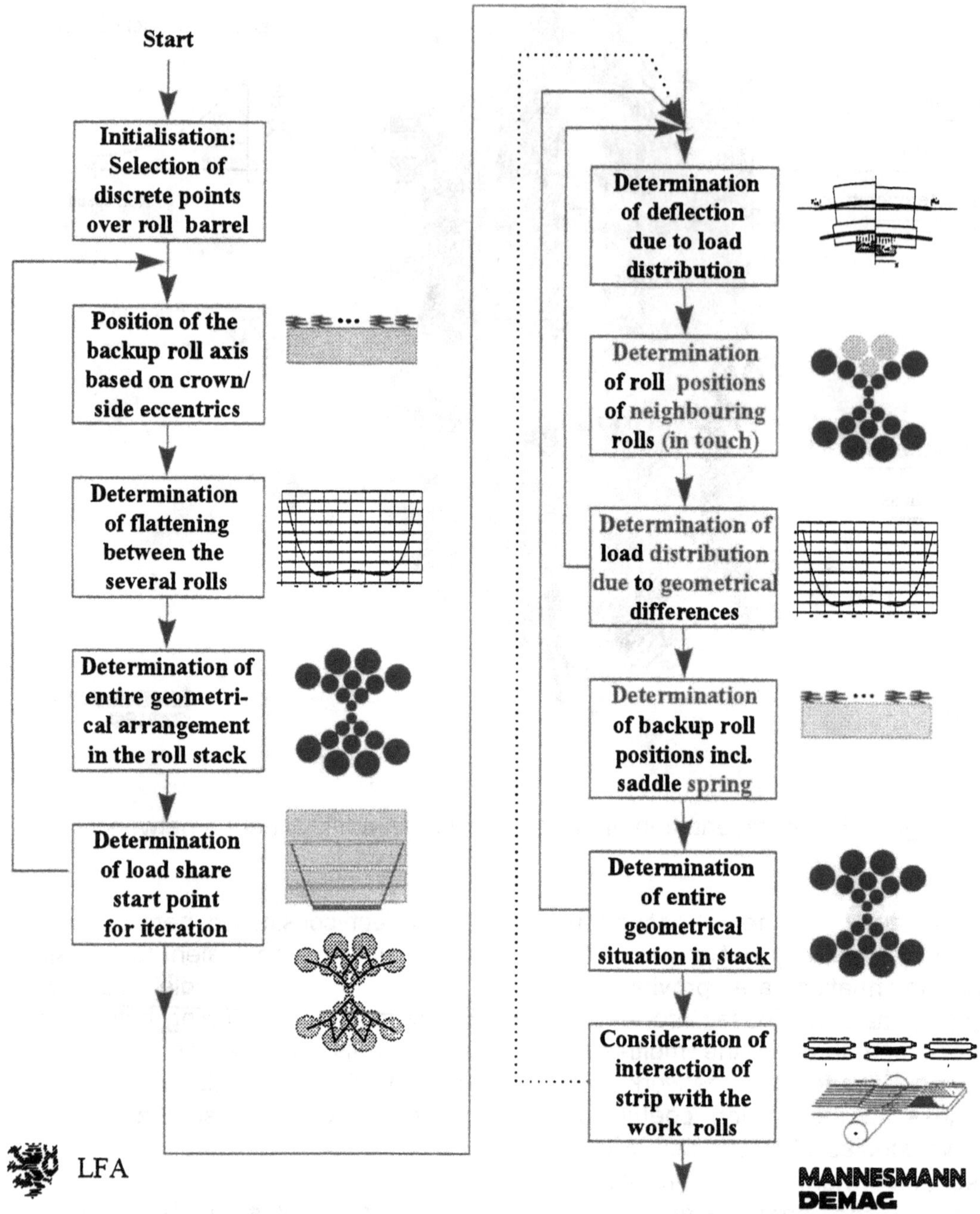

Figure 5 : General concept of model for description of the elastic mill stand behaviour

One requirement placed onto the elastic mill stand model was the ability to cover a variety of different mill configurations, different roll grindings (including the dynamically changing thermal crowns of the work rolls) and different roll material characteristics even broken down to the individual rolls in the stack to be able to cover also process situations, where unusual roll combinations are chosen.

Therefore the description of the elastic mill stand behaviour is based on a numerical solution approach, where the different parameter can be considered and tested. The different effects, such as the flattening between the rolls and likewise between the strip and the work rolls as well as the deflection of the several rolls are derived from multiple iterations. A general overview about the model concept is given in figure 5.

The elastic mill stand model can generally be divided into two parts. An initialisation part (refer to figure 5 - left part) cares on the fast determination of an image about the load share in the roll stack considering flattening and mill spring however neglecting the roll deflection. The deflection is then considered in a more time consuming final part (refer to figure 5 - right part).

Based on the roll separating force, the model first provides a rough image about the load share in the entire roll stack (refer to figure 5 - left part). This initial load share determination already takes into account mill housing spring effects, which are represented at the locations of the saddle segments as well as flattening effects between the rolls. The spring modulus of the mill housing valid at the locations of the saddle segments have initially been determined by FEM calculations and are adapted on-line.

The initial load share determined is taken in the second step as a basis for the iterative determination of the interaction between load distribution, flattening and deflection. The deflection of each roll is derived from the currently determined load distribution. The geometrical differences between two neighbouring rolls are then interpreted as flattening and thus lead to a new load distribution along the contact curve, which finally leads to a new deflection. This iteration is performed until a solution has been reached, where the entire load, the deflection and the flattening of the several rolls match with each other.

Interaction of the strip with the rolls

After the roll gap contour has been determined for a certain load situation in the roll bite, the interaction of the strip with the work rolls is investigated. This is done with the aid of an analytical approach [9,10], which is the solution of the differential equation in the current setpoint, the basic principles of which are explained in the following.

As long as the roll gap contour matches

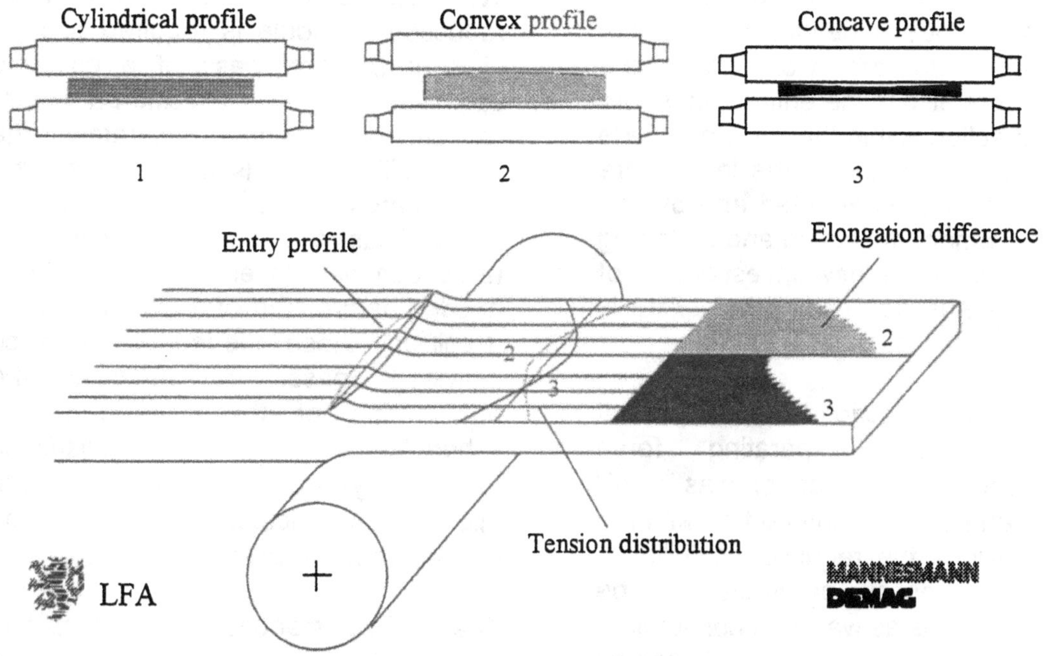

Figure 6 : Thickness profile changes in the roll gap

the incoming strip thickness profile, the thickness reduction appears to be the same throughout the width. Therefore no elongation differences, which are also referred to as flatness errors, are produced (refer to figure 6 - item 1).

Roll gap contours, which differ from the strip thickness profile entering the roll bite, cause local differences in the thickness reduction over the width of the strip (refer to figure 6 - item 2 and 3). The different thickness reductions for their part cause differences in the elongation over the width of the strip, so that certain flatness errors are produced.

Differences in the elongation over the width also cause changes in the local tension applied, which then cause changes in the local pressure. These circumstances apply likewise to flatness errors already approaching the roll bite, which produce local variance in the entry side strip tension. The reduction differences brought about by unmatched roll gap contours also produce local variance in the roll pressure.

Variance in the local roll pressure causes changes in the local flattening of the strip into the work rolls and furthermore also in the entire load share and deflection within the roll stack. Due to these circumstances, the load share in the roll stack is affected and by this also the spring, deflection and flattening effects, so that a new investigation of the entire elastic mill stand behaviour is required (refer to figure 5).

Whereas the described model concept comprising roll separating force calculation, investigation of elastic mill stand behaviour and interaction with the strip provides the required flexibility in terms of process parameter to be considered, it is as well time consuming. Apart from the individual models shown here, also optimisation algorithm form part of the entire system, which require additional computation time. Within the process control computer however, the time frame available for set-up calculations is usually to be kept as short as even possible in order to avoid production delays. Therefore the model concept described requires either adequate hardware or special implementation methods as follows.

On-line implementation concept

The tasks fulfilled by the mathematical models and their dedicated optimisation algorithm running on-line on the process control computer can generally be broken down into the following main items:

- optimisation of pass reduction schedule for the whole coil in advance to rolling
- optimisation of pass specific adjustments (eccentrics, shifting, speed, tension..)
- continuous observation of thermal roll crown and of main and coiler drive status and capabilities
- short and long term adaption

Whereas the determination of the pass reduction schedule is required prior to start of the first pass of a coil, the optimisation of pass specific parameter can only be performed accurately if the actual mill situation is known. This is the case before start of the concerned pass. Thus the several optimisation tasks can be broken down into those needed before start of manufacturing of a coil and those needed prior to each individual pass. In particular the optimisation of the pass reduction schedule is a time consuming functionality due to the fact, that optimisation calculations have to be performed for all planned passes.

These circumstances enable a split of the optimisation tasks, where the time consuming pass schedule determination is performed far in advance to start rolling in a so-called preliminary set-up module (refer to figure 7 - SETPRE) and the less time extensive pass specific optimisations are performed shortly before start of each individual pass in a separate dedicated pass specific module (see figure 7 - SETUP).

information to be considered however causes these investigations to be time consuming too, so that another independent process has been foreseen, which operates after completion of a pass or coil (refer to figure 7 - ADAPT) independent from all other tasks.

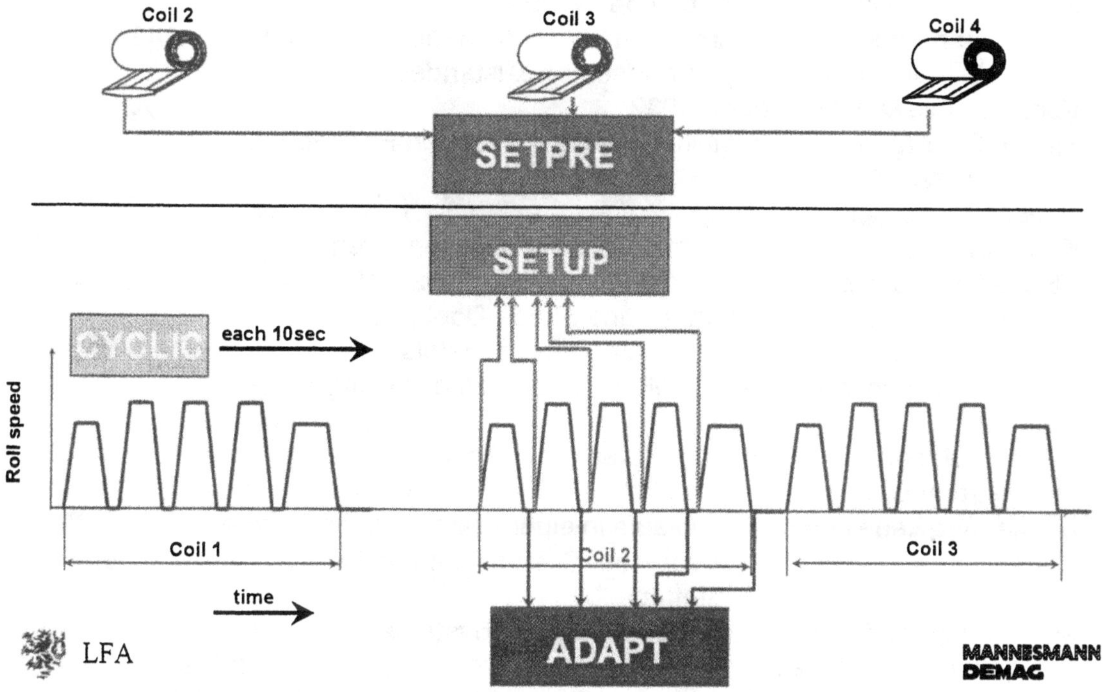

Figure 7 : Principle of task sharing in on-line application

Main and coiler drive status and in particular the thermal expansion of the work rolls change dynamically throughout rolling of a pass and even during mill delay times and thus need also to be continuously investigated. This is performed using a cyclically started module (refer to figure 7 - CYCLIC), which is continuously provided with actual process information rather than with snapshots or summary data only.

Due to the fact, that the amount of reliable measured process information is usually not adequate at a single snapshot, long term adaption calculations can only be performed taking a significant amount of historical data into account. The amount of

Conclusion

The complexity of the mathematical models used to entirely describe and optimise the rolling process provides the capability of taking a great amount of variable process parameter into account. This ensures the process control computer system to be flexible and open for various applications and situations however causes the model calculations to be time consuming.

Due to a special on-line implementation concept, in which the several tasks are broken down to the point in time where they are needed, even standard hardware and operating systems can be used.

References

1. Sendzimir, M. G.: The Sendzimir Cold Strip Mill.
   Journal of Metals 8 (1956) 9, S. 1154-1158
2. Lippmann, H. and Mahrenholtz, O.:
   Plastomechanik der Umformung metallischer Werkstoffe.
   Springer-Verlag, Berlin 1967
3. Karman, T. v.: Beitrag zur Theorie des Walzvorganges.
   Zeitschrift für angewandte Mathematik und Mechanik 5 (1925), S. 139-141
4. Siebel, E. : Die Formgebung im Bildsamen Zustande.
   Verlag Stahleisen, Düsseldorf 1932
5. Siebel, E. : Kräfte und Materialfluß bei der bildsamen Formgebung.
   Stahl u. Eisen, 45 (1925) 37, S. 1563-1566
6. Hensel,A. and Spittel, T. :
   Kraft- und Arbeitsbedarf bildsamer Formgebungsverfahren.
   VEB Deutscher Verlag für Grundstoffindustrie, Leipzig 1978
7. Spittel, M., Spittel, T., Teichert, H. and Skoda-Dopp, U. :
   Material Data Base for Steel and Non-Ferrous Metals,
   Report of Mannesmann Demag MDS Walzwerktechnik, June 1994
8. Sauer, W. :
   Thermal Modelling and Control Strategies in "Advances in Aluminium Rolling"
   Int. Conference of Mannesmann Demag MDS Walzwerktechnik, May 1993, S. 89-95
9. Mielke, R. : Regelung des Bandlaufs in einer Warmbreitbandstraße.
   Ph.D. Thesis, University of Wuppertal, Chair of Automation Techniques, 1992
10. Schneider, A., Kern P. and Steffens, M. :
    Model supported Profile and Flatness Control Systems.
    49° Congresso Internacional de Tecnologia Metalúrgica e de Materiais -
    International Conference 09.-14.10.1994 in São Paulo (Brasil), Vol. 6 S. 49/60

ON-LINE MEASUREMENT SYSTEM FOR THE DEGREE OF ALLOYING

Dalwoo Kim, Choong-Soo Lim and Ki-Jang Oh

Instrumentation Research Team, Research Institute of Industrial Science and Technology,
Pohang, 790-600, Korea

Abstract : Fully automatic on-line optical measurement system for the degree of alloying in galvannealed steel sheet has been developed. In this system, the degree of alloying is measured by detecting the reflectance of galvannealed steel sheet using laser. Since the surface roughness of coated layer scatters the reflected laser beam and reduces the intensity of reflected light, which can be calibrated by measuring the intensity of scattered light in addition to the specularly reflected light. After calibration, the degree of alloying shows the linear relationship with the measured reflectance regardless of surface roughness. According to in-line process experiment, this type of optical measurement technique is reliable to the on-line measurement of the degree of alloying in the process of galvannealing steel strip. *Copyright © 1998 IFAC*

Keywords: On-line control, Steel industry, Optical implementation, Light, Scattering problems

1. INTRODUCTION

Galvannealed steel sheet has good characteristics in welding, painting, and forming, and it is remarkably anticorrosive. All of these properties are strongly affected by the degree of alloying of Zn-coated layer and the control of the alloying is important in this sense. The degree of alloying is the measure of the contents of Fe in Zn-coated layer and is determined by heat treatment after galvannealing process. The on-line measurement is necessary for the control of the heating furnace.

The degree of alloying of iron with zinc is the measure of iron diffused into zinc which is determined by heat treatment after dipping steel strip into a zinc-pot. The degree of alloying in galvannealed steel sheet can be measured by several methods, such as the penetration of γ-ray, laser scattering, x-ray diffraction, and x-ray fluorescence.(Laguitton and Parish, 1977; Honda et. al., 1988) In this study, the degree of alloying has been measured with an optical method by measuring the reflectance of the steel sheet because the reflectance changes as the contents of iron which is diffused into zinc layer. The reflectance of galvannealed steel sheet becomes smaller as the iron diffuses into the zinc layer. Therefore, the changes of the degree of alloying can be measured by detecting the reflectance of coated layer. Our purpose is to develop an optical system which measures the degree of alloying by detecting the reflectance of galvannealed steel sheet. To measure the reflectance of surface, both of the intensities of the whole reflected light and the incident light should be detected and compared. The solid angle formed by reflected light from rough surface is larger than that from smooth surface because of scattering. Therefore, to measure the reflectance of rough surface a larger optical integrator must be used to detect the whole reflected light that propagates divergently. This kind of big optical integrator is not suitable for an on-line measurement system because it can not be positioned closely to the steel sheet which moves with high speed in the production line. An optical measurement system has been developed in which detectors cover only small solid angle instead of measuring the whole reflected light, and the detectors can be installed away from target surface. The light source used in this system is a semiconductor laser and the reflected light within a certain solid angle is detected with a photodiode. The measured reflectances show the linear relationship with the degree of alloying for the negligible variation of surface roughness. As the variation of surface roughness becomes larger, the intensity of the reflected light is affected by the roughness which has no consistent relationship with the degree of alloying. Nevertheless, after calibration of the effect of surface roughness the reflectances show the linear relationship with the degree of alloying.

2. EXPERIMENTS AND RESULTS

The reflectance of galvannealed steel sheet becomes smaller as the iron diffuses into the zinc coated layer. Therefore the changes of the degree of alloying can be measured by detecting the reflectance of coating layer. The purpose of our research is to develop an optical system which measures the degree of alloying by detecting the reflectance of galvannealed steel sheet. To measure the reflectance of surface, all of the reflected lights must be detected and compared with the intensity of incident lights. For the light with some incident angle, the solid angle of reflected light is determined by the surface roughness. The solid angle formed by reflected light from rough surface is larger than that from smooth surface. Therefore, in order to measure the reflectance of rough surface, large optical integrator is necessary. This optical integrator is not suitable for on-line measurement system because this kind of big optical integrator must be installed very closely to the steel sheet which moves with high speed in the production line. For this reason an optical measurement system has been developed which has detectors of small solid angle and can be installed far from target surface.

For the laser light incident with an angle θ, the intensity of the reflected light $I(\theta,\varphi)$ is determined by the product of reflectance $R(\theta)$ and scattering function $S(\theta,\varphi)$, where φ is the detection angle.

$$I(\theta, \varphi) = R(\theta)S(\theta,\varphi). \qquad (1)$$

In equation (1), the reflectance $R(\theta)$ depends only on the degree of alloying for galvannealed steel sheet and $S(\theta,\varphi)$ is determined by surface roughness. With a detector of small solid angle, the measured intensity $I(\theta,\varphi)$ changes with different roughness. So, the effect of surface roughness must be calibrated to find out reflectance $R(\theta)$ from $I(\theta,\varphi)$. According to the scattering theory, the complicated form of scattering function $S(\theta,\varphi)$ can be approximated as a simple form for small incident and scattered angle as follows.(Beckmann and Spizzichino, 1963)

$$S(\theta,\varphi) = \frac{F^2\lambda^2}{4\pi(\cos(\theta)+\cos(\varphi))^2}\left(\frac{T}{\sigma}\right)^2 \exp\left(-\left(\frac{T}{\sigma}\right)^2\right) \qquad (2)$$

In equation (2), F is a geometric parameter determined by the angles θ and φ, σ is root-mean-square surface roughness and T is auto-correlation length of roughness, respectively. In order to calibrate the effect of surface roughness $S(\theta,\varphi)$, we incident a laser beam at a specific angle θ and measure the reflected light both at a specular and at an oblique angle φ. Then, from equation (2), roughness dependant parameters T and σ can be eliminated as equation (3).

$$R(\theta) = C\frac{I(\theta,\theta)}{\ln\left(\frac{I(\theta,\theta)}{I(\theta,\varphi)}\right)} \qquad (3)$$

By using equation (3), pure reflectance can be measured regardless of the surface roughness.

The optical scheme of reflectance measurement is shown in figure 1. the silicon photo detector(P.D. 1) measures the intensity of incident laser light and other photo detectors(P.D. 2, P.D. 3) measure the specular and oblique reflection intensity.

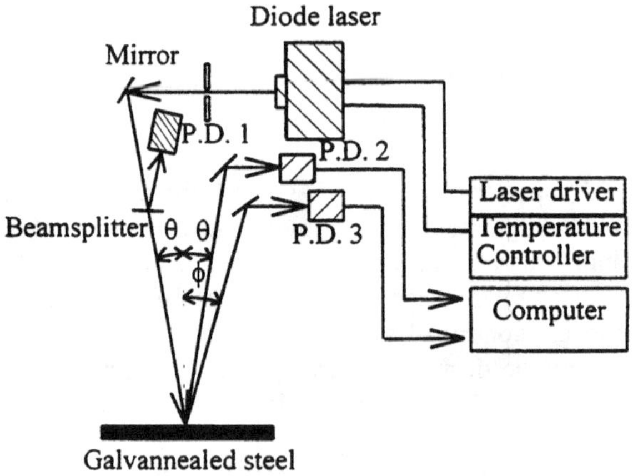

Figure. 1. Optical scheme of reflectance measurement.

The test samples are galvannealed steel sheets produced in Kwangyang Steel Works of POSCO. The degree of alloying was measured firstly by chemical analysis and then the surface roughness was measured. The surface roughnesses of galvannealed steel sheets are ranged between 0.5 - 1.0 μm and have no consistent relationship with the degree of alloying. After measured the intensity of reflected and scattered light, the effect of roughness is calibrated by equation (3). As a result, inversely linear relationship between the degree of alloying and the reflectance has been attained. The measured reflectance is transformed into the degree of alloying by equation (4).

$$\text{degree of alloying} = A \frac{1}{R(\theta)} + Y \qquad (4)$$

In equation (4), A and Y are some fitting constants which are determined by the sensitivity and measuring angles of detectors. These fitting constants can be calculated by comparison of optically measured degree of alloying with chemically measured value. Figure 2 shows the comparison of the degree of alloying measured by an optical method and chemical method. They shows good linear relationship which implies that the optical method can be substituted for conventional chemical method.

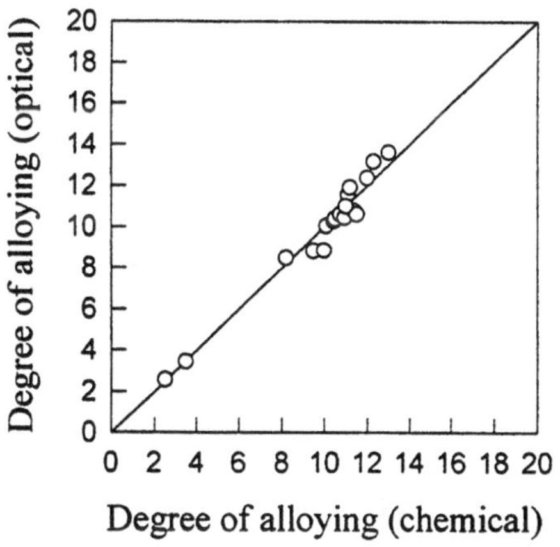

Fig. 2. The comparison of optically measured degree of alloying with chemically measured values.

A fully automatic on-line measurement system which is composed of optical system, linear moving system, and data processor has been designed and constructed. This on-line measurement system has been installed in a continuous galvannealing line of POSCO. The degree of alloying of galvannealed steel sheet in the production line was measured and compared with that of chemically measured one as shown in figure 3.

According to in-line process experiment, this type of optical measurement technique is reliable to the on-line measurement of the degree of alloying in the process of galvannealing steel strip.

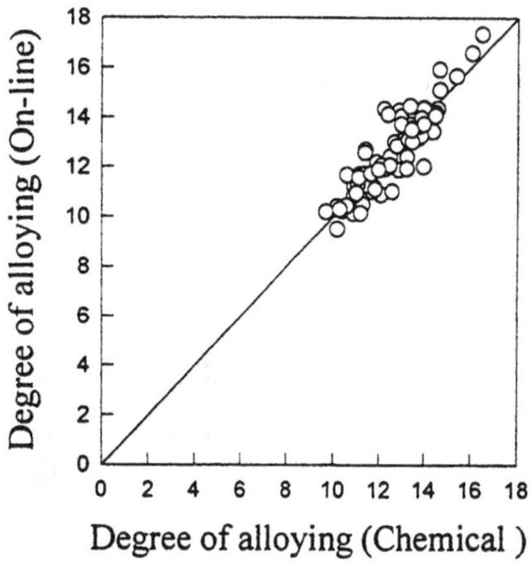

Figure 3. The results of on-line measurement compared with chemical analysis.

3. CONCLUSIONS

A fully automatic on-line optical measurement system for the degree of alloying which calibrates the effect of surface roughness has been developed. This system is mainly composed of an optical system, linear moving system, and data proessor. The optical system moves in the direction parallel to steel sheet to measure the degree of alloying at desired position. The data processor calculates the degree of alloying and displays the data on the monitor screen. The on-line measurement system has been installed in the continuous galvannealing line of POSCO for monitoring the degree of alloying of galvannealed steel sheet. According to in-line process experiment, the degree of alloying is consistent with conventional chemical method. This type of optical measurement technique is reliable to the on-line measurement of the degree of alloying in the process of galvannealing steel strip.

REFERENCES

Beckmann, P., A.Spizzichino (1963). The Scattering of Electromagnetic Waves from Rough Surfaces, Pergamon Press, Oxford.

Honda, T., T.Sakai, Y.Yamamoto, N.Hirayama, M. Yoshida (1988). Current Advances in Materials and Processes, **581**, 1.

Laguitton, D., W.Parish (1977). Anal. Chem.,**49**, 1152.

THE SLAB ACCELERATING COOLING SYSTEM CONTROL MODEL AND ITS APPLICATION STUDY

Shouping Guan, Xiaobo Wang, Tianyou Chai and Xiaogang Wang

Research Center of Automation, Northeastern University 309#
Shenyang 110006 Liaoning P.R.China
Email: yangyx@mail.neu.edu.cn

Abstract: This paper proposes a cooling control model and incorporates it into the application study of laminar cooling control system which is a complicated industrial process. RBF network is used to set up the feedforward model of the control system which is capable of self-learning. Simulation and industrial experiments show that the control model is able to perform effective control of the slab accelerating cooling control process and drive the whole system and slab to meet the desired requirements.

Keywords: neural-network models, multiloop control, feedforward control.

1. INTRODUCTION

The technology of laminar water cooling after the milling is to add a set of cooling equipment between the pyrometer and the straightener which cool the temperature of the slab from 800~900℃ after the final milling to 500~700℃ after the straightening. This cooling process could not only shorten the cooling time of the slab and improve its quality, but also change the metal organization of the slab through cooling rate and therefore improve the strength of slab without weaken its flexibility, which in turn could benefit manufactures a lot economically (Van. et al. 1993, Masahisa. et al, 1978, Hollander, et al, 1971. Moffat. et al, 1985).

The major technical indexes to demonstrate the quality of slab which had gone through cooling process are:

(1) the average temperature of every direction (width, length and thickness);
(2) cooling smoothness;
(3) cooling rate CR;
(4) final cooling temperature FCT.

Because the cooling process is a multi-input, multi-output nonlinear, strongly coupled control process, and the cooling interval is chosen as the length of the slab(5~40m), the sampling temperatures could only be obtained at the entry and exit place. And in the final cooling process, due to the effects of thickness, shape, speed, environment temperature and water temperature on the cooling of slab and inaccuracy of the controlled temperature(800~850 ℃), even for the slabs with same quality, the boundary conditions when entering into the cooling system will be very different. In addition, there exist slab phase changes during the cooing process ($\gamma - F_e \rightarrow \alpha - F_e$), thereby, the cooling system is a complicated process control system. From the slab laminar cooling in 60's to the slab water curtain cooling in 80's, researchers in different countries have made significant contributions to precise cooling control. But a detailed investigation will show that most of them are focused on the improvements of technical aspect (such as various kinds of laminar cooling), and in the aspect of modeling, the process of physical model→simplified model→experiments in lab or operating field→adjustment through expertise→ operation are widely used. Presently, there is no example of building the process control model from the point of control theory, due to the multivariable, strong coupling, nonlinear and time-varying characters of the controlled process. Therefore, using advanced control technology, combining the technical model and artificial intelligent control technology and designing control model for the slab's cooling with self-learning and self-adaptivity are essential aspects in controlling the water cooling process and highly practical.

In order to accomplish that goal, this paper proposes a control strategy which employs advanced control methods to build feedforward (including precomputing model), feedback and self-adaptive control model and incorporate the model into the application study of ANGANG Steel Slab Factory water curtain cooling system through dividing the whole control system into several sub-systems. RBF network is used to obtain the self-adaptive self-learning feedforward model and then is employed in the real water cooing control system. Simulation and industrial experiments show that the slab processed by the water curtain accelerating cooling system could meet the requirements set by manufactures.

2. SYSTEM DESCRIPTION AND ANALYSIS

2.1 Description of laminar water curtain cooling

The laminar flow cooling equipment is installed on the runout table between the roller and straightener, which is 26m long, 4.3m wide, and has 13 groups along the water cooling zone by 2m interval. There are 3 control systems existed in the accelerated cooling system, water-supplying system, runout table's speed system and water curtain system. The capacity of maximum supplying water are $8000m^3/h$. The cooling system can treat the slabs of 10~40mm thickness, 1.4~4.1m width and 15~35m length (see Fig 2.1).

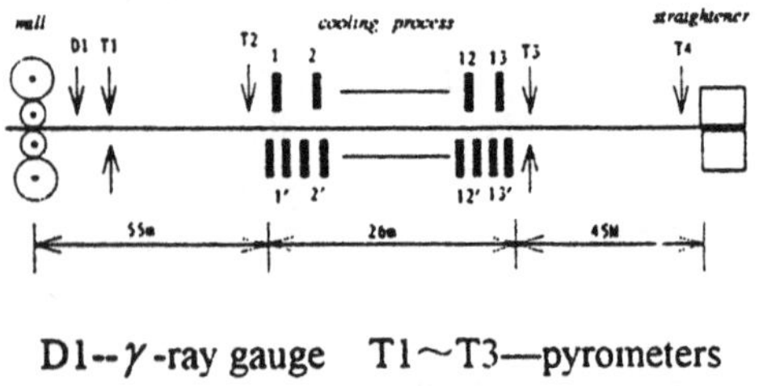

D1--γ-ray gauge T1~T3—pyrometers

Fig.2.1 cooling process

2.2 System analysis

The slab water cooling control system is a complicated industrial process control system, to build the control model, we use centralized control, and the C2000H PLC control system of OMRON co.(Japan). And the control loops are: (1) flow control loop; (2) flow ratio control loop; (3) runout table speed / acceleration control loop; (4) curtain head control loop; (5) cover width control loop; (6) runout table together movement range control loop. So the control problem is using the entry parameters (material of slab, slab temperature, thickness, length, width, water temperature, environment temperature) and the technical targets (final cooling temperature and average cooling temperature) to drive the two temperatures to achieved the target value through properly setting the control variable of each control loop (curtain flow, flow ratio, runout table speed, runout table acceleration, curtain number, distribution of water curtain, curtain head cover width, runout table moving range).

The complicated multivariable control process therefore has been divided into several single variable tracking and adjusting loops and for each control loop, satisfactory performance could be achieved using conventional control strategy. The main goal is how to produce the setpoints of control variables to drive the controlled outputs to meet the desired values and thereby make the system operate at the optimal state as shown in Fig.2.2

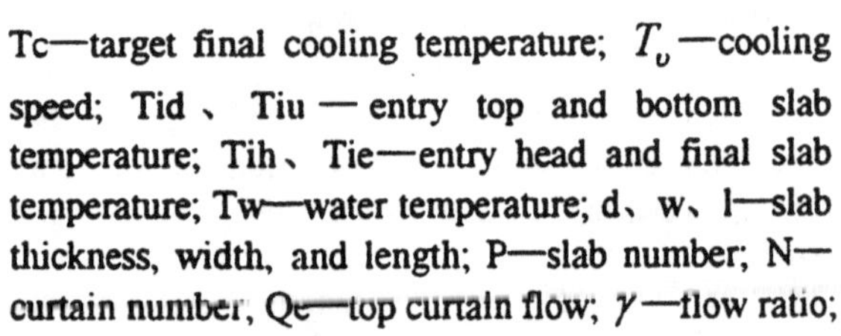

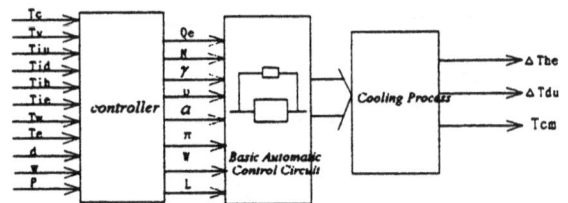

Fig.2.2 Control loop

Tc—target final cooling temperature; T_υ—cooling speed; Tid、Tiu — entry top and bottom slab temperature; Tih、Tie—entry head and final slab temperature; Tw—water temperature; d、w、l—slab thickness, width, and length; P—slab number; N—curtain number, Qe—top curtain flow; γ—flow ratio; υ — runout table speed;. a — runout table acceleration; π — curtain distribution; w — cover width; L — runout table moving range; ΔThe — temperature difference of head and end after the cooling; ΔTdu —temperature difference of top and bottom after the cooling.

3. CONTROL MODEL

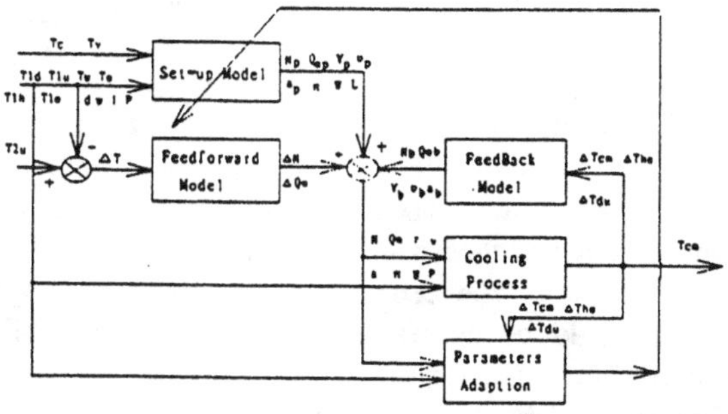

Fig.3.1 The configuration of water curtain cooling control system

As stated above, the control system under discussion is a multi-input multi-output, nonlinear, strong coupled control system, and we could divide it into four sub-systems: set-up calculation, feedforward control, feedback control and model parameters adaptation, as shown in Fig.3.1. Following are the function of each system.

(1) Set-up Model : Because of the need of time of water supply equipment, cooling control equipment and the stable of water curtain, the control value must be pre-computed by employing the set-up model to eliminate the delay. And the set-up is activated every a fixed period of time(5 sec).

(2) Feedforward Model: Before the slab enters into the water curtain, T2 samples the real temperature of the slab and between this temperature and the pre-computed temperature, there exist a difference which is eliminated by the feedforward model. The pre-computation model and this feedforward model together is actually a feedforward controller.

(3) Feedback Model: This model employs the sampling temperature at T3 to adjust the control variable to realize close-loop control.

(4) Parameters Adaptation: After the cooling process is finished for a slab, using the sampling information at T3 and T4 to modify computation model (including the feedforward model) self-adaptively. And the modified model is used to realize close-loop control of the next slab.

4. MODEL SIMPLIFICATION

From Fig.2.2, the cooling process is a multivariable control system, and the variables are strongly coupled, therefore, it is very difficult to use ANN as its control model directly. To combat such difficulty, we divide the control system into several sub-systems to decouple the system and make it easier to employ ANN.

There are five major control variables relating to the control performance, curtain flow Qe, curtain number N, runout table speed υ, flow ratio γ and acceleration a. And π, w and L are decided by the technical model. ΔThe could be adjusted by γ and ΔTdu is eliminated by a, so γ and a are kept as control variable, but N, Qe and υ are not essential control variables and are deserted. From the cooling process, to fix 2 control variables out of 3 could achieve the control goal but because of the flow control, to adjust flow related N and Qe is more convenient than the unrelated υ. In addition, due to the difficulty of obtain precise close-loop control of runout table system, it is not practical to precisely control the speed of runout table. Therefore, after the speed is computed at every thickness, fix it and only when something unusual happens such as the unadequateness of the cooling ability, the speed is adjusted.

The following procedure are adopted to adjust the control variables:

(1) Classify the slab according to the materal and the thickness;

(2) At certain group, produce π, L, w, and υ from technical model;

(3) At certain group, after a and γ are adjusted, it is not necessary to adjust them again in a period of time. Therefore, we could first adjust the acceleration model and flow ratio model and after they are fixed, modify Qe and N model to realize the decoupling of N, a and γ .

(4) According to technical model and physical analysis, the curtain number model is:

$$\Delta N = \Delta N_1 + \Delta N_2 = \frac{\Delta T - \Delta T_0}{\Delta T_0} N_0 + \frac{d - d_0}{d_0} N_0 \qquad (3\text{-}1)$$

(5) Because of the major function the N performs in the cooling process and Qe only realizes the precise adjustment, and it is temperature drop and variation of slab thickness that mainly determines the changes of N. We can use the above equation to determine N and eliminate inaccurate elements of it.

Thus, the ANN is employed to produce Qe and achieve accurate control of the cooling process.

5. RBF CONTROL MODEL

5.1 RBF neural network

The RBF(radial basis function) network (Poggio et al 1990) is a kind of tool which maps the input vector onto multidimensional space, it has the advantage of rapid convergence speed and good approximation ability (Chen,1990), the following Fig.5.1 is its typical structure.

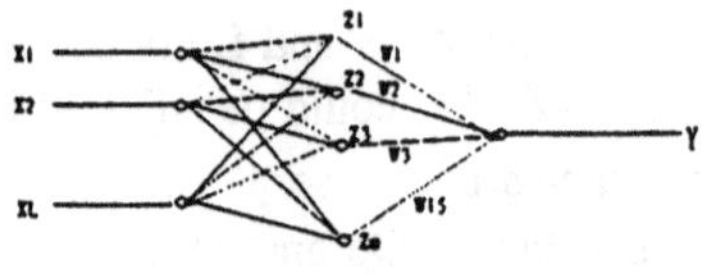

Fig. 5.1 The structure of RBF network

RBF network is a two-layer neural network, its hidden layer process input vector through an activate function and then the output layer compute the linear combination of the hidden layer outputs. The network output is the realization of the mapping: $R^L \rightarrow R$ and is described as follows:

$$Y(X) = \sum_{j=1}^{m} W_j F(\|X - C_j\|) \qquad (5\text{-}1)$$

where $X \in R^l$ is network input vector, $\Phi(\cdot)$ is radial basis function, $\|\cdot\|$ is Euclidean Function, $C_j \in R (1 \le j \le m)$ are the centers, $W_j (1 \le j \le m)$ are weights and m is the number of computing units. And $\Phi(\cdot)$ could be:

a. $\Phi(\cdot) = \exp(-X/2)$; b. $\Phi(\cdot) = X^2 \log(X)$;

c. $\Phi(\cdot) = X^2 + g^2$; d. $\Phi(\cdot) = (X^2 + g^2)^{\frac{1}{2}}$.

In the system, the output of hidden is computed as:

$$z_i = f\left[\sum_{p=1}^{L} \frac{(X_p - C_{ip})^2}{\sigma_{ip}^{\ 2}}\right] \qquad (5\text{-}2)$$

When $f(x) = \exp(-\frac{x}{2})$, (5-2) is changed to

$$z_{ip} = \Phi(\|X_p - C_i\|) = \exp\left[\sum_{p=1}^{L} \frac{(X_p - C_{ip})^2}{2\sigma_{ip}^{\ 2}}\right] \qquad (5\text{-}3)$$

where Z_{ip} is the output of neuron i at the Pth sampling data X_p, C_i are centers, σ_{ip} are widths.

The formula (5-3) shows that through mapping input vector to multidimensional space, the classification is made easier and linearizable. Incorporating the network into our water cooling system, we get

$$y_p = w_0 + \sum_j w_j \Phi(\|X_p - C_j\|) \qquad (5\text{-}4)$$

where y_p is the target output at sampling data P, w_j is the jth weight connecting the hidden layer and the output, w_0 is the threshold of the target unit, j is the number of hidden units.

RBF network could approximate any nonlinear function and is suitable for multivariable function. Its difficulty is the choosing of the centers and as long as they are chosen well, we can achieve terrific approximation. The centers for our water cooling system could be chosen from the real input and output of the control system and using the recursive algorithm, only a small number of centers are needed to obtain good control performance.

In the learning process, unlike BP network, only weights connecting the output layer are adjusted, and therefore it has good memory, which is very important to our water cooling system.

5.2 RBF control model

The simplified structure of RBF controller (MISO) is showed in Fig.5.2.

The dynamic inverse controller showed in Fig.5.2 is realized directly and is used to obtain the control signal through learning the inverse of process model. It could also be viewed as feedforward controller. The learning signal of NN1 is the difference between process output Y and desired output Y_c, and the difficulty of this kind of learning is the inability to decide the error signal $\Delta E = Y_c - Y$ for the modification of $Q_e - \Delta Q_e$, which is the so called Jacobean Problem (Psatis, et al, 1988). Therefore, we need another process simulator NN2 to solve the problem of mapping $\Delta E \rightarrow \Delta U$.

The learning process of RBF controller is actually the identifying process of nonlinear and unknown system. The simplified controller showed in Fig.5.2 is exactly the identification model for unknown water cooling process.

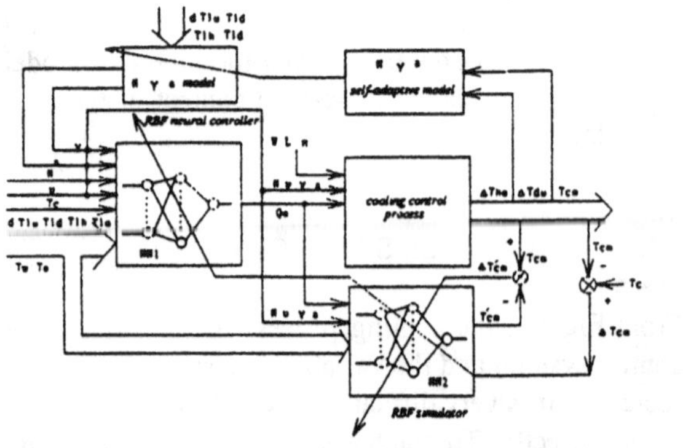

Fig.5.2 The direct controller of water cooling system combining RBF network and simulator

In the identification model, the output of network is:

$$Y = \sum_{i=0}^{m} w_i z_i = Z^T W$$
$$W = \begin{bmatrix} w_0 & w_1 & \cdots\cdots & w_m \end{bmatrix} \qquad (5\text{-}5)$$
$$Z = \begin{bmatrix} z_0 & z_1 & \cdots\cdots & z_m \end{bmatrix}$$

We employ the weighed RLS method as learning algorithm of RBF controller, it is stated as follows:

$$K(n) = P(n-1)Z(n)[\lambda + Z^T(n)P(n-1)Z(n)]^{-1}$$
$$P(n) = \frac{1}{\lambda}[P(n-1) - K(n)Z^T(n)P(n-1)] \qquad (5\text{-}6)$$
$$W(n) = W(n-1) + K(n)[y_d(n) - Z^T(n)W(n-1)]$$

where λ is the forgetting factor and $0 \le \lambda \le 1$.

The performance target is

$$J(n) = \lambda J(n-1) + \frac{1}{2}[y_d(n) - Z^T(n)W(n-1)]^2 \cdot \qquad (5\text{-}7)$$

5.3 The choosing of RBF centers:

We use the n-means clustering algorithm to determine the centers of RBF network. Given initial centers $C_i(0), 1 \le i \le n$, and an initial learning rate for the centers $\alpha_c(0)$, at each sample t the recursive n-means clustering algorithm consists of the following computational steps.

(1) Compute distances and find a minimum distance

$$a_i(t) = \|v(t) - C_i(t-1)\|, 1 \le i \le n$$
$$k = arg[min\{a_i(t), 1 \le i \le n\}] \qquad (5\text{-}8)$$

(2) Update centers and re-compute kth distance

$$C_i(t) = C_i(t-1), 1 \le i \le n, i \ne k$$
$$C_k(t) = C_k(t-1) + \alpha_c(t)(v(t) - C_k(t-1)) \qquad (5\text{-}9)$$
$$\alpha_c(t) = \|v(t) - C_k(t)\|$$

The initial centers are often chosen randomly. The learning rate should be $\alpha_c(t) < 1$, and should slowly decrease to zero. In the present application, $\alpha_c(t)$ is computed according to

$$a_c(t) = a_c(t-1) / (1 + int[t/n])^{\frac{1}{2}}. \qquad (5\text{-}10)$$

The convergence properties of the n-means clustering procedure were studied by MacQueen (1967). The n-means clustering is based on a linear leering rule, thus guaranteeing rapid convergence. It is also an unsupervised procedure using only the network input data. No desired response is required and the procedure will not be affected by the earning rule used for the weights. Notice the similarities between the n-means clustering and Kohonen self-organizing algorithm (Kohonen 1987).

5.4 The classified process of water curtain cooling control system:

Because it is the 10mm～40mm slab that the system has to process, from the above analysis, due to the different thickness, cooling rate, number of cooling section, flow ratio, runout table speed and acceleration are different. So we could classify the slab according to its thickness with the interval of 2mm, and we get 16 groups to realize the off-on control. The control process for each group is the same as has been shown in Fig.5.2, and for changes within 2mm, the model reference self-adaptive control is employed to reduce the input of network and improve control accuracy, and we finally get the off-on RBF control system .

6. SIMULATION RESULTS AND INDUSTRIAL EXPERIMENTS

(1) We get 64 pairs of input-output data(thickness is 20-22mm, entry temperature is 750-850℃) from the working site, as shown in Fig. 6.1(a) (the target final cooling temperature is 600℃) and in Fig6.1(b) (dot line, control variable-flow) to train NN1 and NN2. Number of the centers is chosen as 32, and p=200*eye, $\lambda = 0.99$. After the training, output of RBF network (flow, as shown in dash line in Fig.6.1(b)) and the real controlled output (solid line) could match each other and output of RBF network (temperature, as shown in dot line in Fig.6.1(c)) and the real temperature (solid line) could also match each other.

(2) RBF network is used in on-line control of the cooling process, and when the change of slab thickness is extended (20-26mm), employing control method as shown in Fig.5.2, and entry temperature of each slab as shown in Fig.6.2(a), we could also achieve desired temperature as shown in Fig.6.2(b).

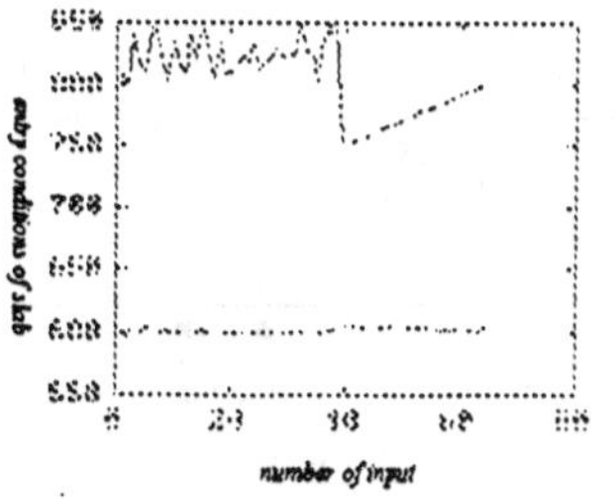

Fig.6.1(a)

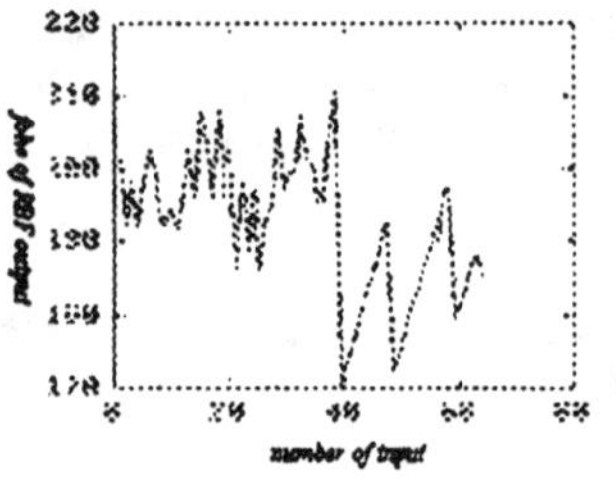

Fig. 6.1(b)

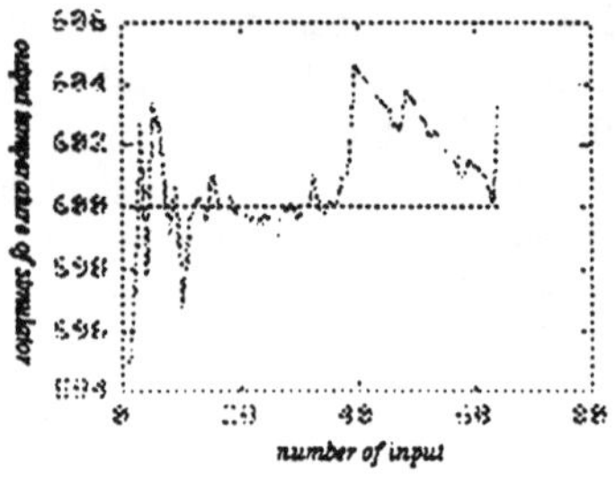

Fg 6.1 (c)

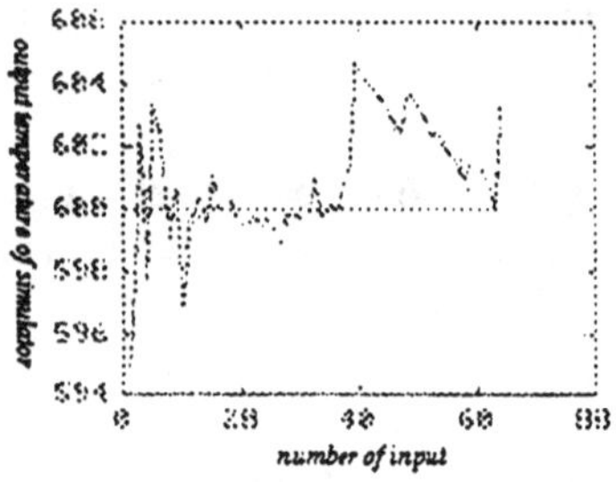

Fig.6.2 (a)

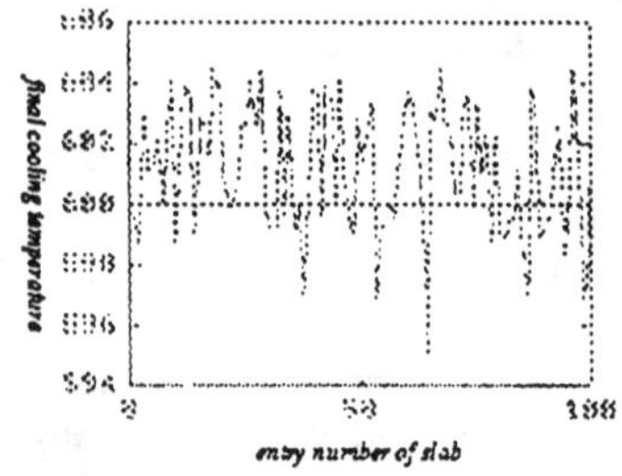

Fig.6.2 (b)

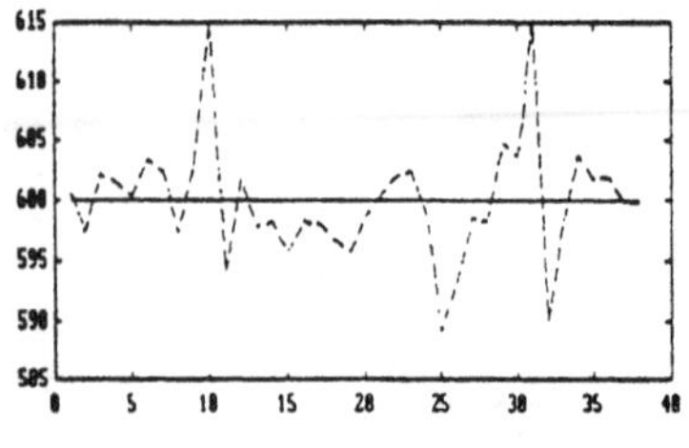

Fig.6.3

(3) Fig.6.3 is the control result when RBF network is combined with the whole cooling control model. It shows that RBF network is capable of self-adaptive adjusting to guarantee stable output of the system and drive the final cooling temperature to satisfactory accuracy. For example, when the 10th slab comes, the entry conditions change, the final temperature could achieve desired target when RBF network is used and the transition period is only 2 slabs.

7. CONCLUSIONS

The simulation and industrial experiments show that it is reliable to employ RBF network as a tool to produce the control variable of the main control loop—curtain flow in the water cooling system. The network is used to combat the difficulty of building control model and our work provides a pioneering example of employing RBF network in the laminar water curtain cooling industrial process. In the mean time, application of the control model improve quality of slab dramatically and bring about economical benefits to ANGANG Steel Slab Factory.

REFERENCES

Chen, S., Billings, S.A. and Grant, P.M., "Nonlinear Systems Identification Using Neural Networks", Int.J.Control, Vol.51, No.6, 1990, pp.1191-1214

Hollander, F.,"Design and control for advanced runout table processing", Iron and Steel Engineer, March, 1971, pp.81~91

Kwasny, S.C. and Faisal, K.A., "Rule-Based Training of Neural Networks", Expert Systems With Applications, Vol.2, 1991, pp.47-58

Masahisa, O.,Tomoki,K., et al,"The computer control of hot strip coiling temperature", Paper for the 7th World Congress of IFAC, June 12~16, 1978, pp.159-166

Moffat, R.W., et al, "Computer control of hot strip coiling temperature with variable flow laminar spray" , AISE Year Book, 1985, pp.474~481

Poggio,T., and Girosi,F.,"Networks for Approximation and learning" IEEE Pro., Vol.4, No.9, 1990, pp1481-1497

Psatis, D., Sideris, A., and Yamamura, A., "A Multilayered Nueral Network Controller", IEEE Control Systems Mgazine, Vol.10, No.3, 1988, PP.44-88

Van Ditzhuijzen G.A.J.M.,"The controlled cooling of hot rolled strip:a combination of physical modelling, control problems and practical adaption"IEEE Trans. on Automatic Control, Vol.28, No.7, 1993, pp.1060~1065

DEVELOPMENT OF A PROFILE MEASURING SYSTEM FOR THE CONDUCTOR ROLL OF ELECTRIC GALVANIZING LINE

Eung Suk Lee, Yong Joon Choi and Ki Nam Baik

Automation Division, RIST, P.O.Box 135, Pohang, Korea, 790-600
eungsuk@risnet.rist.re.kr

Abstract : A surface profile measuring system for the conductor roll of the Electric Galvanizing Line and measurement software is developed in this paper. For the profilemeter, a linear guided control system was designed with laser displacement sensors and developed a 3-dimensional software which is easy to use in a factory. Additionally, the precision position control of the linear guide system was designed using an AC motor with an encoder. The measuring principle of the laser sensor is the optical triangulation method. Two laser sensors were used in order to remove the disturbance and vibration effect on the linear guide system. *Copyright © 1998 IFAC*

Keywords: Profiles, Vibration Measurement, Accuracy, Electric Conductors, Optical Transducers, Measuring Range

1. BACKGROUND OF THE STUDY

In the steel industry, the surface of conductor roll, which is in the Electric Galvanizing Line (EGL) for conducting the steel plate by zinc or nickel anode, must be kept very clean. Otherwise, unexpected surface marks appear on the conducting steel plate. Fig.1 shows a conductor roll used in Pohang Steel Company's EGL (POSCO report).

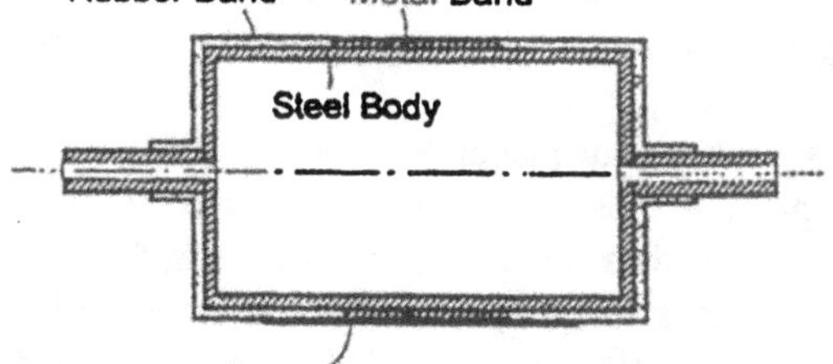

(a) EGL conductor roll

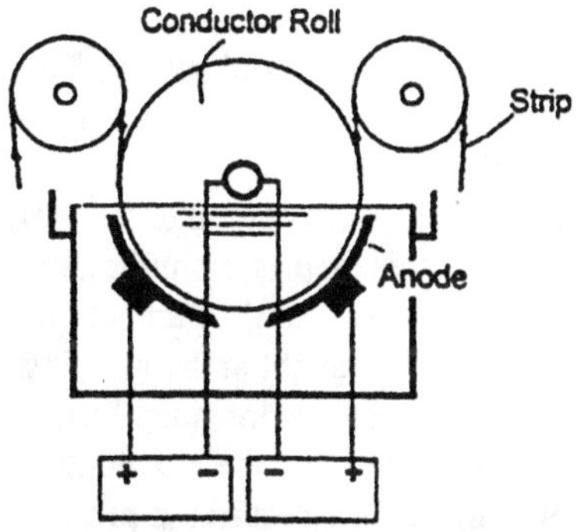

(b) conductor roll configuration

Fig. 1. The shape and configuration of the EGL conductor roll in POSCO.

The conductor roll consists of two parts : a steel band in the middle and polyurethane (rubber) cover on the side of the roll. The rubber parts are necessary not to be conducted on steel roll and to protect the slip between the steel strip and roll. The conductor roll rotates with the strip covering the steel and rubber parts of the roll, which results in band marks on the strip due to the step of the rubber and steel parts. If the part is higher than the rubber, band marks appear and if the rubber is higher than the steel part, arcing marks appear due to the gap between the strip and the steel roll. In the factory, the electric current or production speed is controlled to remove the effects which results in reducing the productivity as shown in Fig.2 (Yuji Acachi and POSCO report). Sometimes anode materials are conducted on the steel roll which results in bad surface quality of the conducting strip.

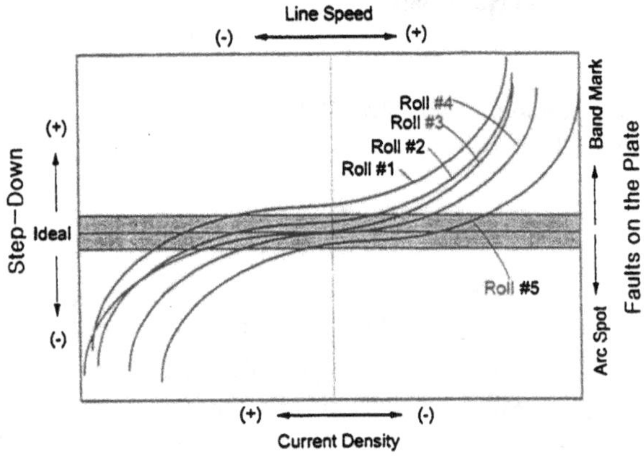

Fig. 2. The relation between current density and productivity with line speed of the EGL.

The object of this study is to develop an on-line conductor roll surface measuring system for the profile and the step between the rubber and steel parts of the roll. The roll temperature at working condition is higher (about 70 °C) than normal ambient conditions at which the roll step is manufactured and adjusted. This results in a different step height between working and manufacturing conditions because of the different thermal coefficients of steel and rubber. Therefore, the roll profile must be measured and maintained at working conditions for adjusting the conducting liquid temperature, the strip tension and the roll cleaning method, etc. The steel step is manufactured 0.6 ~ 0.7 mm higher than the rubber part at the ambient condition to be no-step height at the working condition. The rubber part expands more quickly than the steel part as being warming by the conducting liquid at working conditions. This requires a profile measuring accuracy of less than ±0.01 mm.

In this study, a roll profile measuring system was developed for EGL of POSCO using the laser triangulation displacement sensors. The measuring system was experimented with in the laboratory using a roll simulator with a good result and then tested at POSCO.

2. PROFILE MEASURING SYSTEM USING THREE SENSORS METHOD

A difficulty of this study was to remove the disturbances or vibrations of the roll and measuring system from the measured data at the rotating condition with high measuring accuracy. Fig.3 shows an example to measure the roll diameter in vibration condition (Yuji Acachi). The displacement data by the three sensors are taken in the same time to remove the system vibration effects. If the roll is assumed as a circle, the roll diameter $\triangle R$ can be calculated as Eq.(1). The displacement sensors were water jet type and the measuring accuracy was ±50μm.

$$\Delta R = \frac{\Delta L1 + \Delta L3 - 2 \cdot L2 \cdot \cos\theta}{2(1-\cos\theta)} \qquad (1)$$

where $\triangle L1$, L2 and L3 are the measured sensor displacements.

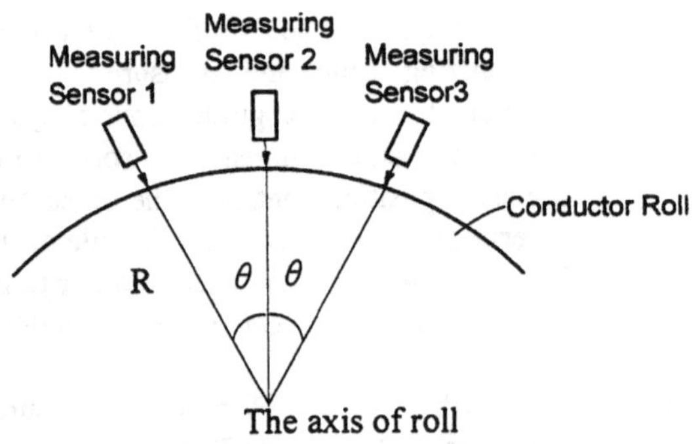

Fig. 3. The method to measure a circular roll diameter.

There are two kinds of vibration mode along the roll axis, parallel motion to the axis and pitching motion, which exist in both the measuring system and the rotating roll as shown Fig.4.

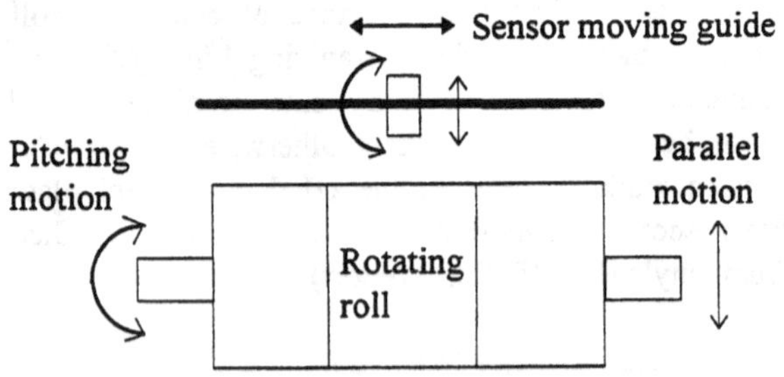

Fig. 4. Two kinds of vibration mode along the roll axis.

Two or three displacement sensors are required to compensate those vibration effects on the measured roll profile data along the roll axis. Fig.5 shows the three sensors method, in which sensors A and C are fixed at both sides of the roll and sensor B is moving for measuring the roll profile. The position of the reference line which is obtained by the two sensors A and C will change from time to time and the sensor B data, roll profile, is plotted on the reference line.

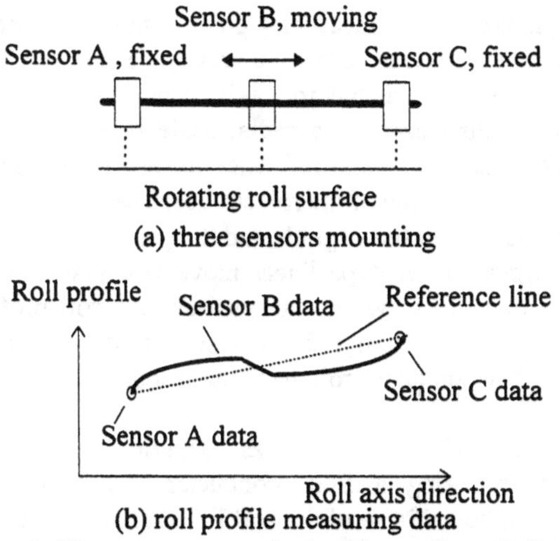

Fig. 5. Three sensors method with a reference line to compensate the vibration effects for measuring the roll profile.

3. TWO SENSORS METHOD FOR COMPENSATING THE VIBRATION COMPONENTS

3.1 Two sensors method

It is possible to use two sensors to remove the vibration components from the measured profile data. Fig.6 explains the two sensor method to remove the vibration components in the measuring system and rotating roll from the measured roll profile. The two sensors move together in steps which should be the same distance as the sensors space. The difference of two sensor values is connected at the end of the previous one and integrated to obtain a profile. To plot a continuous profile, two sensors are to be as close as possible. For the measuring system, we used the laser triangulation micro-distance measuring sensors. The characteristic of the optical triangulation method is a long stand-off distance, a distance between sensor head and measuring surface, which is an advantage to use in the on-line measuring condition. The principle of this method to measure distance 'd' is measuring the distance $\triangle x$ on the optical diode as shown in Fig.7. The distance $\triangle x$ can be represented as Eq.(2), where R_c is the stand of distance and θ is reflection angle to the photo diode (Ernest, 1990).

$$\Delta x = [f \cdot d \cdot \sin\theta]/[\frac{R_c}{\cos^2\theta} + d] \qquad (2)$$

The specification of the laser displacement sensor used for this study was 1μm resolution, 20kHs sampling time and ±8mm measuring range with 50mm stand off distance (Keyence Model LC-2450). For the real environment of the POSCO EGL, the laser head was protected from corrosive fumes of EGL and the air purged for cooling and cleaning.

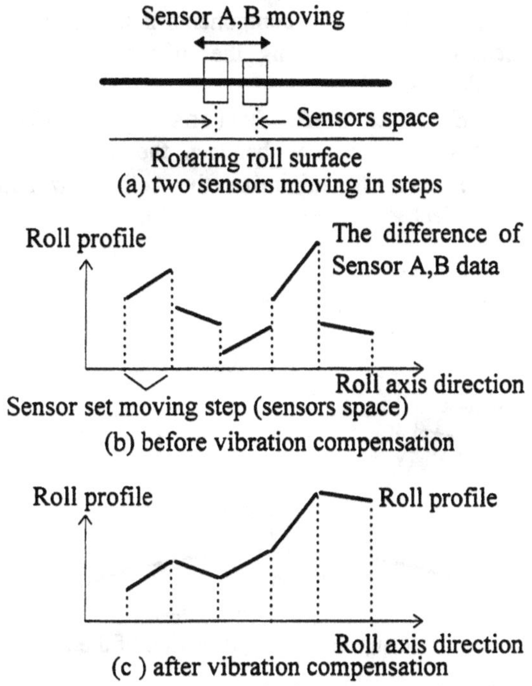

Fig. 6. Two sensors method to remove the vibration components from the measured roll profile.

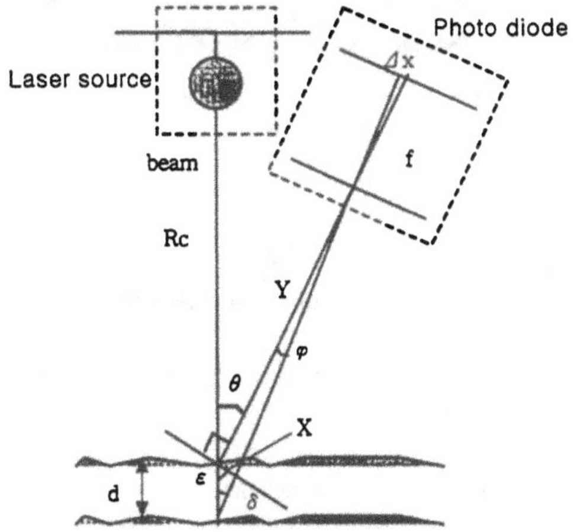

Fig. 7. Irregular reflection type laser triangulation displacement sensor.

The roll profile in the radial direction can also be measured using the same method. Two displacement sensors are set along the roll circular direction of the roll as shown in Fig.8. In this case, we have to take into account the horizontal and vertical vibration of the roll. If the horizontal and vertical vibration components are δx and δy, for sensor A measured error component E(A) is,

$$E(A) \cong [\delta x \cdot \sin\theta,\ \delta y \cdot \cos\theta] \qquad (3)$$

For sensor B, the effect of δx can be ignored when the roll diameter is big because the measuring position is roll center. However, δy is the direct error component in vertical direction of the measuring system, so that the error component E(B) is,

$$E(B) \cong [0,\ \delta y] \qquad (4)$$

Therefore, the total error components E influence on the measuring system using the difference of two sensor values is,

$$E(A-B) \cong [\delta x \cdot \sin\theta,\ \delta y \cdot (\cos\theta - 1)] \quad (5)$$

In this case, the distance between the sensor A,B should be short to minimize the roll horizontal vibration effect.

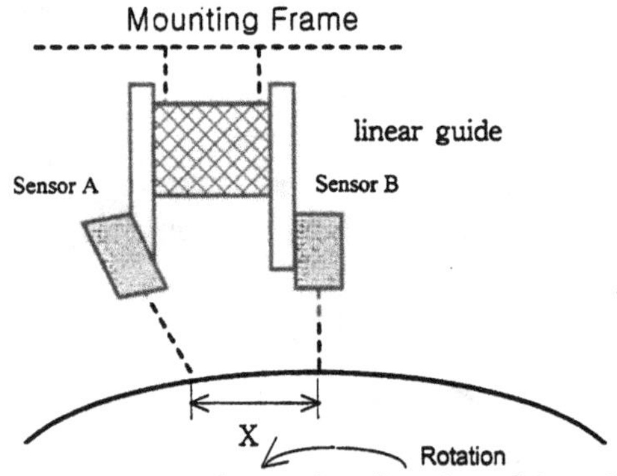

(a) sensors set-up in rotating direction of the roll.

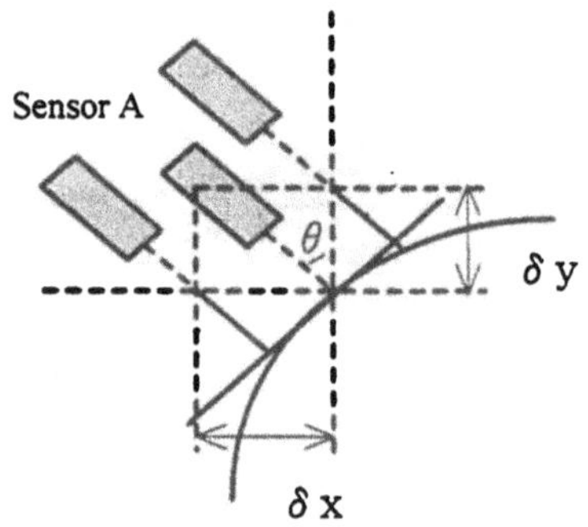

(b) error component for sensor A

Fig. 8. Vibration component for measuring the radial of direction.

3.2 Roll profile measuring system

Fig.9 shows the system developed in this study using two non-contact displacement sensors.

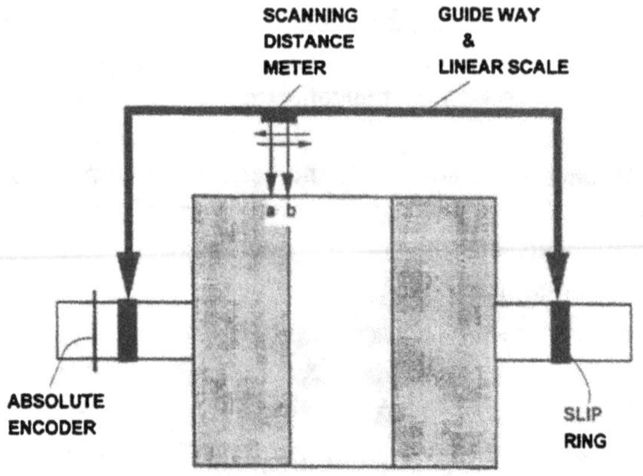

Fig. 9. Two sensors mounting with a linear guide and rotary encoder.

To synchronize the measuring point of the two sensors with the previous point, a rotary encorder(1024 pulse/rev) is set at the roll axis. The rotating angle signal is also used as a radial scale reference with linear scale as an axial reference to display the three dimensional profile of the roll surface. The two sensors are moving on a linear guide and the position accuracy of the geared belt type linear movement system was ±0.05 mm. This system can measure the roll profile and step at maximum 2 meter/sec of strip moving speed. The conductor roll diameter is about 2m.

The measuring system was experimented in a laboratory using a EGL conductor roll simulator, which is the half size of the real EGL roll, as shown in Fig.10. The roll is put into a water tank in which water is heated up to 60°C. The roll consists of two parts, rubber and steel. The roll speed is changed using an inverter with AC motor. We used a geared belt type linear guide to move the sensor in roll axial direction. The positioning accuracy of the guide was less than 0.1mm.

Fig. 10. EGL conductor roll simulator (right side) with profile measuring system

The measuring software was programmed by LabWindows/CVI v3.1 of National Instrument with "C" compiler which works in Windows environment for PC. Fig.11 show an example of roll profile graphic panel of the measuring software with a two or three dimensional display. The step down, the height of rubber part for steel part of the roll, is displayed around the roll in 360°. The average of the step down is calculated and showed in detail. The particular measuring area or position can be enlarged using the zooming function.

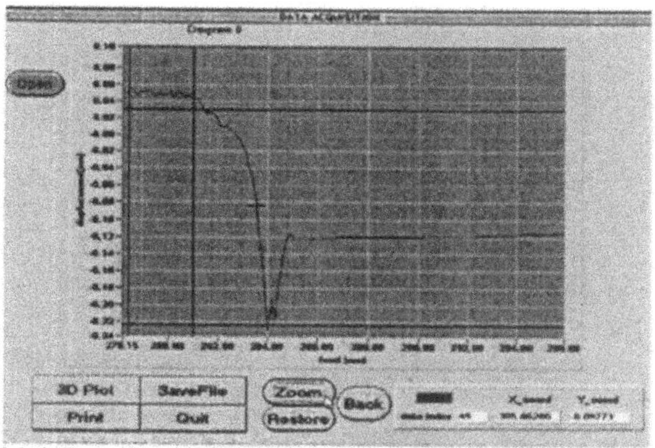

(a) 2-D roll surface profile, left rubber part, right steel part (Y-axis ; step down height)

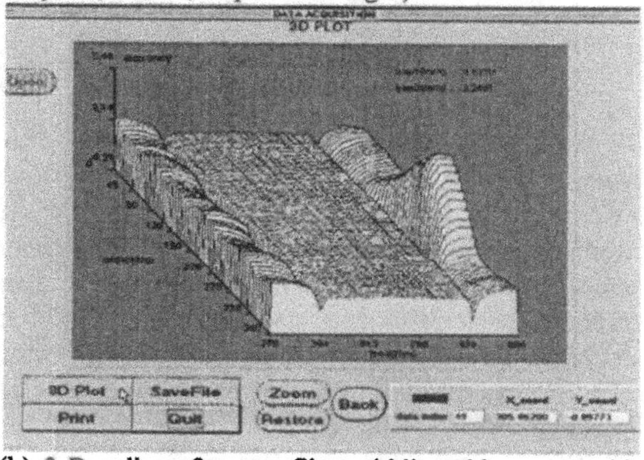

(b) 3-D roll surface profile, middle rubber part, side steel part (Y-axis ; 360° around the roll)

Fig. 11. An example of the surface profile measuring panel with graphic window (X-axis ; axial direction of the roll).

4. EXPERIMENTAL RESULTS

4.1 Results using the roll simulator

The vibration of the measuring system with the roll simulator was measured using a machine analyzer and is showed in Fig.12a, which is very small and can be ignored. Fig.12b shows the vibration measurement with random vibration applied as the similar condition to the EGL conductor roll. To verify the vibration compensation algorithm, the two sensor method is compared with the results using one sensor. Profile measurement using one sensor without the vibration compensation algorithm is shown in Fig.13. In the case of small vibration (Fig.13a), shows that the average step down is 120μm which is the exact value. Fig.13b shows the profile measurement with serious vibration as Fig.12b.

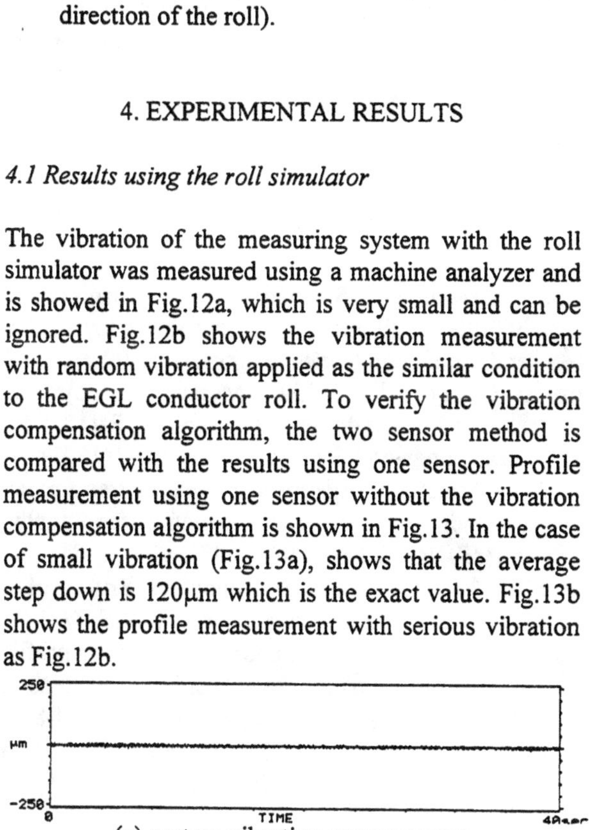

(a) system vibration measurement

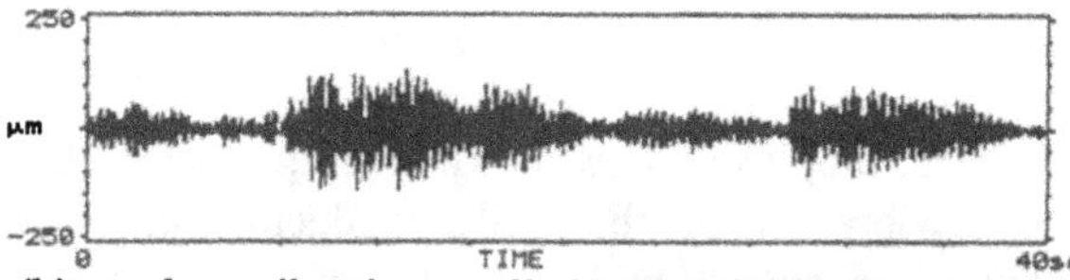

(b) random vibration applied to the measuring system

Fig. 12. Vibration measurement in the horizontal direction at the sensor position of the system using a machine analyzer.

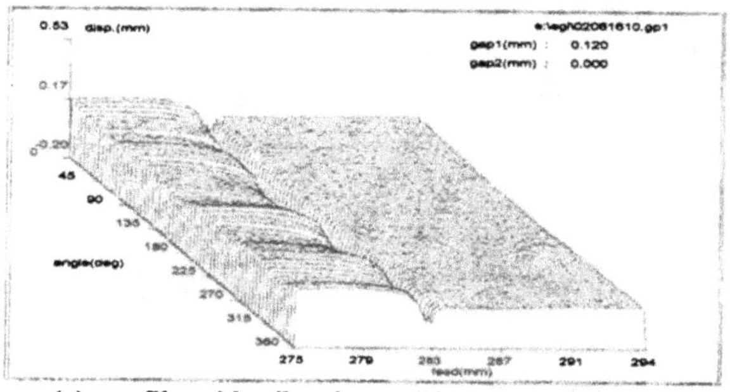

(a) profile with vibration as Fig.2a, step 120μm

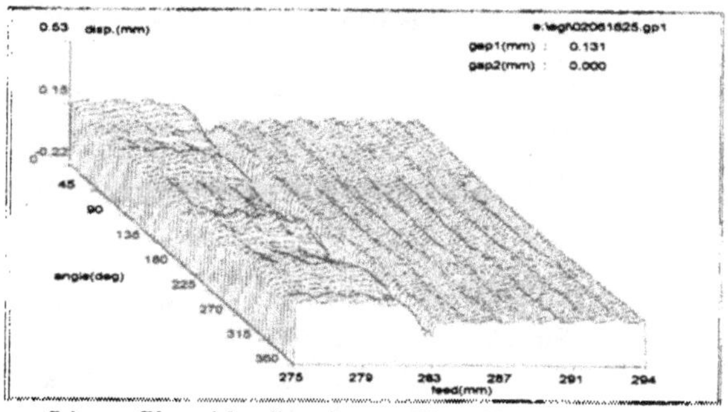

(b) profile with vibration as Fig.12b, step 131μm

Fig. 13. Profile measurement using one sensor with vibration as Fig 12.

The profile measurement using the two sensors method with vibration compensation is shown in Fig.14. The measured average step height was 121μm, which is similar with exact value (Fig.12a). The measured results also show the step down shape in detail without loosing the data between rubber and steel parts of the roll.

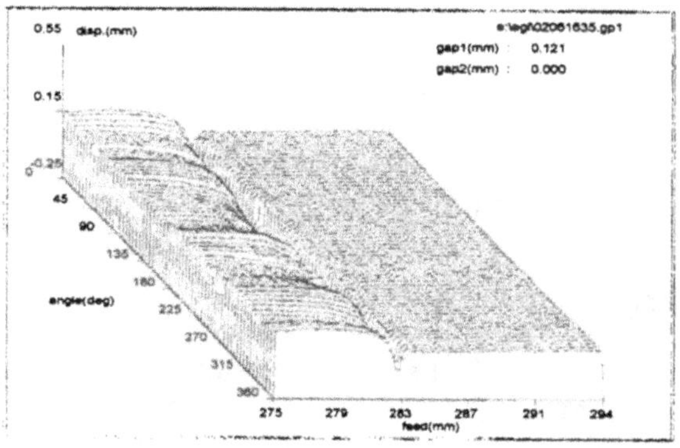

Fig. 14. Profile measurement result using two sensor with vibration compensation algorithm, step 121μm.

4.2 Results for the EGL conductor roll

Before the application of the measuring system, the system vibration of the EGL conductor roll was measured as shown in Fig.14. Therefore, using the system horizontal vibration $\delta x \approx 200$mm and vertical vibration $\delta y \approx 50$mm, from the Eq.(5) the horizontal and vertical error components for the system are,

$$E(A-B) \cong [8.2\mu m,\ 0.2\mu m] \quad (6)$$

where roll diameter= 2440mm, sensor A angle θ=4.7° and sensors A,B space X=100mm. This result shows the profile measuring accuracy of the developed system is about 10μm.

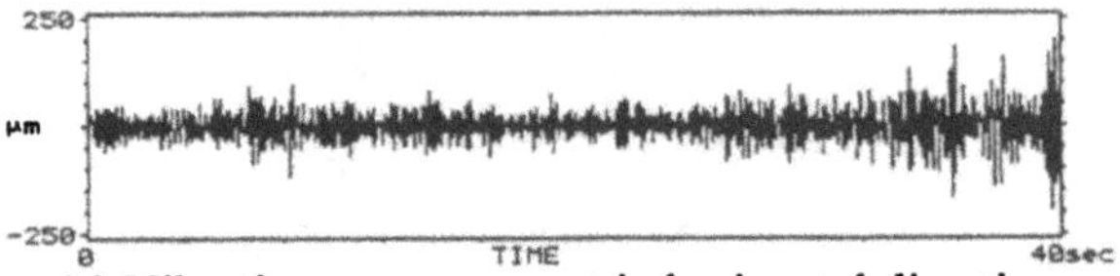

(a) Vibration measurement in horizontal direction

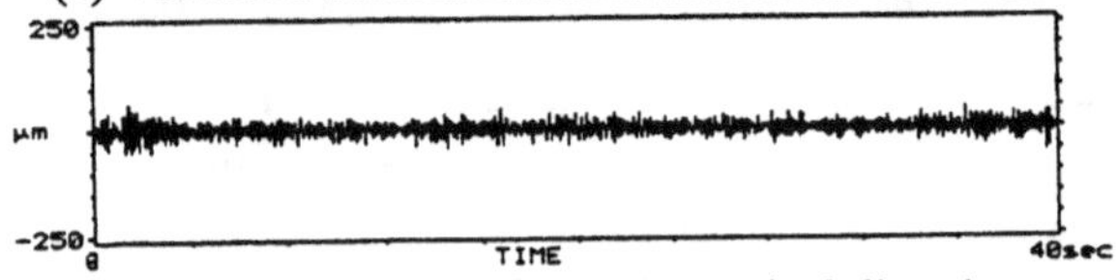

(b) Vibration measurement in vertical direction

Fig. 15. System vibration measurement for the EGL conductor roll.

The step down of the conductor roll was measured as shown in Fig.16. Three dimensional roll profile measured for the EGL conductor roll was also measured as shown in Fig 17. Showing the measured results, the developed system works very well under the bad environmental condition of the EGL.

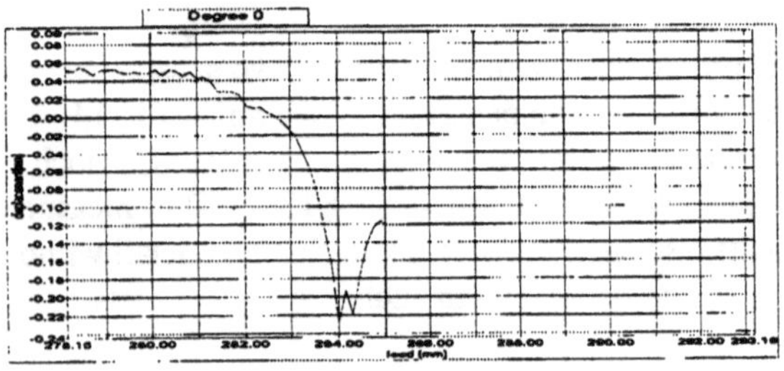

(a) right side of the roll

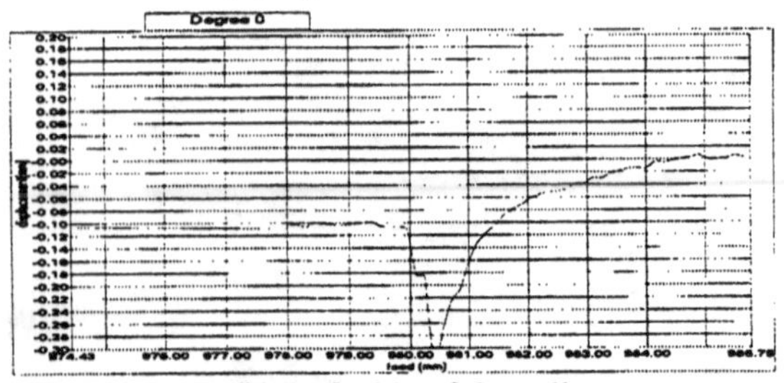

(b) Left side of the roll

Fig. 16. Measurement of step down for EGL conductor roll.

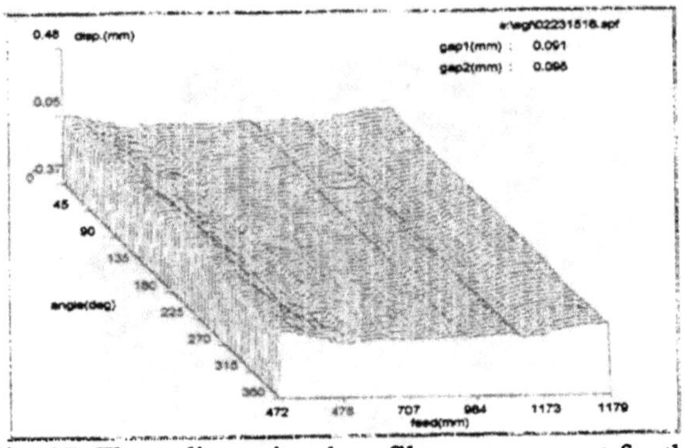

Fig. 17. Three dimensional profile measurement for the EGL conductor roll.

5. CONCLUSIONS

The roll profile measuring system developed in this study worked very well under the serious environmental conditions of EGL, such as the system and rotating roll vibration, etc. The two/three sensors method was useful for compensating the system vibration. The profile measuring accuracy was about 0.01mm. This system can also be used in other kinds of rolls to measure the roll shape and crown, etc. The measuring system will require some modifications, including changing the hardware parts using strong materials to be last in the environment of corrosive fume evaporated from the conducting liquid. Another alternative to this system is to develop a portable system. There are two advantages to the portable system : it would be more economical and it would not be necessary to use with strong materials for the corrosion, because only one portable system is enough for all working rolls with measuring at a necessary roll and time.

REFERENCES

Anand K. Asundi, C. S. Chan and M.R. Sajan (1994). 360-deg profilometry: new techniques for display and acquisition,*Optical Engineering*,Vol.33, No.8

Ernest O. Doebelin (1990). Measurement systems application and design, *McGraw-Hill*, Part 2, Ch2

H.G.Park, Sungwoo.Kim and Juneho Park (1993). 3-Dimensional profile measurement of free-formed surfaces by slit beam scanning topography, *KSME*, Vol.17,No.5, 202-1207

POSCO (1992-1993). EGL Conductor Roll Profile Check Status, *Maintenance Report, POSCO.*

S.H.Bae and H.J.Lee (1995). On the study for profile measurement using Laser displacement sensor, *'95 proceedings of the 10th KACC*, Vol.1, 84-87

Yuji Adachi, et al. (1991). Development of a conductor Roll profile meter for EGL, *CAMP_ISIJ*, Vol. 4

STATISTICAL METHODOLOGY IN PREDICTION OF CAR BODY SHEET DEMAND

Bogusław Filipowicz*, Bogusław Bieda, Bogusław Ochab*****

**AGH-University of Mining and Metallurgy*
Institute of Automatics, Al. Mickiewicza 30, 30-059 Kraków, Poland.
fax/tf.048 012 34 15 68, e-mail:idzi@ia.agh.edu.pl.
***AGH-University of Mining and Metallurgy*
Management Faculty, Al. Mickiewicza 30, 30-059 Kraków, Poland
****Sendzimir Steel Plant, ul. Ujastek 1, 30-969 Kraków, Poland*

Abstract: Automotive industry development causes a permanent increase of demands for various kinds of car body sheets. To satisfy those demands efficiently it is important to realize the significance of the methods of statistical analysis of tendency and development prediction. Model of Box-Jenkins is applied to time-series analysis and decomposition, which are important in car body sheets production. The data come from Sendzimir Steel Plant in Cracow, Poland. Macroeconomic variables such as trends, seasonality and production periodicity, time-series decomposition are used. Method of Box-Jenkins has proved to be one of the best method of time-series analysis.

Keywords: statistical analysis, time series analysis

1. INTRODUCTION

Nowadays firms which work in strongly competitive conditions are obliged to carry out a thorough analysis of their productive capacity, tasks and strategies. A growing competition induces fast implementation of technical innovations. Since market is becoming more and more demanding it is essential not only to produce a new product but to sell it, too.

Production of car body sheets like any other production requires a strategic plan. Such a strategy results from the necessity to conduct a two-fold management optimisation process: economic and productive.

The paper presents the results of time series-analysis concerning car body sheets production at Sendzimir Steel Plant (HTS) in Cracow, Poland. The HTS, the only producer of such sheets in Poland, delivers car body sheets to motorcar factories in Poland and abroad. The Motorcar Factory Daewoo - Poland is one of the main buyers.

The data analysis was made using Box-Jenkins method, see (Brockwell and Davies, 1987). The choice of the method was determined by the objective which was obtaining the most precise prediction (forecasting). The main advantage of this method is the possibility to choose, out of many ARIMA (Auto Regressive Integrated Moving Average) models, the most representative one. The complex methodology, obviously requires the application of software, here a statistical programme STATGRAPHICS ver. 5.0 is applied to data analysis. ARIMA procedure uses Marquard's algorithm for non-linear least square method with a possibility of backward prediction for parameters estimation.

The seasonal ARIMA $(p,d,q,)\times(P,D,Q)$, model of Box-Jenkins has the following formula:

$$\phi(B)\Phi(B^s)(1-B)^d(1-B^s)^{D_s}(X_t-\mu)=\theta(B)\Theta(B^s)a_t \quad (1)$$

see (Hanke and Reitsch, 1981), where $\{X_t, t=1,2,...,n\}$ is the observed series with mean μ, B is the backwards difference operator defined by $BX_t = X_{t-1}$, and $\{a_t\}$ is a white noise process with mean zero and variance σ_a^2. The non-seasonal ARMA components are represented by the polynomials $\phi(B)$ and $\theta(B)$ of degree p and q, respectively, in the backward shift operator, where

$$\phi(B) = 1 - \phi_1 B - \phi_2 B^2 - \ldots - \phi_p B^p \quad (2)$$

$$\theta(B) = 1 - \theta_1 B - \theta_2 B^2 - \ldots - \theta_q B^q \quad (3)$$

The seasonal autoregressive and moving-average components of the model are represented by $\Phi(B^s)$ and $\Theta(B^s)$, polynomials of degree P and Q in B^s, where s is a positive integer denoting the seasonal period of the process. In the Box-Jenkins model, d and D_s are restricted to integer values. The most general seasonal fractional ARMA model allows both d and D_s to be non-integral. The seasonal fractional differencing operator $(1 - B^s)^D$ is defined by the binomial series

$$(1 - B^s)^D = \sum_{j=0}^{\infty} \binom{D}{j} (-B^s)^j = 1 - dB^s - \frac{1}{2} d(1-d) B^{2s} - \ldots \quad (4)$$

where

- p - degree of non-seasonal autoregression component
- d - degree of non-seasonal differentiation component
- q - degree of non-seasonal mobile mean component
- P - degree of seasonal autoregression component
- D - degree of seasonal differentiation component
- Q - degree of seasonal mobile mean component
- s - season length

2. IDENTIFICATION AND ESTIMATION

Table 1 presents the volume of car body sheets production in 1990-1996, in particular months.

Table 1. Volume of car body sheets production in 1990 - 1996

	1990	1991	1992	1993	1994	1995	1996
I	12856	11136	4328	8296	7268	8116	7271
II	10638	10688	5677	9748	5235	8007	8826
III	10960	6478	7174	9543	7011	8013	9899
IV	6569	2358	7149	6431	7832	6921	9718
V	6045	6059	6840	7961	6081	9309	7853
VI	3179	5978	7391	7301	9268	7822	7276
VII	4079	5615	9496	5239	5703	5694	8503
VIII	5663	3539	6022	7514	7525	7812	7923
IX	5366	3581	3429	8319	7603	9376	7424
X	4790	1915	7899	6495	11161	10521	7675
XI	5451	1593	6397	7389	11021	8011	-
XII	6576	2755	6895	5146	10615	5292	-
Total	82172	61695	78697	89382	96323	94894	82368

Time sequence plot of car body sheets production is shown in Fig. 1. The first step in identifying a tentative model is to look at the autocorrelations and partial autocorrelations of the data.

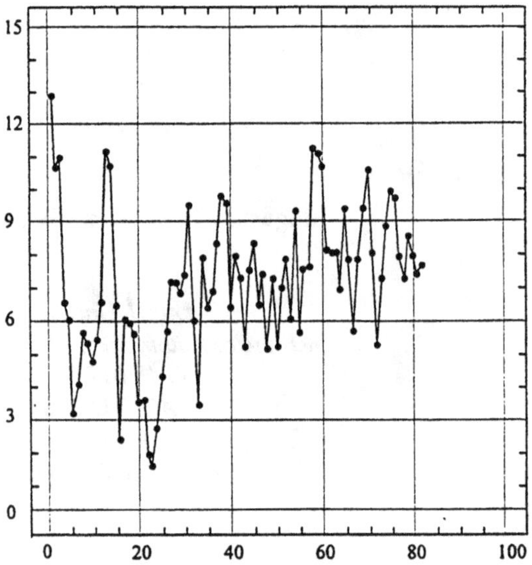

Fig. 1. Time sequence plot

The plots of autocorrelations and partial autocorrelations for the data are shown in Fig. 2 and Fig. 3.

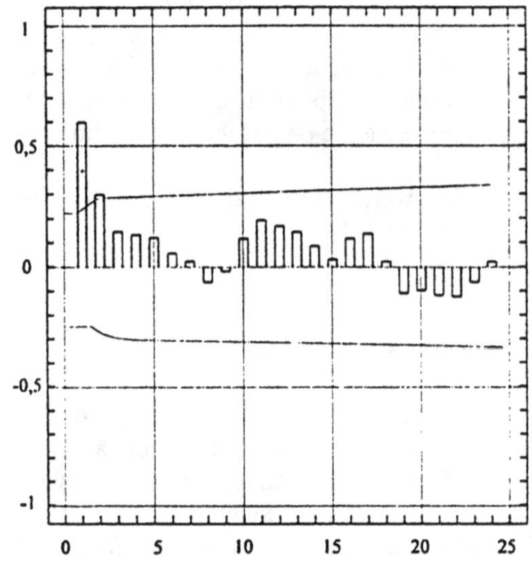

Fig. 2. Autocorrelations plot

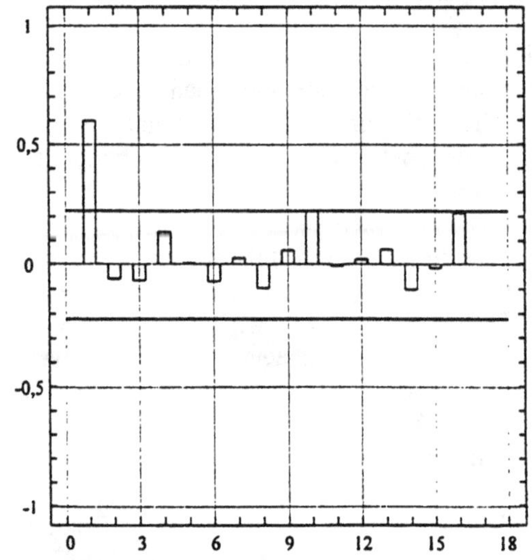

Fig. 3. Partial autocorrelations plot

Table 2 presents estimated` Box-Jenkins model for a certain process.

On the basis of Figures 2 and 3 it can be noticed, that the first 7 autocorrelations appear to be trailing off to zero, which indicates an AR process. Also, the partial autocorrelations drop to zero after the first lag. This indicates that the first order AR, AR(1) process, is presented, see (Hanke and Reitsch, 1981) by:

$$Y_T = B_1 Y_{T-1} + e_T, \qquad (5)$$

where

Y_T - dependent variable

Y_{T-1} - independent variable

B_1 - regression coefficient

e_T - residual term that represents random events not explained by the model.

3. PREDICTION

Prediction consists of basic procedures of times series analysis. They make it possible to decompose particular components of time series and give a possibility to find a model for a given phenomenon.

The results presented in the paper are obtained by Brown's and Winter's exponential smoothing, see (Makridakis and Wheelwright, 1978). Brown's exponential smoothing eliminates the random oscillations from time series.

The results obtained with Brown's procedure are shown in Table 3., but the plot in Fig. 4.

The abbreviations placed in Table 3 are defined as follows:

M.E. - mean error

M.S.E. - mean square error

M.A.E. - mean absolute error

M.A.P.E. - mean absolute percentage error

M.P.E. - mean percentage error

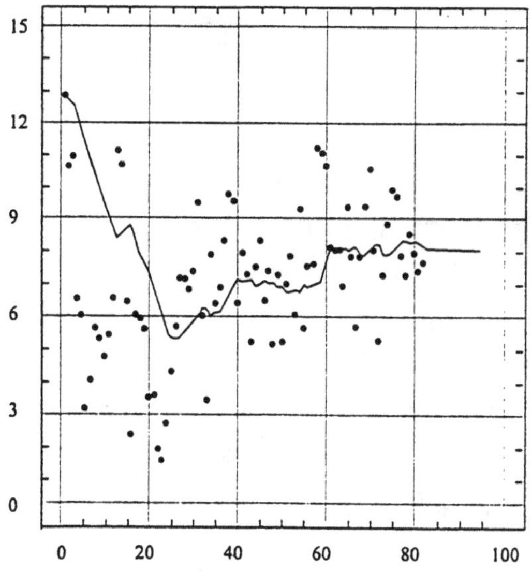

Fig. 4. Brown's seasonal decomposition

Table 2. Estimated parameters of Box-Jenkins model for car body sheets

Estimation begins...				
Initial:		RSS=2.76142E8		b=0.589354 7140.5
Iteration 1:		RSS=2.75399E8		b=0.589675 7299.58
Final:		RSS=2.75292E8	...stopped on criterion 2	
Summary of Fitted Model for: HTSBL.sheets				
Parameter	Estimate	Stnd.error	T-value	P-value
AR(1)	.59368	.09018	6.58307	.00000
MEAN	7364.78532	486.04396	15.15251	.00000
Constant	2992.48509			

Estimated white noise variance=3.44115E6 with 80 degrees of freedom

Estimated white noise standard deviation (std err)=1855.03

Chi-square test statistic on first 20 residual autocorrelations=12.096 with probability of a larger value given white noise=0.842243

Backforecasting: no Number of iterations performed: 2

Table 3. Brown's smoothing

Data:HTSBL.sheets			Percent	100	
Forecast summary	M.E.	M.S.E.	M.A.E	M.A.P.E.	M.P.E.
Simple: 0.1	-592.992	6.74102E6	1973.74	39.84483	-24.6675

Period	Period	Period	Period
83	84	85	86
8052.77	8052.77	8052.77	8052.77
Period	Period	Period	Period
87	88	89	90
8052.77	8052.77	8052.77	8052.77
Period	Period	Period	Period
91	92	93	94
8052.77	8052.77	8052.77	8052.77

Winter's method unlike the Brown's method, can be used to time series with seasonal oscillations, what is a characteristic feature of a given process. Plot of forecast of car body sheets demands according to Winter is shown in Fig. 5.

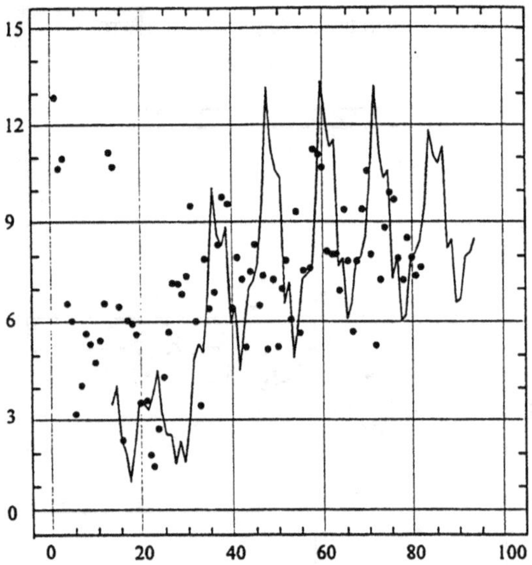

Fig. 5. Winter's forecast plot

Process seasonality according to Winter is shown in Fig. 6.

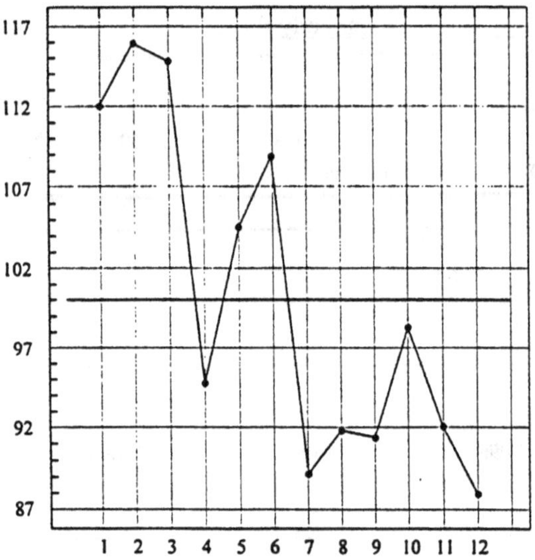

Fig. 6. Winter's seasonality plot

Trend plot obtained in Winter's method is shown in Fig. 7.

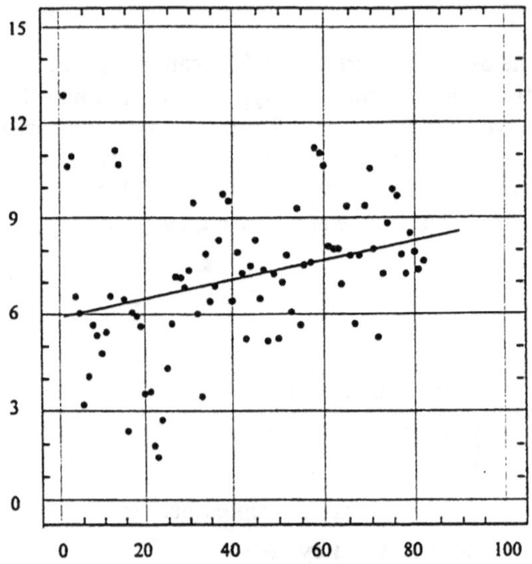

Fig. 7. Trend plot according to Winter's method

4. CONCLUSION

Statistical methods of time series analysis are applied more and more often to management of complex productive systems. The method of Box-Jenkins is particularly useful in the group of statistical methods. The method and statistical programme STATGRPHICS ver. 5.0 are used to analysis of the data concerning the car body sheets production in HTS in Cracow, Poland. Analysing the results obtained it can be noticed, that the significant seasonal fluctuations in monthly production of car body sheets are caused by the clients demands, periodical surveys and routine repairs. The konwledge of the fluctuations can be applied for searching the complementary order book and planning repairs, in a better way.

On the basis of demands forecast in longer time intervals (trend function) and productive capacity of a process line it can be planned an expansion of production volume which is necessary to cover the growing clients' demands.

5. REFERENCES

Brockwell P.J. and A.D. Davies (1987). *Time Series: Theory and Methods*, Springer-Verlag.

Hanke J.E., A.G. Reitsch (1981). *Business forecasting*, Boston.

Makridakis S. and S.C. Wheelwright (1978). *Interactive Forecasting. Univariate and Multivariate Methods.* Holden-Day, USA

DESIGN STUDY FOR CONTINUOUS CASTER TUNDISH WEIGHT CONTROL AND LADLE STEEL FLOW ESTIMATE

R.A. Cockerell* S.F. Graebe** G. Elsley*** S. Crisafulli* G.C. Goodwin****

*CICS Automation Pty Ltd, PO Box 570, Wallsend, NSW 2287, Australia
** OMV AG Refinery, Mannswoerther Strasse 28, A-2320 Schwechat, Austria
*** BHP Steel, Long Products Division, PO Box 196B, Newcastle, NSW 2300, Australia
**** Centre for Integrated Dynamics and Control, University of Newcastle, Callaghan, NSW 2308, Australia

Abstract: This paper describes a study into tundish weight control and steel flow estimation at BHP Steel - Long Products Division, Newcastle, NSW, Australia, performed during late 1995. Following upon a previous project to successfully implement automatic mold level control at this installation, a study was made on how to improve the weight control in the tundish and to also improve the estimate of steel flow exiting the ladle. The project involved plant analysis, controller design, Kalman filtering and on-line implementation. The feasibility study successfully achieved its objectives and resulted in some modifications to the existing control system.
Copyright © 1998 IFAC

Keywords: Continuous steel casting, tundish control, Kalman filtering.

1. INTRODUCTION

Continuous bloom casting is the process of molding molten metal into solid blooms. At the Newcastle Steelworks a ladle transports steel from the steel making process and empties through a ladle slide gate valve (LSGV) into an intermediate reservoir (tundish) which maintains a constant supply of steel to four molds.

Following upon a previous project to successfully implement automatic mold level control at this installation (Graebe *et al.*, 1995), a study was made on how to improve both the weight control in the tundish and the estimate of steel flow exiting the ladle. Particular areas which required attention were: a faster, more accurate (ladle) steel flow estimate signal for use in lead injection; a steadier steel flow to reduce the occurrences of

low flow which cause problems with lead mixing; and, reducing the tundish weight deviation. This paper describes how all three of these major criteria were achieved and have, in fact, been successfully implemented on the bloom caster during the feasibility phase using an advanced control system, UNAC (Crisafulli, 1995).

The existing estimate of steel flow into the tundish is provided by a low pass filtered derivative of the tundish weight with the additive offset of cast flow. This flow estimate is used as a setpoint for the lead injection control. The improved estimate of steel flow into the ladle is based on a Kalman filter. This state estimator estimates the actual tundish weight and inlet steel flow based on the tundish weight measurement, casting flow rate, LSGV pulses and estimated quantities for pulse

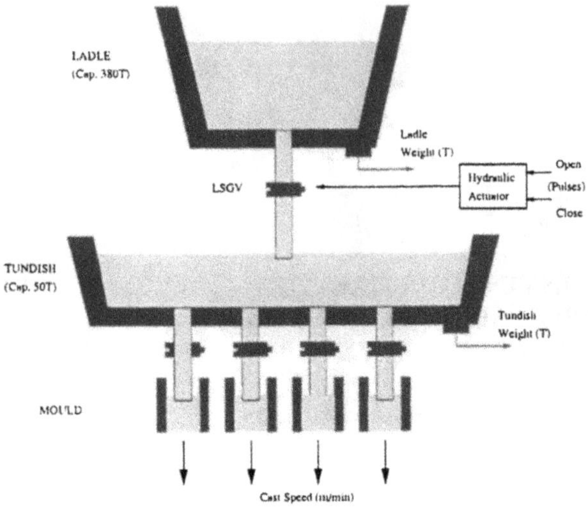

Fig. 1. Bloom Caster Outline.

gains, process noise covariance and measurement noise covariance.

The existing tundish weight controller is a PD feedback controller. The controller output is converted, via pulse width modulation, to provide open and close pulses for the LSGV. The performace of the existing system included poor weight control where oscillations in the order of a one minute period were experienced with significant weight deviations. This resulted in excessive LSGV wear and periods of low steel flow, which is undesirable for lead mixing. Two methods for control of the tundish weight were designed and tested. The first method was a PD controller with a similar structure to the existing controller. An alternate, cascade controller was also implemented which placed an emphasis on balancing outlet and (estimated) inlet steel flows within a tolerable weight range.

A major factor in the success of the study was the use of UNAC as a platform for both the rapid prototyping of controllers, and for data logging and system analysis. UNAC was interfaced to the existing Distributed Control System (DCS) via 4–20mA and relay I/O signals. The estimator and controllers were programmed graphically via block diagrams which could be modified on-line. This allowed the feasibility trials to be performed across two days with no disruption to production.

2. PLANT DESCRIPTION

Figure 1 provides an outline of the caster arrangement. The measurements applicable to tundish weight control are also labelled.

The physics of the tundish weight/level problem are quite straightforward. The tundish is a tank whose rate-of-change of weight is due to any mismatch in the mass flow rates of the incoming and outgoing steel flows. That is,

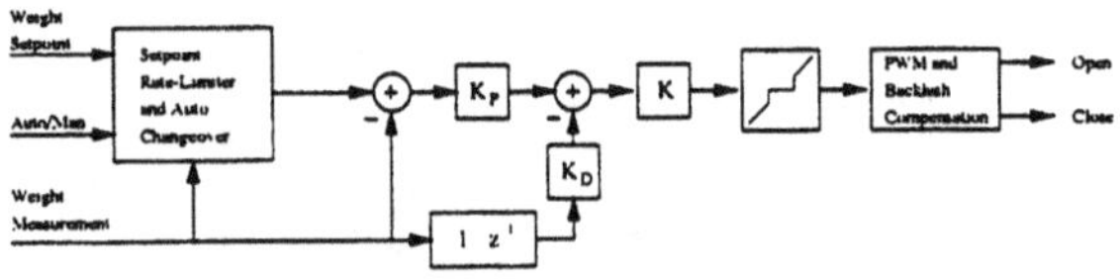

Fig. 2. Existing PD Controller.

$$\frac{d}{dt}w_t = f_i - f_o,$$

where

$f_{i,o}$ = tundish inlet and outlet steel flows,
w_t = tundish weight.

The steel flow into the tundish is controllable with the ladle slide gate valve (LSGV), whilst the outflow is assumed equal to the cast flow of the four caster strands (neglecting small flow variations due to the mold level controller) which is inferred by the measurements of casting speed and the mass per unit length of the cast steel.

Plant limitations which affect both the steel flow estimate and tundish weight control system are:

- coarse quantisation (1T) of the ladle weight measurement and errors upon ladle changeover (there are two ladles on a rotating turret) renders it unusable for short term flow estimation;
- quantisation of the tundish weigher (0.1T) is reasonable for weight control but limits the bandwidth of the steel flow estimator;
- LSGV hydraulics has non-linear effects such as friction and backlash, variability between different ladles, and (originally) too coarse a resolution in minimum control pulses; and,
- there is no position feedback or limit switches on the LSGV.

The weight control system is currently implemented in the DCS computer. A simplified block diagram is shown in Figure 2.

This controller is a Proportional-plus-Derivative (PD) feedback controller where the derivative action is active only on the tundish weight measurement. The controller output is passed through an output deadzone before the control signal is converted, via Pulse Width Modulation (PWM), to provide open and close pulses for the LSGV. Backlash compensation is provided in the form of an extended ON time for the first pulse, in either direction, after a change in pulse train direction. (If the integrating valve behaviour is considered as part of the control system, then overall it is essentially PI control of the tundish weight/level. That is, the PI controller is being implemented in velocity form within the DCS with the valve performing the accumulating function.)

The performance of the existing system included poor weight control where oscillations in the order of a one minute period were experienced with significant weight deviations. This resulted in

- excessive LSGV wear; and,
- periods of low steel flow which is undesirable for lead mixing.

The current estimate of steel flow into the tundish, $\widehat{f}_i$, is provided by a low pass filtered derivative of the measured tundish weight, w_t^m, with the additive offset of cast flow, f_c. The overall transfer function is

$$\widehat{f}_i = \frac{1}{(10s+1)}\frac{60s}{(6s+1)}w_t^m + f_c,$$

where the flow rates are measured in T/min. This flow estimate is used as a setpoint for the lead injection control. The low pass filtering is required to reduce noise problems due to differentiating the quantised measurement, however a time lag is consequently introduced.

3. IMPROVED LADLE STEEL FLOW ESTIMATE

The proposed estimate of steel flow into the ladle is based on a Kalman filter. This state estimator estimates the actual tundish weight and inlet steel flow based on the tundish weight measurement, outlet steel flow, LSGV pulses and estimated quantities for LSGV pulse gains, process noise covariance and measurement noise covariance. The noise covariances form the filter tuning parameters.

One of the advantages of this method over the existing estimator is the use of the LSGV pulses to improve the response time. These signals essentially provide, in a control framework, a feedforward type action to assist the feedback corrective action based on the rate-of-change of weight. To further improve the usefulness of these pulse signals, a backlash compensation system was implemented in UNAC to subtract a user-specified pulse width (i.e. the estimated backlash effect) from a pulse which is in the reverse direction to the previous open or close pulse.

Simplified plant equations are based on the tundish tank and the control pulses to the LSGV:

$$\frac{d}{dt}w_t(t) = f_i(t) - f_o(t)$$
$$\frac{d}{dt}f_i(t) = K_oP_o(t) - K_cP_c(t)$$

where

$w_t(t)$ = tundish weight (T),
$f_{i,o}(t)$ = inlet and outlet steel flows (T/sec),
$P_{o,c}(t)$ = LSGV open & close pulses,
$K_{o,c}$ = open & close pulse gains (T/sec^2).

Thus, the state space plant model is

$$\frac{d}{dt}\begin{bmatrix} f_i \\ w_t \end{bmatrix} = \overbrace{\begin{bmatrix} 0 & 0 \\ 1 & 0 \end{bmatrix}}^{A}\begin{bmatrix} f_i \\ w_t \end{bmatrix} + \underbrace{\begin{bmatrix} 0 & 1 \\ -1 & 0 \end{bmatrix}}_{B}\begin{bmatrix} f_o \\ K_oP_o - K_cP_c \end{bmatrix} + \begin{bmatrix} v_1 \\ v_2 \end{bmatrix}$$

$$w_t^m = \underbrace{\begin{bmatrix} 0 & 1 \end{bmatrix}}_{C}\begin{bmatrix} f_i \\ w_t \end{bmatrix} + n.$$

The state observer used has the general form (Middleton and Goodwin, 1990)

$$\frac{d}{dt}\widehat{x} = A\widehat{x} + Bu + H(y - C\widehat{x}),$$

which, in this case, corresponds to

$$\frac{d}{dt}\begin{bmatrix} \widehat{f}_i \\ \widehat{w}_t \end{bmatrix} = \begin{bmatrix} 0 & 0 \\ 1 & 0 \end{bmatrix}\begin{bmatrix} \widehat{f}_i \\ \widehat{w}_t \end{bmatrix} + \begin{bmatrix} 0 & 1 \\ -1 & 0 \end{bmatrix}\begin{bmatrix} f_c \\ \widehat{K}_oP_o - \widehat{K}_cP_c \end{bmatrix} + \begin{bmatrix} h_1 \\ h_2 \end{bmatrix}\left(w_t^m - \begin{bmatrix} 0 & 1 \end{bmatrix}\begin{bmatrix} \widehat{f}_i \\ \widehat{w}_t \end{bmatrix}\right),$$

where the observer gain vector, H, can be calculated for optimal performance. For example, if v & n are independent white noise processes with known spectral densities, then the Kalman filter gains can be calculated by solving a Riccati equation.

Essentially there are two tuning effects. Firstly, the ratio of estimates of the spectral densities of the measurement noise, n, to process noise, v, provides a measure of how much "notice" the filter should take of the measurement signal vs. the model output. Secondly, the ratio of the states' process noise, (v_1, v_2), spectral densities provides an indication of the relative "confidence" of each part of the model. As is the case with many practical applications, the statistical properties of these noise terms are unknown.

Alternatively, the state observer can be expressed in transfer function form

$$\widehat{x} \triangleq T_1u + T_2y,$$

where

$$T_1(s) = (sI - A + HC)^{-1}B$$
$$T_2(s) = (sI - A + HC)^{-1}H.$$

The resulting transfer functions are

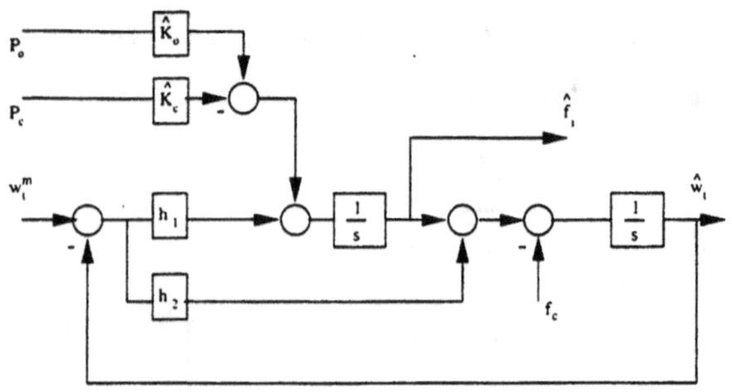

Fig. 3. State Estimator.

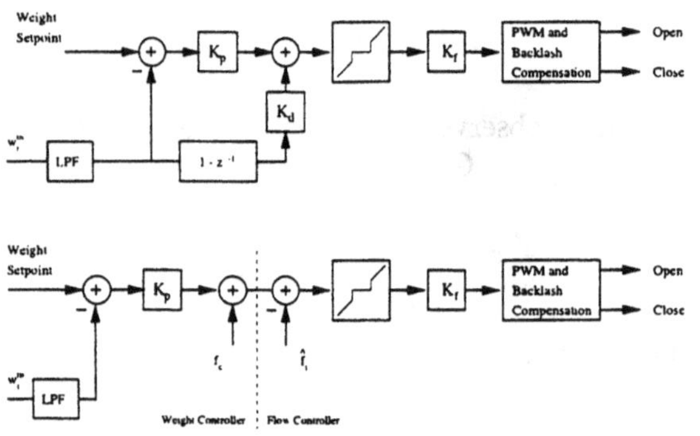

Fig. 4. Proposed Control Schemes.

$$\widehat{f}_{\mathrm{i}} = \frac{h_1 f_{\mathrm{c}} + h_1 s w_{\mathrm{t}}^{\mathrm{m}} + (s + h_2)(\widehat{K}_{\mathrm{o}} P_{\mathrm{o}} - \widehat{K}_{\mathrm{c}} P_{\mathrm{c}})}{s^2 + h_2 s + h_1}$$

$$\widehat{w}_{\mathrm{t}} = \frac{-s f_{\mathrm{c}} + (h_2 s + h_1) w_{\mathrm{t}}^{\mathrm{m}} + (\widehat{K}_{\mathrm{o}} P_{\mathrm{o}} - \widehat{K}_{\mathrm{c}} P_{\mathrm{c}})}{s^2 + h_2 s + h_1}.$$

In this form it was found easier to select the gains h_1 & h_2 based on the desired filter frequency responses. A schematic representation for the estimator is provided in Figure 3 which shows that there is essentially a PI feedback controller with the I-term augmented by a feedforward correction due to LSGV pulses.

Values of h_1 & h_2 were easily selected by experimentation using logged data since this was an open loop filtering problem. The estimated valve gains and backlash times were also determined from logged plant data.

4. PROPOSED TUNDISH WEIGHT CONTROLLER

Two methods for control of the tundish weight have been designed and tested. Figure 4 shows a block diagram summary of the controllers. The first method is a PD controller with a similar structure to the existing DCS controller discussed in Section 2. An alternate controller (preferred) was also implemented which placed an emphasis on balancing inlet and outlet steel flows within a tolerable weight range. Further details of these controllers is provided in the following sections.

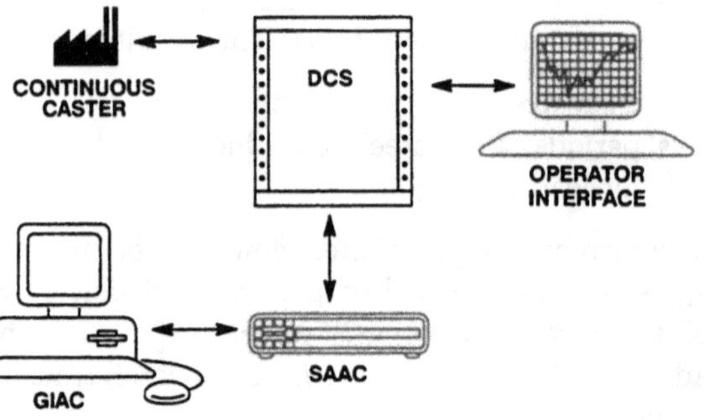

Fig. 5. Outline of Implementation Arrangement.

The output of either controller was passed through an identical Pulse Width Modulation (PWM) and backlash compensation circuit.

4.1 *PD Control*

This is a standard PD controller which is basically identical to the existing DCS controller structure. It was designed for both comparison and fallback purposes.

4.2 *Cascade Weight/Flow Control*

This scheme is a variation on the PD control aimed to provide better disturbance rejection to cast speed disturbances and to also improve regulation performance.

The inner loop of the controller is a P-only flow controller which uses the ladle steel flow estimate as a soft-sensored measurement. The aim of this inner loop is to regulate a desired inlet steel flow. (Recall that the valve performs integrating action on the controller output pulses.) The setpoint for the inner loop is the current cast flow in addition to the output of a P-only feedback weight controller. Therefore, if the tundish weight is at the required setpoint, then the inner loop will have the cast flow as the demanded inlet steel flow. Note that immediate action is taken for casting speed changes before any significant weight disturbance occurs, unlike the PD schemes. (Note, integral action for the outer loop would have to be considered if there were significant errors in the cast flow measurement.)

5. IMPLEMENTATION AND RESULTS

5.1 *Implementation Platform*

Implementation of the controller was performed using a portable UNAC control system which was interfaced to the existing DCS. Figure 5 shows an outline of the implementation arrangement.

UNAC is a platform ideally suited to prototyping controllers as it enables changes in both parameters and structure to be made rapidly. UNAC comprises of two main components. GIAC is a software package executing on a workstation (in this case an IBM-PC/AT compatible laptop) which enables a control system to be designed in block diagram form, then tested using a plant simulation. Once the design is ready it is downloaded to the implementation platform, SAAC, which executes the control system using a real-time operating system and interfaces to the an external DCS/PLC or directly to field I/O. The downloading operation is automatic and does not involve any code compilation, linking, etc. GIAC can then be used as an engineering workstation to display trends, log data, analyse signal frequencies, alter controller parameters, etc. If a change in controller structure is required (e.g. add a filter to a noisy input), then the schematic can be quickly modified in GIAC and reloaded to SAAC.

Some of the benefits of using an implementation method like this are:

- minimal changes need to be made to the existing control system;
- advanced control techniques can be made available quickly and cheaply; and,
- operators retain their familiar interface to the control system.

The interface to the DCS consisted of 4-20mA and 1-5V signals, a digital input for selecting UNAC control. No signals were required to be interfaced directly to field equipment. The plant operator was provided with a function for selecting between UNAC and DCS control.

5.2 *Steel Flow Estimation*

There is a significant improvement in the response of the Kalman filter based steel flow estimator over the existing estimator. This is largely due to the incorporation of the LSGV control action into the model.

The second plot of Figure 6 shows the results of the new steel flow estimator on logged data. Note the sharp edges on the estimate corresponding to the control pulses and the corrective action of the weight measurement which adjusts the flow estimate for model errors (i.e. valve gain settings) after the true effect of the control action is determined.

The stepped nature of the flow estimate is what is expected for the true ladle flow and would allow the lead injection rig to more accurately control the lead mixing. Also, a more accurate flow estimate allows situations of low flows to be dealt with faster.

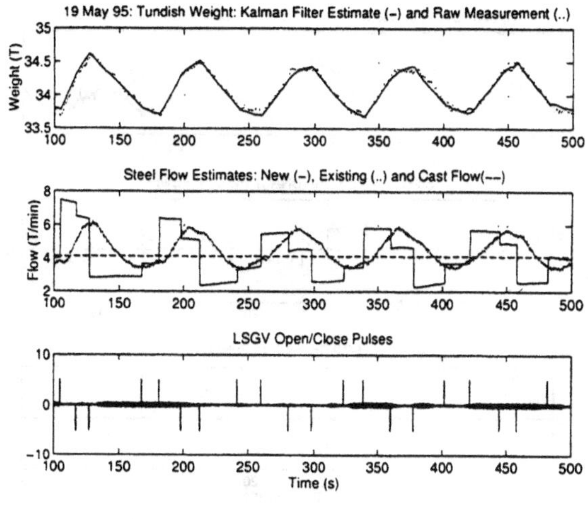

Fig. 6. Steel Flow Estimation.

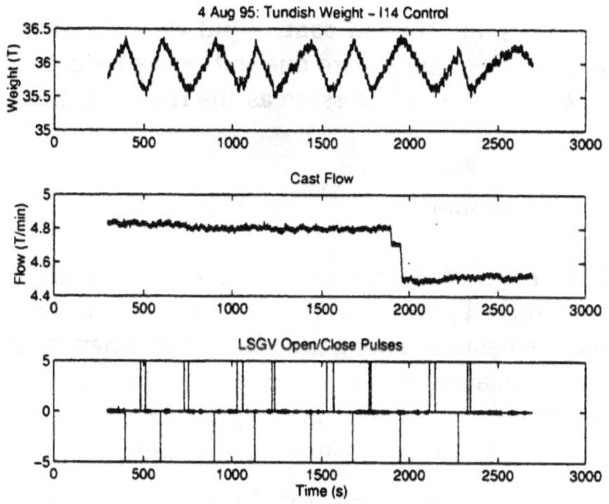

Fig. 7. Existing Tundish Weight Control.

With no field measurement available, it is not possible to exactly verify the accuracy of either steel flow estimate. However, the close matching of the weight estimate (i.e. the other state from the Kalman filter) and weight measurement, shown in the upper plot of Figure 6, supports the accuracy of its associated flow estimate.

5.3 *Existing PD Control*

The first, logged weight control trials took place in August 1995. Figure 7 shows the performance of the existing DCS controller. A steady cycle period of about 4 minutes and ±0.4T deviation exists for the duration of the heat, during which time there was only one major change in casting speed. (The UNAC/PD controller has comparable design and performance.)

5.4 *UNAC/Cascade Control*

Figure 8 provide the results of the trialling of the cascade controller. Note the closeness in inlet/outlet steel flows and the associated reduced

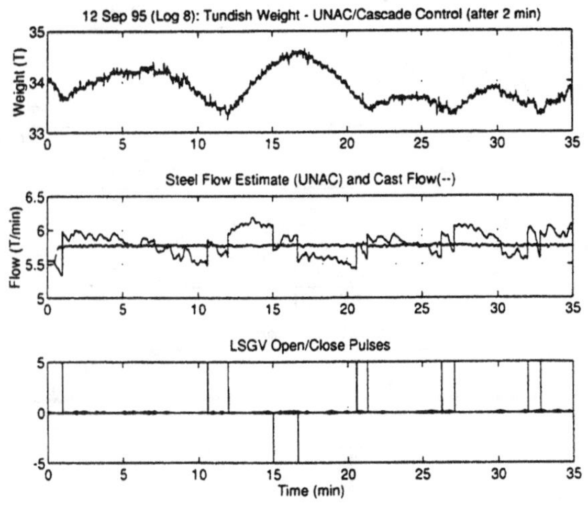

Fig. 8. UNAC/Cascade Control.

frequency in required control pulses. (The more frequent open pulses are due to the the reduction in ferrostatic head pressure as the ladle empties.)

5.5 *Discussion*

The design study has successfully achieved all of the required goals. This paper has described those achievements including descriptions of a new steel flow estimator and tundish weight controller. The design has been tested on the plant using UNAC, and very good results have been achieved.

A quantitative comparison between the performance of the existing and UNAC tundish weight controllers is avoided because of the limited number of logs taken of both systems. However, the results of this section show the performance which the UNAC system achieved.

The new steel flow estimator has been shown to definitely provide a more accurate (via a faster response time) value than the existing scheme. This is partly due to the incorporation of LSGV pulses in the estimator structure.

The improved control and estimation schemes were not adopted, even though the results were good, due to improvements made to the existing system during the study period.

6. CONCLUSION

This paper has described a study undertaken into tundish weight control and steel flow estimation. The study showed how to improve the weight control in the tundish by incorporating an (improved) estimate of steel flow exiting the ladle based upon a Kalman filter. The implementation was performed using the UNAC control system and provided very good results compared with the existing system.

7. REFERENCES

Crisafulli, S. (1995). UNAC: A control development and implementation system. *Australian Journal of Instrumentation and Control* **10**(2), 19–20.

Graebe, S.F., G.C. Goodwin and G. Elsley (1995). Control design and implementation in continuous steel casting. *IEEE Control Systems* **15**(4), 64–71.

Middleton, R.H. and G.C. Goodwin (1990). *Digital Control and Estimation - A Unified Approach.* Prentice-Hall.

AUTHOR INDEX

www.ingramcontent.com/pod-product-compliance
Lightning Source LLC
Chambersburg PA
CBHW072058220326
41598CB00068BA/4451